T0138360

DREAMERS,
VISIONARIES,
AND
REVOLUTIONARIES
IN THE
LIFE SCIENCES

Dreamers, Visionaries, and Revolutionaries in the Life Sciences

EDITED BY OREN HARMAN
AND MICHAEL R. DIETRICH

THE UNIVERSITY OF CHICAGO PRESS

Chicago and London

The University of Chicago Press, Chicago 60637

The University of Chicago Press, Ltd., London

© 2018 by The University of Chicago

All rights reserved. No part of this book may be used or reproduced in any manner whatsoever without written permission, except in the case of brief quotations in critical articles and reviews. For more information, contact the University of Chicago Press, 1427 E. 60th St., Chicago, IL 60637.

Published 2018

Printed in the United States of America

27 26 25 24 23 22 21 20 19 18   1 2 3 4 5

ISBN-13: 978-0-226-56987-1 (cloth)

ISBN-13: 978-0-226-56990-1 (paper)

ISBN-13: 978-0-226-57007-5 (e-book)

DOI: https://doi.org/10.7208/chicago/9780226570075.001.0001

Library of Congress Cataloging-in-Publication Data

Names: Harman, Oren Solomon, editor. | Dietrich, Michael R., editor.

Title: Dreamers, visionaries, and revolutionaries in the life sciences / edited by Oren
    Harman and Michael R. Dietrich.

Description: Chicago ; London : The University of Chicago Press, 2018. | Includes
    bibliographical references and index.

Identifiers: LCCN 2017057789 | ISBN 9780226569871 (cloth : alk. paper) | ISBN 9780226569901
    (pbk. : alk. paper) | ISBN 9780226570075 (e-book)

Subjects: LCSH: Biologists—Biography. | Scientists—Biography. | Biology—Biography. |
    Science—Biography.

Classification: LCC QH26.D74 2018 | DDC 570.92—dc23

LC record available at https://lccn.loc.gov/2017057789

♾ This paper meets the requirements of

ANSI/NISO Z39.48-1992 (Permanence of Paper).

To Sam Silverstein, a dreamer and a friend

—O. H.

To Dick Lewontin, a revolutionary in the best way

—M. R. D.

CONTENTS

OREN HARMAN & MICHAEL R. DIETRICH

INTRODUCTION PERCHANCE TO DREAM
Fostering Novelty in the Life Sciences

INTRODUCTION

Biology isn't always kind to its dreamers. The status of "visionary" is often granted in retrospect and usually when that vision has already enjoyed some modicum of success. Even then, the moniker is often loaded. Lynn Margulis's obituaries, to take one example, heralded her as a visionary, and rightly so. Her advocacy of endosymbiosis as a form of evolutionary innovation altered foundational principles of evolutionary change, and her pursuit of "unconventional" ideas is legendary.[1] Yet even in death, the paleontologist and geologist Andrew Knoll remembered her as "a fountain of ideas—fertile, original, inspiring, contentious, and unedited." As a person, "Lynn could infuriate her colleagues, but at least one of her proposals changed the way we think about life."[2] Infuriating but original, inspiring but contentious: Is the price of scientific novelty always so steep, the acknowledgment of innovation always so hard fought?

Dreamers, Visionaries, and Revolutionaries in the Life Sciences explores biologists who had grand ideas that went beyond the "run of the mill" science of their peers. They each espoused theories, practices, or applications of science that were visionary, sometimes fantastical or even quixotic, but always challenging, and even threatening and destabilizing. Our goal is to understand the conditions that fostered such scientists as they advanced genuine novelty, the challenges and imaginations that, from the nineteenth century and forward, helped to shape modern biology.

Dreamers, Visionaries, and Revolutionaries in the Life Sciences completes our trilogy that began with *Rebels, Mavericks, and Heretics in Biology* (Yale, 2008) and continued with *Outsider Scientists: Routes to Innovation in Biology* (Chicago, 2013). Some of the scientists we feature here might be considered "rebels," since they went against widely accepted tenets in their field, and some were undoubtedly "outsiders," in that they weren't trained as biologists. But "dreamers" deserve a category of their own. Uniquely, the last part of our trilogy concerns the conditions that fostered innovations in biological theories, methods, and practices. This is admittedly a more difficult category to define precisely compared to rebels, who attacked certain well-articulated icons, and outsiders, who came into biology from other fields and made a difference. What distinguishes the innovation of the "dreamer" from that of any other creative biologist is that a "dreamer's" innovation is genuinely novel for their field. But what does that mean? We claim that "dreamers" imagine and articulate newness: theories, models, methods, practices, or applications of science that represent changes of kind, not just degree. These changes are not revisions or refinements of an existing feature of a field; they are original additions that extend beyond what had constituted the boundary of the field. Dreamers are therefore often perceived as radical, visionary, even revolutionary, and they excite reactions spanning the gamut from ridicule to amazement, disbelief to ire.

While we include the category of revolutionaries in our analysis, innovations need not be revolutionary in the sense intended by the philosopher of science Thomas Kuhn. Historians of biology have seldom been fond of Kuhn's historiographic commitment to scientific revolutions, and for good reason: biology, it seems, just doesn't work the way physics does.[3] Because Kuhn required that revolutions arise from a crisis in problem solving, dissent and the creation of novelties were only justified if the status quo demanded an alternative. But this seems too circumscribed to describe much of the history of the life sciences and, in our view, is an unnecessarily conservative approach to the production of novelty.[4] As the chapters in this book reveal, the wellsprings of deep change are pluralistic: there are many more paths to novelty in biological research than imagined by Kuhn and his followers.

Seeking to expand the ambit of novelty, we join a community of scholars, including Hans-Jörg Rheinberger, Lindley Darden, Peter Galison, Paul Thagard, and others, who have been describing forms of scientific change and creativity that do not conform to Kuhnian strictures.[5] And while a post-Kuhnian consensus remains elusive, new frameworks for imagining innovation are beginning to be accepted. Conceptual change, to take only one form of change, has been framed in terms of extension, replacement, elimination, and reorganization, yet none of these actions need be revolutionary in any Kuhnian sense.[6] *Dreamers, Visionaries, and Revolutionaries in the Life Sciences* offers an original contribution to the literature on innovation and change within biology, and science more generally, because it focuses specifically on the many different ways that novelty, not just incremental change, has been generated in different fields of the life sciences.[7]

The broad array of figures profiled in this collection allows us to explore a diverse set of approaches to novelty in the life sciences and the conditions that fostered and supported this profound form of scientific creativity. Here is a thematic microhistory that asks, What allowed dreamers, visionaries, and revolutionaries in biology to imagine and follow a transformative path?[8]

What aspects of their personality, their training, their collaborations, their institutional support, and the social and scientific dynamics of their field carved the space where they could dream of novel approaches and theories in exquisitely new ways, inviting dreamers to champion them within a sometimes unaccommodating, or beguiled, scientific landscape?[9]

Whether the interventions of our dreamers were imagined by way of a fresh extrapolation from known facts or by challenging the relevance and centrality of consensus assumptions; whether they are a result of asking questions for which no clear path to answers yet exists, or by moving the focus of a field to an entirely new area, their essential feature is the unsettling of accepted views and practice, the transcending of an unmarked boundary. But ours is not the view that all revolutionary interventions necessarily produce a perceptual shift or lead by fiat to an intellectual incommensurability. Rather we focus on the circumstances that allow novel alternatives, whether or not ultimately adopted by the community, to be created in biology.[10]

Dreamers are never divorced from their communities. Being a dreamer is a relative calling, something that comes about in reaction to a given scientific ecology. Based on the distinction we make above, dreamers represent a very small fraction of scientists, whether due to their daring, blindness, disregard for accepted foundations, or unique and rare ability to think like children (Jane Goodall), a water creature (Rachel Carson), or even a bug (Eugene Rosenberg). The impact of this minority can be transformative, though this is not always the case. Whatever the consequences of the vision, what is of particular interest is how, situated in their institutional context, intellectual milieu, and historical time, certain individuals are able to extricate their thinking from the accepted science of their peers, introducing novel alternative approaches to important problems. "To sleep, perchance to dream, there's the rub," Hamlet laments, tortured by the fear that even in death there will be no escape from his troubles. But dreamers and visionaries cast aside apprehensions, willingly stepping beyond the frontier of their communities into new territory, where it is not at all clear that they will be followed.[11]

Dreamers, Visionaries, and Revolutionaries in the Life Sciences offers a comparative analysis of historically significant novelties in the life sciences, whether they were enshrined within the realm of scientific consensus, discarded, or remain pushing at the gate. How do different historical contexts, institutional circumstances, and the state of research at a particular time allow such novelties to occur in a scientific community? This is what we seek to understand. As with our earlier books in this trilogy, rather than merely producing a set of disjointed sketches, our goal is to assemble a body of evidence that provides insight into the collective historical role of a distinct category of life scientists. The biographical format, far from ushering a "great-man" approach through the back door, is marshaled as *ancilla historiae* for contextualized social and intellectual histories of science. Scientists, even the rare reclusives among them, never work in a vacuum. Dreamlike and visionary thinking always occurs and is negotiated within and around specific communal circumstances.[12]

Yet it is fitting to raise a caveat. The categories of "dreamers" and "visionaries," even "revolutionaries," a skeptic might offer, are too diffuse to be of any value to the student of science. After all, there are many ways to be a dreamer, visionary, or revolutionary: How can all such ways be subsumed under ecumenical titles, oblivious in their generality to the many nuances that necessarily obtain? Our reply to the skeptic is meant to disarm: We agree, the categories are general. But their theoretical usefulness in science is not diminished but enriched by the broad range of interventions that they cover. Here's why.

Dreamers, Visionaries, and Revolutionaries in the Life Sciences is not about thinkers who succeeded in convincing the scientific community, or necessarily brought us closer to genuine understanding in our inexorable march toward some form of absolute "truth." Such a suggestion not only constricts the categories, but betrays a triumphalist account of the march of knowledge. In a very real sense, we may never be able to grasp absolute truth and must make a point of reminding ourselves that much of what we call "true" today will most likely be "false" tomorrow. Science is our loyal, tested method to erase the truths of yesterday as we deepen our own limited understanding. And the point of "dreamers" is not to pick out winners in retrospect.

The primary criterion for inclusion in the book, rather, is that the subject illustrates a meaningful dimension of novel scientific thinking, whether by being unafraid to think in dramatically bold, even fantastic, ways about an important problem, or by choosing an unorthodox, even unrecognized, problem to begin with, often turning the usual thinking on its head. Being a dreamer, revolutionary, or visionary does not mean being someone who necessarily saw "the truth" ahead of everyone else. The label describes someone who was able to advance an alternative that was fundamentally distinct from those of others working in the same field, whether "right" or "wrong." It is the blend of personal, intellectual, institutional, and historical circumstances that allowed for striking out on such paths that is of interest. And the wider the category, the richer and more telling the circumstances.

In juxtaposition to accounts that focus on innovations within science under the rubric of conceptual change, we have therefore cast our net more broadly to capture innovative social applications of science as well as new forms of managing and funding science. Indeed, not all the dreamers in *Dreamers* were scientists. We have consciously chosen visionaries from a broad range of fields, with a broad set of agendas, in order to provide as rich an analysis of the phenomenon and its importance in the history of biology as possible. As we believe the different test cases demonstrate, dreaming has been a significant motor of change in modern biology. Imposing rigid, narrowly defined categories on the agents of scientific novelty can only obfuscate both the nuances and communalities of the phenomenon. If we are to begin to understand the role of dreamers, visionaries, and revolutionaries in the sociology of science and in the advancement of knowledge, we do better to define our terms as widely as possible, allowing patterns and themes to emerge not from above but from below and within.

BOOK STRUCTURE

The chapters that follow are arranged in rough chronological order, beginning in the nineteenth century and dipping into the twenty-first, but with heavy representation from twentieth-century efforts. To bring these chapters into conversation with each other, we have ordered them in six parts by topic: (1) "The Evolutionists," (2) "The Medicalists," (3) "The Molecularists," (4) "The Ecologists," (5) "The Ethologists," and (6) "The Systematizers." Rather than comprehensive, the mix of subjects within each part is emblematic and contrastive. No historian would claim that Jean-Baptiste Lamarck, Ernst Haeckel, and Peter Kropotkin stand in for every visionary in the history of evolutionary biology. But taken together, Richard Burkhardt's analysis of Lamarck's dream for an expansive zoological system, Robert Richards's reflections on Haeckel's tragi-romantic vision of the phylogenetic tree, and Oren Harman's examination of Kropotkin's twin advocacy of mutual aid in nature and in politics, provide insight into different dimensions of evolutionary thought in the eighteenth, nineteenth, and early twentieth centuries, illustrating some of the various routes of transcending quotidian thinking that were possible within this field.

Mary Lasker, Jonas Salk, and Mina Bissell are the subjects of the second part of this book. As biomedical researchers, Salk and Bissell pursued groundbreaking research on polio and cancer, respectively. Charlotte DeCroes Jacobs's account of Salk's research highlights not only his novel approach to developing a polio vaccine but the sharp resistance his ideas encountered within the biomedical community—indeed, so strong was this resistance that Salk tested his polio vaccine in secret. Anya Plutynski's chapter analyzes Bissell's idea of "dynamic reciprocity" between cells and their environment, showing how Bissell's approach pushed cancer researchers to appreciate the significant role of cellular microenvironments in disease formation and progression. In contrast, Kirsten Gardner's profile of Mary Lasker calls our attention, not to a cancer researcher, but to a deft organizer and public trumpet who successfully advocated for federal biomedical funding and the creation of the National Institutes of Health. The polio researcher excited tempers; the cancer researcher opened eyes; and the advocate opened hearts and pocketbooks. All three were dreamers who saw well beyond the visible horizon.

The transformative impact of molecular biology is captured in part 3 with chapters on W. Ford Doolittle by Maureen O'Malley, Margaret Dayhoff by Bruno Strasser, and George Church by Luis Campos. Dayhoff's effort to create the first database of molecular sequences arguably set the stage for modern bioinformatics and eventually provided the empirical basis from which Doolittle would apply his pluralistic approach to the evolution of introns, selfish DNA, and the tree of life itself. Building on the foundations of modern molecular biotechnology, George Church seems to revel in the fantastic, whether dreaming of bringing mammoths or Neanderthals back from extinction.

In part 4, the rise and influence of the environmental movement are brought to life by Janet Browne's account of Rachel Carson's appeals to the public and Michael Dietrich and Laura Lovett's description of John Todd's assembled ecosystems. Carson brought science to the public by presenting a morally compelling case against pesticides, specifically DDT. Todd's own research on DDT contributed to his radicalization and eventual departure from academia to create "living technologies" for sustainable food production and later bioremediation. In juxtaposition to these socially oriented efforts, Philippe Huneman's analysis of Steven Hubbell reveals a radical reframing of ecology in probabilistic terms. Hubbell's neutral ecology adapted the neutral theory of molecular evolution to create a novel theoretical edifice that directly challenged accepted theories of biodiversity and biogeography.

Behavioral dimensions of the life sciences are represented in part 5 with chapters by Dale Peterson on Jane Goodall, Rick Grush on Francis Crick, and Mark Borrello on David Sloan Wilson. Crick and Wilson are not usually considered ethologists, but we include them under this heading because they both addressed dimensions of behavior, and this grouping casts a new light on their work. While the mysteries of animal cognition and human consciousness alike were considered simply too hard to crack by many contemporaries, Goodall's untutored, direct, and personal observations of the chimpanzees of the Gombe National Park stand in contrast to Crick's theoretical and intellectual approach, and both left their mark. Wilson's more recent application of evolutionary thinking to improve the quality of life in

Binghamton, New York, builds on a tradition ranging from social Darwinism to sociobiology, but Borrello sees his approach to city design as a reimagining of social psychology in Darwinian terms, a dream too naive and historically blind to stand a hope for any great success.

The systematizers assembled in part 6 represent those who sought comprehensive systems of understanding, whether the developmental systems of plant and animal forms described by D'Arcy Thompson at the beginning of the twentieth century, James Lovelock's system of planetary interactions that comprised his Gaia hypothesis of the 1970s, or Eugene Rosenberg and Ilana Zilber-Rosenberg's contemporary and comprehensive account of symbiosis embodied in their hologenome theory. Tim Horder's chapter on Thompson lays out the conditions that fostered Thompson as a comprehensive if idiosyncratic intellectual who brought together classics and biology in a groundbreaking treatise on biological form, one that would not survive the test of time. Sébastien Dutreuil's chapter on Lovelock reveals how Lovelock's exit from academia granted him the space to explore geochemistry on a new scale, winning followers but also many detractors, while Ehud Lamm's chapter shows how the grand theorizing of the Rosenberg's was born within an academic setting—first from being stuck with a difficult empirical problem in the coral seas, second from borrowing creatively from other disciplines, and third—by thinking like a bug. Despite their obvious differences, all three illustrate a common principle: advancing a new vision entails daring leaps of the imagination.

We acknowledge that in any collection such as this, there are bound to be dreamers who "got away" so to speak. One is reminded of the Frenchman Jacques du Vaucanson, who built the first android, arguing in Lucretius's footsteps that life was nothing more than material mechanics. Or of Marie Stopes's post–Great War dream of "reproductive rationality," memorialized (and somewhat trivialized) in the children's rhyme: "Jeanie, Jeanie, full of hopes / Read a book by Marie Stopes / But, to judge from her condition / She must have read the wrong edition." An account of Joshua Lederberg's imagining of unimaginable life forms on other planets as he went about creating the field of exobiology could have been included, as could one of the twenty-first century neuroscientist Sebastian Seung's dreaming of rivers and riverbanks as visual metaphors to describe the architecture of consciousness. To a degree, the chapters in this book are the idiosyncratic result of our effort to match historians and subjects. Our initial list of contributors and subjects extended well beyond the chapters included here and encompassed scientists such as Sarah Hrdy, Charles Davenport, Gerald Edelman, Craig

Venter, Eva Jablonka, and many others. We expect that still others could have fallen under the rubric we present here, and hope that *Dreamers, Visionaries, and Revolutionaries in the Life Sciences* will touch off a more far-reaching conversation.

THE ANALYSIS OF DREAMS

What, then, have we learned?

To begin with, guiding our thinking has been the widely shared observation that dreamers in science need not always themselves be scientists, nor is science, obviously, confined to academic pursuits. Biology includes scientific intellectual advancements alongside developments regarding the ways in which science is managed, institutionalized, supported and applied. We have therefore fittingly considered "dreams" that are not necessarily constitutive of scientific theory and practice, but extend to visions of how science can be applied socially and how science itself may be structured. Mary Lasker, for instance, had a transformative impact on cancer research, not as a cancer biologist, but as a formidable advocate, organizer, and fund-raiser. John Todd's living machines represent important innovations in applied ecology that were realized outside of academic science as parts of nonprofit and for-profit businesses. Living machines were assemblages conceptually and biologically, but Todd's dream was in their transformative application to real-world problems of waste treatment.

Some of the dreamers profiled here fostered profound transformations, such as Rachel Carson's reimagining of humankind's relationship to the environment. To be sure, Jonas Salk's dream, as well as Jane Goodall's, had huge subsequent impacts on those who would suffer from polio and on our approach to animal cognition, respectively. But not all the dreamers and visionaries featured in this collection have been success stories, as we've already mentioned. D'Arcy Thompson's imagining of development from the point of view of allometric transformations was a causal theory that has not been borne out by later research, though it has exercised important influences. James Lovelock and his Gaia hypothesis, considered to this day "kooky" and downright wrong by many scientists, nevertheless provided a radically alternative perspective on the planet and its "homeostasis," challenging interlocutors to define their basic terms more precisely. Francis Crick's correlates of consciousness have been superseded by subsequent developments in brain science. David Sloan Wilson's advocacy for group selection in evolutionary theory remains far from being accepted; indeed, Sloan Wilson's fight to adopt lessons from group selection theory to better human environments is a battle which many leading evolutionists and sociologists consider obsessive, even wrong-headed. Jean-Baptiste Lamarck's ideas were not championed by biologists in his own day, but they were enthusiastically revived (albeit selectively) in the decades after the publication of Darwin's *Origin of Species*, only to wane again in the twentieth century, being called both empirically and theoretically injurious, though, *mutatis mutandis*, this too is changing.

Alongside the success/failure axis with respect to dreamers, there resides a further continuum, in fact somewhat closer to a dichotomy. While many dreamers marshaled their dreams in productive, positively transformative ways, there is a dark side to visionary dreaming. Many ethicists take exception to George Church's dreams of regenesis and de-extinction, arguing that this particular biologist's dreams will turn out to be all of our nightmares. More often, a particular vision, while innovative at first, can throw a field into disarray, or down an unproductive alley. An argument can be made that D'Arcy Thompson's advocacy of geometry at the expense of biochemistry and morphology, or Haeckel's romantic embryology, which overshadowed his own experimental embryology, or, conversely, Francis Crick's reduction of consciousness at the expense of a more holistic approach, serve as historical examples of misleading visions, albeit judged from our own modern perspective.

Regardless of judgments of their impact and influence, the interventions almost always fall into one category: combinatorial novelties that arise from the original and creative integration or assembly of ideas and practices from multiple fields—in other words, from making daring, often surprising connections between hitherto separate spheres. Complete novelties that owe little to thinking either in the field in question or related fields are in fact strikingly rare, if they exist at all, and represent an overly romantic view of novelty. Mary Lasker's advocacy for federal funding structures and a nationally coordinated fight against one disease called cancer (rather than many) redrew the existing biopolitical landscape, undoubtedly. Jonas Salk's attempt to rid the world of what seemed like an intractable children's disease, polio, was as fantastic in his day as is George Church's promise to use the tools of synthetic biology to create new life forms in the twenty-first century. Both men, and Lasker too, adamantly refused to be constrained by their contemporaries' understandings of the "proper" limits of vaccine research, synthetic life research, and the boundaries of the laboratory, respectively. Jane Goodall is another dreamer who was not afraid to look at a problem from an entirely new angle, perhaps unknowingly: by retaining a naive faith in untutored but systematic and direct observation of animal behavior, Goodall had a profound impact on the received wisdom of our behavioral, and moral, origins. Goodall, Church, Salk, and Lasker each achieved dramatic impacts, but even their transformative interventions were rooted in existing knowledge, practice, and technology, and in the history of their fields. No novelty ever arises *ex nihilo*.

Pure originality is at best rare, and in any case the definition is contested: What precisely does it mean? An honest reply would be that a precise definition is hard to provide, if not impossible (if you don't believe us, close your eyes and try to imagine a completely new color). Novelty seems more often than not to consist in the combination or rearrangement or recasting of materials drawn from disparate sources. Many of the figures represented in these pages pulled together ideas and practices from multiple fields, some scientific, some not, to create genuinely original innovations within their area of study: Church with regenesis, Salk with a dead virus for a vaccine, Goodall with her anthropomorphic approach, and Lasker taking medicine to the media and to government are examples. But there are more. Contra to Tennyson's dictum, and just about everyone else's assumption, the Russian anarchist Prince Peter Kropotkin painted a picture of evolution as a game of cooperation and trust rather than cut-throat competition. Novel as it was to evolution, Harman demonstrates that Kropotkin's cooperative vision was deeply influenced by his political thinking as an avowed anarchist. In his words, Kropotkin's "dreams of social justice and natural order grew in tandem, reinforcing and buttressing each other rather than deterministically producing each other."

Margaret Dayhoff's dream of integrating computation, molecular information, and molecular biology, especially in the form of molecular systematics, likewise brought together fields that failed to see connections between them, even if all were in science. Although James Lovelock's Gaia hypothesis would take on a political and social life, his exhortation to think of our planet as a living, breathing, organic system was grounded in his development of technologies to measure atmospheric chemicals and then integrated with the help of biologists such as Lynn Margulis with ideas regarding ecosystems. Ilana Zilber-Rosenberg and Eugene Rosenberg similarly build on Margulis's ideas, this time her idea of a symbiotic holobiont, to reconceive ourselves as made up of far more than just our own cells. Describing the sources of their novel, even fantastic, theorizing, Ehud Lamm writes, "A unique amalgam of mischievousness, bug-thinking, work on nutrition, a study of sociology, and a leap of imagination had all played their role." While it is true that the chain of thinking that lead the Rosenbergs to the hologenome started with a failure of their experimental system, their innovation was not rooted in an intractable anomaly but in a carefully fostered creative amalgam.

THE IMPETUS TO DREAM

Dreamers stand apart from their peers, but, of course, they are embedded in institutions, cultures, and history. Such structures have certain features that play a role in fostering innovation, and although they are not universally shared, dreamers have often sought out institutions and cultures that allowed for interdisciplinary movement of people and ideas. Notably, for some this meant leaving academia altogether, or turning instead to the public. Others sought to surround themselves with young, less constrained collaborators in order to directly foster creativity. And while some dreamers proved almost willfully tone deaf to their times, others revealed a special sensitivity to the tunes of politics, picking out hardly audible rhythms or otherwise riding on growing symphonic crests as they offered social change through their science.

Lamarck, for example, as Burkhardt notes, "had not been in the field of invertebrate zoology for long—not quite six years—when he first began elaborating his ideas on organic evolution." Being new to the field combined with Lamarck's "long-established habit, for better or worse, of building broad, explanatory edifices," which he knew would not be appreciated by his peers, but which led to adventurous and innovative theorizing. In declining health and convinced that he could do little to sway his critics, Lamarck felt that he had nothing to lose. His visionary behavior was not rooted in any stubborn persistence, but in the freedom granted by institutional security for an established scientist entering a new field where he was relatively unfettered by its traditions. The same could be said of Francis Crick's move to tackle the hard problem of consciousness after a much-lauded career as one of the founders of molecular biology. Rick Grush finds the source of Crick's innovation in his "ability to step back from the way a phenomenon is currently being approached by the experts in a field, and see the phenomenon itself (to whatever degree this is possible)." Secure in his standing and place at the Salk Institute, Crick could safely ignore previous research on consciousness, bringing a set of mathematical tools from signal processing and optimal filtering with which to gain a revolutionary, if ultimately superseded, understanding.

While moving into a new field is a well-traveled route toward innovation, as we have shown in *Outsider Scientists*, a number of the dreamers we consider here were well established in their fields and brought interdisciplinary innovation to their work by structuring generative collaborations. W. Ford Doolittle, for instance, reinforced his interest in taking a philosophical perspective on science by bringing philosophers of biology into his lab group. Mina Bissell actively sought out cross-disciplinary collaboration both on principle and as a research strategy. Plutynski also shows how Bissell recruited female students, in particular, "who may not have found a home in a (then) more prestigious lab, because they were pregnant, had young children, or were non-native English speakers." Bissell welcomed them into her lab and cultivated relationships with "women who were similarly brilliant, ambitious, and also mothers, as well as scientists"; this secure and supportive space fostered fresh, out-of-the-box thinking from young lab members. Outside of academia, a large part of Mary Lasker's success is attributed to her network of collaborators and contacts. In Kirsten Gardner's words, "She surrounded herself with allies who offered clever suggestions and necessary votes, sympathetic friends who championed her cause, and useful supporters who often held influential political seats or board seats." While not directly a source of Lasker's grand, dreamlike ideas, the community she created around herself was an essential element of her success because it provided the structure to articulate and realize her vision of biomedical research and funding.

A third group of dreamers either left academia altogether or did not enter to begin with. In the less "disciplined" spaces they created for themselves, they were not completely independent of the scientific community but certainly less constrained. Jane Goodall's move from Louis Leakey's secretary to his field biologist charts a path autonomous from academia and speaks directly to her unique ability to pursue field observation in a manner that was unparalleled at the time (as well as, perhaps, to Leakey's own unorthodox genius). John Todd's decision to leave his position at the Woods Hole Oceanographic Institute and begin to form both for-profit and nonprofit corporations was an important step toward the realization of his dream to transform how we live, how we use water, and how we process waste. Academia had undoubtedly been a crucial incubator for his ideas, but he realized that corporations were the only efficient means to see his ideas put into practice in the world. James Lovelock's decision to leave his academic appointment similarly freed him to innovate with fewer restraints and to take his message to the public more directly. Without the filter—and impediment—of scientific journals, his research reached a different audience and took on a different tenor that would eventually also undercut his scientific credibility among many who remained in the academic fold. Margaret Dayhoff, for her part, pursued her *Atlas* of sequences at the National Biomedical Research Foundation. Founded by Robert S. Ledley as a nonprofit research institute to promote computerization, the NBRF supported Dayhoff's dream but burdened it with financial demands that pitted the *Atlas* against a scientific ethos valuing free distribution of information. Finally, Jonas Salk, when the scientific establishment rejected his dead polio vaccine in favor of a rival's attenuated live one, turned above the heads of his colleagues to the public: their dimes had paid for his research, he felt, and so it was their right to take sides. Only after he died, and thanks to public pressure induced by media coverage he himself had instigated, was Salk's safer vaccine eventually vindicated.

Perhaps not surprisingly, a good number of dreamers and visionaries proved unusually sensitive to the cultural cadences of their times, making the argument for seeing innovation as inextricably linked to forms of social politics. The expatriate Kropotkin may have arrived back in Russia too late, following the revolution, but his championing of mutual aid in nature as well as political anarchy resonate with the unrest of the times and are prescient in retrospect. Before him, Haeckel had captured the romantic winds of a generation, offering a tragic, as well as inspiring, vision of nature in his phylogenetic tree and biogenetics—a vision that arguably caught on, especially in Germany, as a sign of the times. So too can Rachel Carson's, and later James Todd's, environmentalism be viewed as proactive fashionings of political passions already stirring beneath the surface in an increasingly environmentally conscious West. Carson is often credited with jump-starting the movement, although she probably would not have begun her crusade had she not first heard the rumblings that would help it catch on. George Church, for his part, remains a fascinatingly emblematic character: sniffing out the ambiguous modern taste for taking control of creation—the disgust as well as the attraction—he faithfully represents the times, exciting both outrage and titillation.

THE DREAMER

Sir Peter Medawar once wrote, "Among scientists are collectors, classifiers, and compulsive tidiers-up; many are detectives by temperament and many are explorers; some are artists and others are artisans."[13] Dreamers, visionaries, and revolutionaries, too, function in different ways and are seldom cut from the same cloth.[14] Haeckel's view of the natural world may seem as romantic as James Lovelock's approach to the living Earth, but it is hard to imagine more opposing personalities and styles of scientific practice— the first idealistic and expressive, the second empirical and hard-nosed. Likewise, Kropotkin's and Goodall's descriptive approach contrasts starkly with Stephen Hubbell's technical method, and once more with W. Ford Doolittle's philosophical bent. Both Jonas Salk and Mina Bissell sought to alleviate human suffering, but the latter's environmental approach was almost precisely opposite to the former's internalist, reductionist one. Nor would the two share a common path: one would be ridiculed as a Don Quixote, haughty and even worse—unoriginal; the other was celebrated early on as a brilliant visionary.

Divided by a century, Peter Kropotkin and David Sloan Wilson both stressed the role of cooperation in evolution, but one was led to anarchism in Russia, while the other embraced the capitalism of small-town America. D'Arcy Thompson saw patterns in life—that was his vision. George Church dreams of fantastic, new genetic chimeras while simultaneously harking back to a lost, Neanderthal past. He does not want to recognize patterns so much as to create new ones. Ilana Zilber-Rosenberg and Eugene Rosenberg, by way of comparison, look within organisms rather than between them. Their dream seems to admonish us to figure out how species come together to form the chimera of individuality before we begin inventing new species by crosses. Visionaries and dreamers come in all shapes and sizes. One person's dream may be another's mundane reality—even their most freakish nightmare.

Still, the dreamers assembled here often share certain traits, despite all their differences. One common thread is a willingness to question orthodoxy, sometimes rooted in a deep-seated skepticism. A second is an abiding persistence or loyalty to their ideas or perspectives, and a third is the capacity to seek out institutions and situations that will allow them to pursue their vision. Maureen O'Malley described Ford Doolittle's research agenda as being "driven by a commitment to questioning orthodoxy, by examining its assumptions and implicit limitations." Doolittle, in fact, has "led a double life: as a scientist and as an 'embedded' philosopher who has challenged assumptions ('deconstructively') and offered novel and sometimes left-field explanations ('playful speculation')." George Church is likewise described by Luis Campos as deliberately seeking to question authority—in his case, indeed, to shock an establishment that he himself has paradoxically come to embody. When this skepticism is combined with confidence or persistence, the ground has been laid for the dreamer to undertake the tough task of championing his or her novel views. Jonas Salk, for instance, is described as "rarely express[ing] self-doubt," exhibiting "a fierce tenacity," and "repeating his views over and over"; Charlotte DeCroes Jacobs states that "many found this doggedness exhausting." Mina Bissell is also characterized by her "persistence," though by her "curiosity, and willingness to ask questions," as well. According to Anya Plutynski, "they were what enabled her to challenge mainstream views of cancer causation in her research, and launch a more expansive view." Robert Richards's account of Ernst Haeckel's dream of nature's perfection reflects a similarly dogged and obsessive desire to articulate as fully as possible the tenet of Darwinian evolution. In the process of becoming Darwin's German champion, Haeckel formulated his biogenetic law, powerfully combined Darwinism and monism, and made biological communication profoundly visual. Even though charges of fraud and scandal followed his efforts, Haeckel was unwavering, perhaps due to an almost complete conflation between nature and love.

Many of the dreamers that we know best were also extraordinarily good communicators. It may be that this ability to articulate a compelling vision of transformative change is not a feature of dreamers *per se*, but of those dreamers who are successful at capturing both scientific and public attention. But such success can be double-edged, as public acclaim may be perceived to supply unearned scientific credibility. Charlotte Jacobs writes of Jonas Salk that "scientists accused him of crossing the line of acceptable academic behavior by soliciting media attention." Salk's ability to create a relationship with the media and use it to gain public support and acclaim is taken by Jacobs as a central motivator of scientific resentment. In her words, "one cannot underestimate the role of jealousy, a repercussion experienced by many successful visionaries." Lovelock, Goodall, Carson, Todd, and even Kropotkin took their cases to the public with great effect but not with the intent of gaining credibility as much as creating social change.

It is striking that many of the dreamers in this volume have professed a faith in science as a social force, and others held deep political convictions that they did not separate from their science. The intertwining of the social and scientific is a recurring thematic, whether in Kropotkin's dream of cooperative nature extending to human societies, David Sloan Wilson's vision of a Darwinian city, John Todd's assembled ecosystems for sustainable water treatment, or Jane Goodall's dream of more amicable relations between animals and humans. Radical novelty in the life sciences does not have to promise social transformation, but when it does, it may be doubly potent. Dreaming, as the saying goes, is really a form of planning.

FINAL THOUGHTS

The comedian Steven Wright once claimed to be a peripheral visionary: "I could see the future but only way off to the side." Sometimes dreamers do see the future ahead of us; sometimes they seek to create that future, and sometimes they exhibit surprising blindness. Almost always, they take disparate quantities—different disciplines, or a new technology, or application—and tie them together in ways that none before them have done. However different our dreamers may have been from each other, their various narratives show that novelty need not arrive to extricate a scientific community from a dead end, necessarily, or otherwise represent advances in the social relations of science. Dreamers, visionaries, and revolutionaries have challenged the foundations of the life sciences, and they have called for new applications, for better or worse. They have flourished and continue to do so within the academy and without, often fashioning their own environments to foster creativity. Serving as both a challenge to and a reflection of their times, dreamers have been obsessive and annoying while at the same time original, compelling, and inspiring. Dreamers are no joke. Like the ever-cleansing bite of humor and the merciful winds of change, may they persist for the benefit of us all.

NOTES

1. Dorian Sagan, ed., *Lynn Margulis: The Life and Legacy of a Scientific Rebel* (White River Junction, VT: Chelsea Green, 2012); Satish Kumar and Freddie Whitefield, eds., *Visionaries: The 20ᵗʰ Century's 100 Most Inspirational Leaders* (White River Junction, VT: Chelsea Green, 2006).

2. Andrew H. Knoll, "Lynn Margulis, 1938–2011," *Proceedings of the National Academy of Sciences* 109 (2012): 1022.

3. Paul Hoyningen-Huene, *Reconstructing Scientific Revolutions: Thomas S. Kuhn's Philosophy of Science* (Chicago: University of Chicago Press, 1993), 174, 233. See also L. Soler, H. Sankey, and P. Hoyningen-Huene, eds., *Rethinking Scientific Change and Theory Comparison: Stabilities, Ruptures, Incommensurabilities?* (Dordrecht: Springer, 2008), especially Thomas Nickles, "Disruptive Scientific Change," 349–77; Ernst Mayr, "Do Thomas Kuhn's Scientific Revolutions Take Place?," in *What Makes Biology Unique? Considerations on the Autonomy of a Scientific Discipline* (Cambridge: Cambridge University Press, 2004), 159–69; John C. Greene, "The Kuhnian Paradigm and the Darwinian Revolution in Natural History," in *Perspectives in the History of Science and Technology*, ed. D. H. D. Roller (Norman: University of Oklahoma Press, 1971) , 3–15; and Peter Godfrey-Smith, "Is It a Revolution?" (essay review of E. Jablonka and M. Lamb, *Evolution in Four Dimensions*), *Biology and Philosophy* 22, no. 3 (2007): 429–37.

4. For an account of piecemeal change in biology, see Sylvia Culp and Philip Kitcher, "Theory Structure and Theory Change in Contemporary Molecular Biology," *British Journal for the Philosophy of Science* 40 (1989): 459–83.

5. Hans-Jörg Rheinberger, *On Historicizing Epistemology* (Palo Alto, CA: Stanford University Press, 2010). Lindley Darden offers an alternative account of theory change in genetics that extends beyond Kuhn in *Theory Change in Science: Strategies from Mendelian Genetics* (New York: Oxford University Press, 1991). Peter Galison updates (and unravels) Kuhn in *Image and Logic: A Material Culture of Microphysics* (Chicago: University of Chicago Press, 1997). See also Paul Thagard, *Conceptual Revolutions* (Princeton, NJ: Princeton University Press, 1992); and W. Patrick McCray, *The Visioneers: How a Group of Elite Scientists Pursued Space Colonies, Nanotechnologies, and a Limitless Future* (Princeton, NJ: Princeton University Press, 2016).

6. Thagard, *Conceptual Revolutions*; Alan C. Love, ed., *Conceptual Change in Biology* (Dordrecht: Springer, 2015).

7. Michael North, *Novelty: A History of the New* (Chicago: University of Chicago Press, 2013).

8. A kindred project is Andrew Pickering, *The Cybernetic Brain: Sketches of Another Future* (Chicago: University of Chicago Press, 2011). On microhistory, see Jill Lepore, "Historians Who Love Too Much: Reflections on Microhistory and Biography," *Journal of American History* 88 (2001): 129–44; Soraya de Chadarevian, "Microstudies versus Big Picture Accounts?," *Studies in History and Philosophy of Biological and Biomedical Sciences* 40 (2009): 13–19.

9. Note that we are not asking what circumstances contributed to the acceptance of their ideas. Certainly not all imagined innovations will gain acceptance within a scientific community, but even a novel proposal that is ultimately rejected by most can become an object of extended discussion, refinement, and testing.

10. David Hull offered a model of innovation by evolution instead of revolution in *Science as a Process: An Evolutionary Account of the Social and Conceptual Development of Science* (Chicago: University of Chicago Press, 1988). While we are using the language of revolution to refer to significant innovations, we are open to thinking broadly about the processes of scientific innovation.

11. As Joseph Rouse notes, disagreement is common within a field. Rather than view a field or a paradigm as monolithic, he suggests that a field provides common ground, a set of common beliefs that act to create a sense of shared community. See his "Kuhn's Philosophy of Scientific Practice," in *Thomas Kuhn*, ed. Thomas Nickles (Cambridge: Cambridge University Press, 2003), 109. Kuhn also recognized that consensus was not necessary and reframed part of his analysis of disagreements regarding theory choice in terms of judgments based on shared values. See Thomas Kuhn, "Objectivity, Value Judgement and Theory Choice," in *The Essential Tension* (Chicago: University of Chicago Press, 1977). For subsequent work, see Helen Longino, *Science as Social Knowledge: Values and Objectivity in Scientific Inquiry* (Princeton, NJ: Princeton University Press, 1990). On communities and social order, see Barry Barnes, "Thomas Kuhn and the Problem of Social Order in Science," in Nickles, *Thomas Kuhn*, 122–41.

12. K. Brad Wray, *Kuhn's Evolutionary Social Epistemology* (Cambridge: Cambridge University Press, 2011).

13. Peter B. Medawar, "Hypothesis and Imagination," in *The Art of the Soluble* (London: Methuen, 1967), 132.

14. Despite the many differences between them, the men and women of *Dreamers, Visionaries, and Revolutionaries in the Life Sciences* may describe a shared persona or collective cultural identity—a mask that they may have worn, to borrow Daston's metaphor (Lorraine Daston and H. Otto Sibum, "Introduction: Scientific Personae and Their Histories," *Science in Context* 16, no. 1/2 [2003]: 1–8). Focusing exclusively on the persona of the "dreamer," however, would shift our gaze to the manner in which each scientist navigated his or her identity and place within a community in the wake of a novel proposal—to the reception and interpretation of their dreams and themselves as dreamers, rather than to the conditions that made those dreams possible. That said, some of the figures we consider here did seem to embrace and cultivate a persona of a visionary, such as George Church, James Lovelock, and James Todd.

Part I: The Evolutionists

Figure 1.1.
Jean-Baptiste Lamarck. 1907 heliogravure based
on original oil painting by Charles Thévenin,
c. 1802. Image from *Archives du Muséum
d'Histoire Naturelle*, 6th series, 6 (1930).

RICHARD W. BURKHARDT JR.

JEAN-BAPTISTE LAMARCK
BIOLOGICAL VISIONARY

In his capacity as "perpetual secretary" of the Academy of Sciences in Paris for nearly three decades, the distinguished comparative anatomist Georges Cuvier prepared the eulogies of thirty-nine different scientists. The very last was of the zoologist Jean-Baptiste Lamarck, Cuvier's former colleague at both the Academy of Sciences and the Museum of Natural History in Paris. As it happened, Cuvier died before he could deliver his *éloge* of Lamarck to the academy, but the piece was later read in his stead and published. It caused a stir because Cuvier had chosen to single out Lamarck's career as an object lesson in how science should *not* be done. Cuvier's obituary of Lamarck stands today as one of the cruelest scientific "eulogies" ever written.[1]

Cuvier readily acknowledged that Lamarck had made important contributions to science through his work on animal and plant classification. What Cuvier could not abide was Lamarck's penchant, as Cuvier represented it, for building grand, explanatory edifices on wholly imaginary bases. These edifices, Cuvier said, were like the enchanted castles one reads about in old novels—they vanished in thin air as soon as the magic talismans on which they depended were broken. Maintaining that it would be a benefit to science if, instead of simply commenting on Lamarck's positive contributions to science, he also examined the ways in which Lamarck had strayed from science's true path, Cuvier offered a reconstruction of the "genealogy" of Lamarck's misguided theories, showing how Lamarck's system-building zeal was manifested in his quixotic efforts to overthrow the chemistry of Lavoisier, his various meteorological and geological ideas, and his theory of the origin and successive development of life in all its many forms. Of this last theory Cuvier wrote that such a theory might amuse a poet or stimulate more system-building by a metaphysician, "but it could not sustain for a moment the examination of anyone who has dissected a hand, an internal organ, or merely a feather."[2]

In dismissing Lamarck's theory of organic evolution (and the idea of species mutability in general), Cuvier left himself, at least in this particular regard, on the wrong side of history. This is particularly ironic, given that Cuvier was the leading zoologist of the first third of the century. He transformed zoological classification with his comparative studies of animal organ systems. His pathbreaking work in vertebrate paleontology focused attention on the history of life on earth in a way that was unprecedented. The case can indeed be made that Cuvier's contributions to zoology and vertebrate paleontology ultimately helped make an evolutionary interpretation of life more plausible, notwithstanding his own ardent rejection of the idea of species change. But our primary goal here is not to contrast Lamarck's and Cuvier's respective contributions to biology. Our aim instead is to consider Lamarck's own style of thinking and how it helped him produce a broad, bold, novel vision of how the different forms of life came into existence. Half a century before Darwin, Lamarck maintained that nature had begun with the very simplest forms of life and then from these had successively developed, over time, all the others, from the tiniest monad all the way up to human beings. It was a vision well suited to a man who identified himself not just as a naturalist but as a naturalist-philosopher, and it was of profound importance for biology.

Born in 1744, Lamarck as a young man had wanted to be a soldier, but an injury during the Seven Year's War ended his military career.[3] He proceeded to Paris, where he eventually made a name for himself as a botanist. The Comte de Buffon provided him an affiliation with the King's Garden (Jardin du Roi). In 1793, at the height of the French Revolution, and when Lamarck was nearly forty-nine years old, the King's Garden was reconstituted as the National Museum of Natural History. The transformation of the institution transformed Lamarck's career as well. He was appointed professor of the "insects, worms, and microscopic animals"—the group of animals we now call the invertebrates, thanks to Lamarck himself. He readily embraced the two primary responsibilities of his new position: teaching an annual course on the invertebrates and bringing order to the museum's invertebrate collection. He also served the museum ably in a series of administrative capacities, thereby demonstrating that he was no idle dreamer when it came to performing the institutional tasks required of him.

Had Lamarck remained a botanist, he most likely would never have come to believe in organic mutability, but taking on the new professorship opened new horizons for him. Identifying, ordering, and attempting to make sense of the remarkable array of different invertebrate forms was one part of it. Having to *teach* students about these animals and explain to students the importance of invertebrate zoology was another. Years later he recounted that when he was first given the job he felt he had been put in charge of the less interesting part of zoology: "It in effect seemed to me that there was more advantage and greater interest to be stimulated in the demonstration of the characters, ways of life, and habits of the lion than those of the earthworm."[4] What Lamarck concluded not long after he began teaching was that the study of the invertebrates afforded insights that were truly profound. He was impressed by the way invertebrates manifested the intimate relationship between animal faculties and animal organization. He furthermore concluded that the way the invertebrate classes could be arranged in a series of increasing complexity corresponded to the actual course nature had taken in bringing all the different forms of life into existence.

The turn of the century was an immensely fertile period of Lamarck's career as a biologist, a period of breakthrough biological dreaming on his part, with a bold theory of organic change as the primary product. It bears underscoring that Lamarck initially made his new ideas on organic mutability known by presenting them *to his students, not to his colleagues.* He offered the earliest glimpses of his new thinking in the lecture that introduced his course on invertebrate zoology at the museum in 1800. He presented the first broad overview of his general explanation of organic diversity in his opening lecture of 1802. In his introductory lecture of 1803, he zeroed in on the species question. His students (a highly diverse lot in terms of their age, provenance, and career intentions) were signing up for his course in these years in appreciable numbers (132 in 1802, 71 in 1803).[5] Knowing they would be struck by the novelty and singularity of his new views, he asked them to suspend their judgment until they had the chance to consider carefully these views and all the facts related to them. "It is on you yourselves," he urged, "that I call to pronounce on this great subject, when you have sufficiently examined and followed all the facts that bear on it."[6]

Lamarck's invitation to his students represented the flip side of how he anticipated his colleagues would treat the same ideas. He had already seen his colleagues at the Academy of Sciences turn a cold shoulder to the chemical and meteorological memoirs he presented to them. He concluded they were too invested in their own ideas and reputations to greet new ideas with an open mind.[7]

It may give us some appreciation for Lamarck's dissonant feelings in this period if we consider his two primary professional activities on the day he first offered a broad overview of his theory of organic change. The day was the seventeenth of May 1802 (the twenty-seventh of *floréal,* year ten, by the Republican calendar). Lamarck's course on invertebrate zoology was to meet for the first time that year, at half past noon. The lecture he gave on this occasion stands in retrospect as a milestone in the history of biology. Later that evening Lamarck met with his fellow professors/administrators for the museum's weekly administrative assembly. There he reported on the latest developments at the museum's menagerie, which he had been charged with overseeing since the previous July. His chief announcement was the happy news that the museum's female elephant had completely recovered from the worrisome digestive problems that had been ailing her. Ironically, while we know that Lamarck's colleagues greeted this news with satisfaction (and elected to send a letter of thanks to the veterinarian who helped cure the elephant), we know nothing about how Lamarck's students responded to the remarkable lecture he gave them earlier in the day. Also ironic is that Lamarck most likely said nothing to his fellow professors about what he had told his students earlier in the day. The weekly assemblies were devoted to handling the museum's business, not discussing scientific theory, and Lamarck in any case would not have expected his colleagues to welcome a new theory from him.[8]

Lamarck was keen, nonetheless, that the new views he had presented to his students *not be misrepresented*. He decided he should get his lecture into print quickly and that he should accompany it with additional comments so that he would be better understood. Before he knew it (as he later explained), he had a small book on his hands without ever having intended to write it. A mere two months after delivering the opening lecture of his course, he presented the museum with a copy of his new book, *Researches on the Organization of Living Bodies*. In addition to the original lecture, the book included a long second section on what he called "direct generation" (i.e., "spontaneous generation") and the effects of moving fluids on living bodies, plus "some considerations relative to man," an "appendix" on the meaning of the word *species*, and a section entitled, "Researches on the Nervous Fluid: Preliminary Considerations."[9]

Clearly Lamarck was not someone who deliberated for long before letting his ideas be known. That said, his *Recherches* of 1802 do not represent the whole sweep of his intellectual ambitions at the time. He had hoped to lay the groundwork for a new science he was naming "*biologie*," conceived as but one part of a three-part "terrestrial physics" also involving meteorology and "hydrogeology." His *Hydrogéologie* had already appeared, but he was putting his "*biologie*" aside, he explained, because of his other scientific responsibilities and his increasing ill health.[10] A month later he used the excuse of ill health again when he asked the museum to relieve him of the responsibility of overseeing the menagerie.[11]

Over the next decade and a half, poor health notwithstanding, Lamarck elaborated more fully his understanding of how the diversity of life arose. He did so most notably in his *Zoological Philosophy* of 1809 and in the 1815 introduction to his *Natural History of the Invertebrates*. The subtitle of his *Zoological Philosophy* is additional evidence of the breadth of his explanatory goals. He was offering *Considerations relative to the natural history of animals; to the*

*diversity of their organization and the faculties they derive from it; to the physi-
cal causes that maintain life in them and give rise to the movements they execute;
finally, to those [physical causes] that produce sensation in some [animals] and
intelligence in those that are endowed with it.*[12]

Here we confine our attention to the basic elements of Lamarck's thoughts
on organic mutability. The first glimpse Lamarck provided of his new thinking, in 1800, was his conclusion that the standard view of the relation between
animal structures and animal habits needed to be turned on his head. Habits,
he said, were responsible for the form of the body and its parts, not the other
way around. The webbed feet of ducks and geese, the curved claws of perch-
ing birds, the long legs and necks of shore birds—all these, he explained,
were not there because the species had been originally created that way but
were instead the consequence of long-maintained habits. By Lamarck's account, changes in environmental circumstances induced animals to adopt
new habits. These new habits led over time to changes in structure and the
acquisition of new faculties, "and little by little nature has arrived at the state
where we see her now."[13]

In the broader picture Lamarck set forth in 1802, the environmental influence on habits and structures took second place to the general tendency to
increased complexity. He saw the process beginning with the "direct generation" from nonliving matter of the very simplest forms of life. These forms
became increasingly complex as the result of the physical action of subtle
fluids (primarily caloric [heat] and electricity) coursing through them. Subtle
fluids and then ponderable fluids, carving out channels and organs in living forms, constituted the basic cause of the increasing complexity displayed
in the general series of animal classes. Departures from the general series
were the result of the influence of diverse environments. Environmental constraints were the reason why species, unlike the general "masses" of organization represented by the animal classes, could not be aligned in a single chain
of being. The species formed "lateral ramifications" around the masses. As
Lamarck explained, all that nature had required in bringing all the different
forms of life into existence were "direct generation," the constructive action of fluids in motion, an infinite number of diverse and favorable circumstances, and an immensity of time.[14]

Subsequently, in 1809, and again in 1815, Lamarck continued to present
a theory of organic change featuring two different factors. In his *Zoological
Philosophy* he wrote, "The state in which we now see all the animals is, on the
one hand, the product of the increasing composition of organization, which
tends to form a regular gradation, and, on the other hand, that of the influences of a multitude of very different circumstances that continually tend to

destroy the regularity in the gradation of the increasing composition of organization."[15]

Lamarck's two-part theory was in essence his explanation of why the diverse forms of life exhibited the broad pattern they did when one properly arranged them. In that regard explaining species change looks rather like a subsidiary project for Lamarck, but he knew he had to address it head on. Thus in 1803, having introduced his general theory to his students the previous year, he made the focus of his introductory lecture of 1803 "that great question of natural history, what is the *species* among living things?"[16]

Lamarck recognized that he was up against religious as well as scientific orthodoxy in his belief that species changed. He proceeded to argue that the presumed stability of species was only an appearance, a result of failing to appreciate the immense amount of time nature had at her disposal when it came to bringing about change. As for the belief that species were as old as nature and the result of an initial creation by "the supreme author of all things," Lamarck readily acknowledged that "nothing exists except by the wish of the supreme author of all things," but he proposed that it was only through observation that one could identify "the mode that [the supreme author] was pleased to follow in this regard."[17]

Lamarck was confident that his theory could handle even the most problematic cases. "Would one dare carry the spirit of system-building [*l'esprit de système*] so far," he asked,

> as to say that it is nature, by herself, that created this astonishing diversity of means, ruses, cunning, precautions, and patience of which the *industry* of animals offers us so many examples? Is not what we observe with the class of insects a thousand times greater than what is necessary to make us perceive that the limits of nature's power in no way allow her to produce so many marvels herself? And to force the most obstinate philosopher to recognize that here the will of the supreme author of all things has been necessary and has been alone sufficient to bring into existence so many admirable things?

Lamarck was not daunted. Nature, he allowed, had the faculty to produce all the marvelous examples of instincts and industry in animals—and *reason* as well. There was nothing irreligious about this view, he claimed, for it was in the supreme author's power to *will* that nature have this faculty. Furthermore, there was no reason to admire less the power of the "supreme cause of everything" if it pleased it to operate in this way rather than having to occupy itself with all the details of every particular creation as well as all the changes these creations would then undergo over time.[18]

The reader may be surprised at this point not only by the boldness of La-marck's theorizing but also that we have not yet mentioned the particular idea with which Lamarck's name is now most commonly associated—the idea of the inheritance of acquired characters. That idea was indeed an es-sential part of Lamarck's explanation of organic change (even if Lamarck did not use the phrase, which was yet to be coined). Interestingly enough, how-ever, this was not an idea that Lamarck identified as an intellectual advance of his own making. It was not, in and of itself, an example of Lamarck thinking or dreaming big. He portrayed it as a law of nature "so striking, so much at-tested by the facts, that there is no observer who has been unable to convince himself of its reality."[19]

Lamarck encapsulated it in his *Zoological Philosophy* in the form of two "laws":

First Law: In every animal that has not reached the end of its development, the more frequent and sustained use of any organ will strengthen this or-gan little by little, develop it, enlarge it, and give to it a power proportion-ate to the duration of its use; while the constant disuse of such an organ will insensibly weaken it, deteriorate it, progressively diminish its facul-ties, and finally cause it to disappear.

Second Law: All that nature has caused individuals to gain or lose by the influence of the circumstances to which their race has been exposed for a long time, and, consequently, by the influence of a predominant use or constant disuse of an organ or part, is conserved through generation in the new individuals descending from them, provided that these acquired changes are common to the two sexes or to those which have produced these new individuals.[20]

It was true that other thinkers in Lamarck's day believed in the inheritance of acquired characters (and despite what some modern biology texts suggest, Darwin believed in it too). One of Lamarck's contemporaries who endorsed the idea strongly was Frédéric Cuvier, Georges Cuvier's younger brother.[21] But none of these other authors had supposed as Lamarck did that the inheritance of acquired characters could proceed far enough as to produce new species. If Lamarck did not claim to have originated the concept of the inheritance of acquired characters, he did, however, claim to have been the first to recognize "the importance of this law and the light it sheds on the causes that have led to the astonishing diversity of animals." Recognizing the significance of this law was more important to him, he said, than the "classes, orders, many genera, and great quantity of species" that he had made known to science.[22]

Lamarck's willingness to go beyond where other naturalists feared to tread was manifested in another feature of his theorizing, his discussion of how an apelike creature might come to walk and talk and resemble a human. He introduced the subject in his *Researches* of 1802 and returned to it in 1809 in his *Zoological Philosophy*. "If man were only distinguished from animals by way of his organization, it would be easy to show that the characters used to form a separate family of man and his varieties are all a product of ancient changes in his actions and of habits he took up which have become particular to the individuals of his species."[23] Lamarck speculated that if some quadrumanous species, like the orangutan of Angola (i.e., the chimpanzee), had come down from the trees and acquired the habit of walking, it would, over time, have come to stand upright. And if it no longer used its teeth as weapons but only for chewing, its facial angle would have become broader, and its muzzle shorter. As its societies grew larger, its needs would also have expanded, leading its members to communicate by signs and eventually by articulate sounds. The creature would thus have evolved the ability to speak.

Lamarck's discussion included an explanation for the gap between the most advanced form of life and its closest competitors. The gap was produced and maintained, he explained, by the highest form of life driving its competitors away to environments unsuited to the development of the sorts of habits they would have needed in order to rise further. Thus one could not anticipate finding an apelike creature actually in the process of becoming human.

In emphasizing here the scope of Lamarck's vision, we have not reviewed all the details of his argument and the different kinds of evidence he cited in his support. That evidence included changes in animals under domestication, vestigial organs, the blindness of moles, the geographical variation of species, and more. However, his evidence was hardly conclusive, and persuading others of the validity of his views was not as easy as persuading himself. Nonetheless, his example was not without influence, and his ideas were not forgotten. It is hard to believe that the animus in Cuvier's eulogy of Lamarck was not stimulated in part by Cuvier's displeasure to find that at the beginning of the 1830s the idea of species mutability was attracting new advocates.[24] Cuvier had behaved as if he hoped that by treating Lamarck's ideas as not worthy of public consideration, yet ridiculing them in private, they might simply go away. Lamarck's two-factor theory of organic change may not have gained any major disciples, but at the time of Lamarck's death, the idea of species mutability seemed to be coming into play once again, and not everyone was prepared to grant Cuvier full authority on the matter.[25]

Cuvier, we admit, was not wrong in maintaining that Lamarck's theory-building in biology was akin to his unsuccessful theory-building in other fields. These various constructions indeed bore a family resemblance in style and scope. However, Lamarck's evolutionary views did not depend on his intellectual constructions in these other areas. Cuvier's effort to damn Lamarck's evolutionary thinking through guilt by association ultimately foundered on the fact that Lamarck, in his idea of the successive production of living things by natural processes, had arrived at a vision of truly fundamental importance for biology. Cuvier, the severe, precise, well-connected legislator of zoology, could not imagine that Lamarck, the loner and dreamer, was actually on to something big. Lamarck knew it, however. He told his students in 1806 that coming to understand how living bodies had come progressively into existence represented the unveiling of "the greatest of nature's secrets."[26]

In thinking of how Lamarck's example relates to the general theme of dreamers or visionaries in science, a number of features of his case call out for attention. Let us first consider Lamarck's age. Lamarck was already fifty-six years old in 1800 when he first offered a glimpse of his revolutionary new views on life. There is no evidence, furthermore, that he had waited for any great length of time before this happened. One tends to think of great conceptual breakthroughs in science being made primarily by younger scientists. However, a change in discipline, a move to a new field, might in some instances, at least to some degree, parallel the experience of a young scientist just beginning his or her career. Lamarck had not been in the field of invertebrate zoology for long—not quite six years—when he first began elaborating his ideas on organic evolution.

For that matter, the field of invertebrate zoology was itself new when Lamarck entered it. With their remarkably diverse forms and extraordinary faculties, the invertebrates taken as a whole represented virgin territory for a thinker seeking to understand nature's broad patterns and principles. There were naturalists who knew more about certain areas within the field than he did (Lamarck's assistant, Pierre Latreille, for one, was a greater authority in entomology). It was also the case that Georges Cuvier's comparative anatomical researches in the mid-1790s had served to pioneer a rethinking of the classification of that part of the animal kingdom that Linnaeus had left under the unsatisfactory rubric of "insects and worms." But the job of overseeing the whole field in effect became Lamarck's own, thanks to his position at the museum and to his assiduous efforts to make sense of these animals in all their diversity.

In addition, Lamarck's impetus to think big was boosted by certain specific dimensions of the field. When he was first appointed to his professorship at the museum, the only part of invertebrate zoology in which he had any expertise was in conchology—he was an ardent collector of shells. His detailed

Figure 1.2. Jean-Baptiste Lamarck. *Galerie des naturalistes* de J. Pizzetta (Paris: A. Hennuyer, 1893).

knowledge of shells gave him the confidence at the end of the century to speak to the increasingly urgent question of how to explain the differences between fossil and living shellfish forms. Doubting that living representatives of most of the fossil forms would one day be discovered, but unwilling to believe that a global catastrophe, as Cuvier was proposing for the vertebrates, had wiped out an entire fauna, Lamarck opted for a third alternative: the idea that the forms of the past had been transformed into the forms of the present.[27]

Was this the context in which Lamarck first came to the idea of species change? The evidence does not allow us to know for certain. It is clear that two additional considerations were crucial for the broader picture of organic change that Lamarck developed. One was his success in arranging the different classes of invertebrates in a single series of increasing complexity, which encouraged him to think that this series represented the very path nature had followed in bringing all the different forms of life into existence. Equally important for him were his consideration of what constituted the nature of life in the very simplest forms that exhibited it and his conclusion that the simplest forms of life could be produced from nonliving material by spontaneous or "direct" generation. The conceptual implications of this, as he told his students in 1802, were profound: "Once the difficult step [of admitting direct generation] is made, no important obstacle stands in the way of our being able to recognize the origin and order of the different productions of nature."[28]

We have additionally suggested that Lamarck's visionary thinking drew inspiration and energy from his teaching. His charge gave him the impetus to ponder the importance of the invertebrates for understanding the laws of nature. It furthermore gave him a ready-made forum for expounding his views and an audience that he hoped might be open-minded enough to appreciate him.[29]

The above factors, related specifically to Lamarck's professorship, helped fuel the burst of fruitful dreaming that characterized his biological thought at the beginning of the new century. Added to these were three more features of his career trajectory that we have already mentioned: (1) his long-established habit, for better or worse, of building broad, explanatory edifices; (2) his prior experience with his fellow scientists, which led him to believe there was nothing he could do to make them any more receptive to his broad theorizing; and (3) his declining health and advancing age, which made him suspect that his time and energy as a productive scientist were likely running out. These factors combined to urge him on in his visionary behavior.

Lamarck would not have thought of himself as a dreamer. He would have found that characterization too pejorative, too suggestive of a scientist whose views lacked a factual basis. But he did consider his own way of thinking to be special. He saw his intellectual aspirations and powers of thought rising above those of the average naturalist content to occupy himself with identifying and distinguishing species, and he strongly objected to the claim that the accumulation of more facts was science's only sure way forward. He saw himself bringing together "large" facts that had been recently discovered and using them "to discover unknown truths."[30]

Of Lamarck's conclusions, the most significant for biology was his bold new view that "the simplest of [nature's] productions have successively given rise to all the others"[31]—including all their diverse faculties, and including man. Lamarck set forth this view in no uncertain terms. He did not succeed in convincing his contemporaries of the whole of that vision. Indeed, few of them were attracted to it even in part. Yet the intellectual terrain could never be the same afterward. Lamarck's vision made things different, albeit not necessarily easier, for those who came after him. Charles Darwin wrote in the back of his copy of volume 1 of Lamarck's *Natural History of the Invertebrates*, "It is doubtful whether Lamarck has done more good by awakening subject, or harm by writing so much with so few facts."[32] However, just a few years after Darwin published the *Origin of Species* (and thereby reshaped the intellectual terrain again), Charles Lyell, in a letter to Darwin, expressed his belated admiration for the strength of Lamarck's arguments, given the period in which Lamarck was writing, and in particular for the way Lamarck had gone "the whole orang."[33] Lyell's comment is a good way to think of Lamarck as a biological visionary who did not flinch in his thinking about what a broad theory of organic change ultimately had to encompass—even if the theory Lamarck formulated differs in fundamental ways from present-day views of how organic change has taken place and is continuing to do so.

FURTHER READING

Burkhardt, Richard W., Jr. *The Spirit of System: Lamarck and Evolutionary Biology*. 1977. Reprint, Cambridge MA: Harvard University Press, 1995.
———. "Lamarck, Evolution, and the Inheritance of Acquired Characters." *Genetics* 194 (2013): 793–805.
Corsi, Pietro. *The Age of Lamarck: Evolutionary Theories in France, 1790–1830*. Berkeley: University of California Press, 1988. Revised edition, *Lamarck: Genèse et enjeux du transformisme, 1770–1830* (Paris: CNRS Éditions, 2001).
———. "Before Darwin: Transformist Concepts in European Natural History." *Journal of the History of Biology* 38 (2005): 67–83.
Lamarck, Jean-Baptiste. *Zoological Philosophy*. Chicago: University of Chicago Press, 1984.
Laurent, Goulven, ed. *Jean-Baptiste Lamarck, 1744–1829*. Paris: Éditions du CTHS, 1997.

NOTES

1. Georges Cuvier, "Éloge de M. Lamarck," in *Recueil des éloges historiques*, 3 vols. (Paris: Firmin Didot 1861), 3:179–210.

2. Cuvier, "Éloge de Lamarck," 3:200. All translations in the present paper are the author's own.

3. The primary biography of Lamarck remains Marcel Landrieu, *Lamarck, le fondateur du transformisme* (Paris: Société zoologique de France, 1909). For assessments of Lamarck by historians of science, see especially Richard W. Burkhardt Jr., *The Spirit of System: Lamarck and Evolutionary Biology* (1977; repr. Cambridge, MA: Harvard Uni-

versity Press, 1995); Pietro Corsi, *The Age of Lamarck: Evolutionary Theories in France, 1790–1830* (Berkeley: University of California Press, 1988); and Goulven Laurent, ed., *Jean-Baptiste Lamarck, 1744–1829* (Paris: Editions du CTHS, 1997). Most of Lamarck's theoretical writings, plus additional bibliographical information on Lamarck, are to be found on the invaluable website, "Jean-Baptiste Lamarck: Works and Heritage," at www.lamarckcnrs.fr.

4. Jean-Baptiste Lamarck, "Discours d'ouverture pour le course de 1816," in *Inédits de Lamarck*, ed. Max Vachon, Georges Rousseau, and Yves Laissus (Paris: Masson, 1972), 28.

5. Max Vachon, "Lamarck Professeur," in *Lamarck et son temps: Lamarck et notre temps* (Paris: J. Vrin, 1981), 248. Pietro Corsi and others have undertaken to identify all the students inscribed on the registers for Lamarck's courses. For the state of this project, see "Lamarck," www.lamarckcnrs.fr.

6. Jean-Baptiste Lamarck, "Discours d'ouverture d'un cours de zoologie, prononce en prairial an XI . . . sur la question, qu'est-ce que l'espèce parmi les corps vivans?," in *Discours d'ouverture des cours de zoologie donnés dans le Muséum d'Histoire Naturelle (AN VIII, AN X, AN XI et 1806)*, ed. A. Giard (Paris, 1907), 92. Phrasing this slightly differently in 1806, Lamarck told his students that, insofar as they were beginners in the study of nature, they should not let themselves be swayed by authorities one way or the other with respect to his views (121).

7. Burkhardt, *Spirit of System*, 40–45.

8. Meeting of 27 *floréal an* 10, "Procès-Verbaux des Assemblées des professeurs," Archives Nationales de France, AJ/15/103, 57. Hereafter this series is cited as AN, with the appropriate document reference.

9. Jean-Baptiste Lamarck, *Recherches sur l'organisation des corps vivans* (Paris: Maillard, 1802).

10. Pietro Corsi suggests that the reason Lamarck shelved his *"biologie"* was not so much his ill health but instead the increasingly conservative political climate that made Lamarck's materialistic and atheistic tendencies more dangerous for him. See Pietro Corsi, *"Biologie,"* in *Lamarck, philosophe de la nature*, ed. Pietro Corsi, Jean Gayon, Gabriel Gohau, and Stéphane Tirard (Paris: Presses Universitaires de France), 37–64.

11. His request was made on 5 *fructidor an* 12; see AN, AJ/15/103, 128.

12. Jean-Baptiste Lamarck, *Philosophie zoologique*, 2 vols. (Paris: Dentu, 1809); *Histoire naturelle des animaux sans vertèbres*, 7 vols. (Paris: Déterville, 1815–1822).

13. Jean-Baptiste Lamarck, *Système des animaux sans vertèbres* (Paris: Déterville 1801), 15.

14. Lamarck, *Recherches*.

15. Lamarck, *Philosophie zoologique*, 1:221.

16. Lamarck, "Discours d'ouverture, an XI," in *Discours d'ouverture des cours de zoologie*, 85–105.

17. Lamarck, "Discours d'ouverture, an XI," 94–95.

18. Lamarck, "Discours d'ouverture, an XI," 100–101.

19. Lamarck, *Histoire naturelle*, 1:200. For more on Lamarck and this concept, see Richard W. Burkhardt Jr., "Lamarck, Evolution, and the Inheritance of Acquired Characters," *Genetics* 194 (2013): 793–805.

20. Lamarck, *Philosophie zoologique*, 1:235.

21. See Richard W. Burkhardt Jr., "Lamarck, Cuvier, and Darwin on Animal Behavior and Acquired Characters," in *Transformations of Lamarckism: From Subtle Fluids to Molecular Biology*, ed. Eva Jablonka and Snait Gissis (Cambridge, MA: MIT Press, 2011), 33–44.

22. Lamarck, *Histoire naturelle*, 1:191.

23. Lamarck, *Philosophie zoologique*, 1:349–57.

24. See Pietro Corsi, "Before Darwin: Transformist Concepts in European Natural History," *Journal of the History of Biology* 38 (2005): 67–83.

25. See Burkhardt, *Spirit of System*, 199–200.

26. Lamarck, "Discours d'ouverture de 1806," in *Discours d'ouverture des cours de zoologie*, 120.

27. See Burkhardt, *Spirit of System*, chap. 5.

28. Lamarck, *Recherches*, 121–22.

29. In publishing his *Philosophie zoologique*, Lamarck explained that it was his experience in teaching that had shown him how useful such a book would be for zoology at its present stage of development (i).

30. Lamarck, "Avertissement," in *Recherches*, iv.

31. Lamarck, *Recherches*, 38.

32. Darwin's annotation appears on the next to last page of his copy of volume 1 of the second edition of Lamarck's *Histoire Naturelle*. The copy is in the University Library, Cambridge.

33. Katherine M. Lyell, ed., *Life, Letters and Journals of Sir Charles Lyell, Bart* (London: John Murray, 1881), 2:365.

ROBERT J. RICHARDS

ERNST HAECKEL

A DREAM TRANSFORMED

2

Ernst Haeckel (1834–1919) was Darwin's foremost champion, not only in Germany but throughout the world.[1] More people at the turn of the century learned of evolutionary theory from his pen than from any other source, including Darwin's own writings. His book *Die Welträthsel* (Riddles of the world, 1899) sold more than four hundred thousand copies before the First World War, and was translated into most of the known and many of the unknown languages of the world, including Esperanto. The great historian of biology Erik Nordenskiöld, writing in the first decades of the twentieth century, judged that Haeckel's *Natürliche Schöpfungsgeschichte* (Natural history of creation, 1868), which went through twelve editions during his lifetime, was the world's introduction to Darwinian evolutionary theory.[2]

In addition to being an extraordinary research scientist, Haeckel was an artist of considerable accomplishment. He filled his twenty or so technical monographs and several popular books with his own illustrations, which furnished much of the persuasive power of his monographs. Darwin's *Origin of Species* included only one generic line drawing. Haeckel's illustrations, however, would ignite a controversy among professional biologists and religious objectors that still smolders even today. If one sought the source of the longstanding enmity between evolutionary thinkers and the religiously orthodox, one could start and finish with Ernst Haeckel. Unlike Darwin, he relentlessly attacked what he took as religious superstition, especially when religion became mixed with biology. His attitude was reciprocated by the Church.

Haeckel's heterodox reputation did not, however, dampen the enthusiasm of his students. He drew to his small university outpost in Jena, in the center of the German lands, some of the best biologists of the next generation, including the brothers Oskar (1849–1922) and Richard Hertwig (1850–1937), Wilhelm Roux (1850–1924), and Hans Driesch (1867–1941), all of whom made their marks by the turn of the century.

Haeckel's legacy is palpable even today. He introduced into biology many concepts that remain viable, including the idea that the nucleus of the cell contains the hereditary material, as well as the concepts of phylogeny, ontogeny, and ecology. He was among the first to use the graphic device of the evolutionary tree and made it a fixture of evolutionary explanation. That device,

35

Figure 2.1. Ernst Haeckel (*seated*) and his assistant Nikolai Miklucho on the way to the Canary Islands in 1866. Haeckel had just visited Darwin in the village of Downe. Courtesy of Ernst-Haeckel-Haus, Jena.

along with Haeckel's emphasis on phylogenetic development, shifted the orientation of evolutionary studies dramatically, especially as newly discovered species had to be fitted into proliferating branches of the tree of life. It is likely that more newly discovered organisms have his name attached than that of any other biologist.[3] Haeckel popularized the idea of the missing link between man and the apes, and his protégé Eugene Dubois (1858–1940) found its remains in Java—that is, the first fossils of *Homo erectus*.

We owe to Haeckel the currency of the biogenetic law that ontogeny recapitulates phylogeny. The principle states that the embryo goes through the same morphological stages as the phylum went through in its evolutionary descent. The human embryo, for instance, begins life as a one-celled creature, just as we suppose life began in the sea as a single reproducing cell; then the embryo takes on the form of an invertebrate, then something like a fish, then a primitive mammal, a primate, and finally a distinctive human form. Though Darwin quite early on had embraced the recapitulation hypothesis, he was emboldened by the confirmation his friend offered. Haeckel's theory of the mind-brain relationship had a measured impact on Darwin's own views about the evolution of human beings.

Haeckel's innovations in evolutionary theory completely altered the texture of the discipline, from his use of tree diagrams to his introduction of novel concepts and his elevation of the principle of recapitulation as both crucial evidence for evolution and as a proximate cause of phylogenetic continuity. He laid the empirical foundations of evolutionary transitions and introduced experimental procedures to secure those foundations (see below). But if one sought the lasting impact of this revolutionary thinker, it would be found in a more abstract contribution that has had extensive concrete effects on cultural life up to the present time: he argued with telling consequence that human beings were completely natural animals. Darwin only suggested this and usually side-stepped the issue. Haeckel made it his central argument, especially as against theologically minded scientists and spiritually engrossed philosophers. Haeckel almost single-handedly excavated one of the deepest fissures of our contemporary intellectual life. And all because he had a dream.

HAECKEL'S DREAM

While working on his medical dissertation, the young Haeckel had a dream. It was of a "true German child of the forest, with blue eyes and blond hair and a lively natural intelligence, a clear understanding, and a budding imagination."[4] That dream was embodied in Anna Sethe (1835–1864), his first cousin. Haeckel had met her only occasionally at family gatherings; but in 1857, when she and her mother moved to the outskirts of Berlin, where Haeckel was working on his dissertation, they became inseparable. Their growing love turned into an engagement, though one with an indefinite promise of marriage, delayed until Haeckel had the security of a job. In letters that recounted his waking dream, the poetically inclined medical student would recall to Anna recent excursions, for example, one in which they made their way through the forest to a mountain stream, where they lay down on a mossy bank:

And your sighing breath, your warm cheek on mine announced to me at every blissful second that sweet unspeakable happiness that I held in my arms, close and sure, so that I might never lose it. Then we lay on my good old plaid, placed on the natural bed of the forest, upholstered with dry beech leaves . . . O, Anna, those were moments I will never, never forget, moments of the greatest human happiness . . . One forgets heaven and earth, past and future, and lives purely and completely in the present. Here Faust himself could exclaim, "Tarry a while, you are so beautiful," so he might secure the moment which sadly only too quickly dissolves.[5]

Though Haeckel desperately wanted to marry his cousin, the practice of medicine would not be the financial route, since he could not abide the thought of dealing with patients. He received an invitation for habilitation research from Carl Gegenbaur (1826–1903) at Jena, who had been an acquaintance at medical school in Würzberg. He and Gegenbaur were to have traveled together to southern Italy, but when his mentor was advanced to ordinarius professor (roughly full professor in the American system), obligations kept the older naturalist at the university. Haeckel decided to travel alone. He had no secure idea for a research subject; he simply hoped some marine organism would capture his imagination. But he also saw in this travel an opportunity to indulge his growing passion for artistic development—and his talent as a painter had begun to bloom. Thus, the Italian journey would also be one of *Bildung*, of cultural and artistic formation.

At the end of January 1859, Haeckel left Berlin, traveling down the Italian boot, lingering in Florence and Rome, where he took in the museums and art galleries, but finally arriving in Naples at the end of March to begin serious work on his research. In the morning, after a swim in the bay, he would inspect the catches the fishermen brought ashore from this marine Eldorado. But in that wealth he could not find his way, and grew increasingly unhappy, a tale of frustration that he detailed in his many letters to Anna. In June, no longer able to stomach the city, he packed up his sketch pads and paints, and escaped to the island of Ischia, just across the bay. There he fell in with another German, the poet and painter Hermann Allmers (1821–1902), who would become his lifelong friend. They tramped through the island, visiting the ruins of past civilizations while glorying in the natural beauty of the landscape. They indulged each other's interests, Allmers inquiring about botany and marine biology, while Haeckel gave himself over to "the misty distances of a dreamy poetry." Haeckel wrote Anna that his new friend "struck a responsive chord in me, has awakened feeling and effort that I believed had already completely died; and in a sense, he has given me back to myself."[6] In August,

Haeckel and his new friend set sail for Capri, where they would indulge in the bohemian life of hiking through the countryside, bathing in small lakes, and painting. Haeckel felt the temptation to abandon his academic pursuits and to give himself over completely to his artistic desires—except that he kept returning in imagination to that dream, a life with Anna, which steeled him to achieve his professional goal.

The urge to abandon an academic career was born, partly at least, out of frustration in finding the right kind of organism to study amid the mountains of beautiful and astounding animals that fishermen had brought up from the waters of the bay: siphonophores, petropods, heteropods, medusae, sponges, and more—some whole families of creatures never before classified or described. Haeckel finally hit on one group of animals, virtually infinite in profusion yet remaining almost unknown to the biologist—the radiolarians. These one-celled animals are the size of a pinhead and secrete an exoskeleton of silica. About 20% of the muck lying at the bottom of the oceans consists of their skeletons. Their species are distinguished by the unusual geometries the skeletons exhibit. Haeckel became entranced by their beauty, and in the illustrations for his prize-winning monograph, their images grow to the size of the sun compared to the Earth of these very small creatures. When Darwin received the gift of Haeckel's two-volume *Die Radiolarien*, he declared them to be "the most magnificent works which I have ever seen."[7]

Haeckel returned to Berlin in April 1860 and began the writing of his habilitation, the German equivalent of a second thesis. He finished the work in 1861 and had it rendered into Latin, still the required language for this kind of academic exercise. He continued his study of the radiolarians, ultimately turning his research into the book that Darwin so much admired, a giant two volumes on the classification and description of the many species, genera, families, and orders of those organisms he had newly discovered and those few already known. The first volume of 570 pages was accompanied by an atlas of thirty-five copper-etched plates, some brilliantly colored to show the inner cellular capsule surrounded by the skeletal frame. Haeckel said he attempted a natural system, rather than a Linnaean artificial system, because of the extraordinary book he had read while preparing his specimens, *Über die Entstehung der Arten im Thier- und Pflanzen-Reich durch natürliche Züchtung* by the English naturalist Charles Darwin.[8] What especially excited the young disciple was the opportunity the translator, Georg Bronn, had brought to the fore. Bronn had added an appendix to his translation of the *Origin of Species*, in which he argued that the Englishman had offered a theory merely of the possibility of descent. Haeckel thought his own research demonstrated the reality of descent, at least in the case of the radiolarians.

Haeckel received an invitation from Gegenbaur to become his assistant at Jena and to serve there as Privatdozent (a lecturer paid by students) and he readily accepted. During his teaching and research duties, he worked feverishly on his radiolarian book. With its publication in spring of 1862, he was offered an advancement to the permanent position of extraordinarius professor in the medical faculty. He immediately wrote Anna to boast of his elevation to the "Archducal-Saxonish-Weimarish-Colburgish-Altenburgish-Meiningenish Extraordinary Professor."[9] The advancement solidified his financial situation, though an added contribution from his father did help. Now the dream could become reality—a life with the "loveliest, and purest maiden soul." Anna promised "everything that science cannot give."[10] Haeckel and Anna were married in Berlin on 18 August 1862.

From the time Haeckel first read the *Origin of Species*, his devotion to the Darwinian theory almost rivaled his affection for Anna— so much so that she had taken to referring to him as "her German Darwin-husband."[11] Through his publications and lecturing, Haeckel's support for Darwin became better known to the public, and he was invited to lecture on Darwinian theory at the first plenary session of the Society of German Natural Scientists and Physicians, which met in the Prussian city of Stettin during September 1863.[12] As judged by the reporter from the *Stettiner Zeitung*, Haeckel's "exciting lecture" met with "a huge applause."[13]

Anna had accompanied her husband to Stettin and undoubtedly gloried in the adulation he received. When they returned to Jena, he plunged into further study and application of Darwinian theory. He wrote Allmers just before Christmas that "I am now convinced that a great future lies before this theory and that it will slowly but surely loose us from the bonds of a great and far-reaching prejudice. For this reason, I shall dedicate my whole life and efforts to it."[14]

In his pursuit of Darwinian theory, Haeckel expected to continue receiving the loving support of Anna. But shortly after writing Allmers, in late January of 1864, Anna suffered a severe attack of pleurisy, which lasted through the first part of February. She recovered, but then in mid-February she again became ill with severe abdominal pains. In the evening of February fifteenth, the pains became acute. The next day, Haeckel's thirtieth birthday and the day he received word that his radiolarian book had been awarded a significant prize, his beloved wife of eighteen months died. Haeckel became mad with grief, falling unconscious and remaining in bed for some eight days in partial delirium. The dream of this "German Darwin-husband" had evanesced in an acrid vapor. His parents thought he might commit suicide and arranged to have him travel to Nice to attempt some kind of recovery. From the southern coast of France, he wrote them a searing letter:

> The last eight days have passed painfully. The Mediterranean, which I so love, has effected at least a part of the healing cure for which I hoped. I have become much quieter and begin to find myself in an unchanging pain, though I don't know how I shall bear it in the long run. . . . You conclude . . . that man is intended for a higher, godlike development, while I hold that from so deficient and contradictory a creation as man, a personal progressive development after death is not probable; more likely is a progressive development of the species on the whole, as Darwinian theory already has proposed it. . . . Mephisto has it right: "Everything that arises and has value comes to nothing."[15]

Thereafter on his birthday, Haeckel harbored thoughts of suicide. In 1899, he wrote a dear friend, an Anna reincarnated, "Thursday, 16 February, is my sixty-fifth birthday, for me the saddest anniversary of the year, since on this same day in 1864, I lost my most beloved and irreplaceable first wife. On this sad day, I am lost."[16] After thirty-five years, the memory of great happiness oppressed by great sorrow had remained vivid.

While recovering along the shore of the Mediterranean, Haeckel chanced to notice in a tidal pool a medusa—a jellyfish—the tendrils of which reminded him of Anna's golden braids. He named it in memory of his wife, *Mitrocoma Annae*—Anna's headband. Next to the illustration that would later appear in *Das System der Medusen* (1879), Haeckel wrote, "I name this species, the princess of the Eucopiden, as a memorial to my unforgettable dear wife, Anna Sethe. If I have succeeded, during my earthly pilgrimage in accomplishing something for natural science and humanity, I owe the greatest part to the ennobling influence of this gifted wife, who was torn from me through sudden death in 1864."[17] Haeckel wrote this while married to his apparently forgettable second wife, Agnes. The living dream of Anna and happiness were supplanted by Darwinian theory and a strident determination to combat biological orthodoxy and religious superstition—at least the pattern of evidence, which I relate in the remaining part of this essay, suggests this transformation.

Haeckel returned from France in the summer of 1864 to a home empty of the joy it once knew. He threw himself into his classes, brief diversionary travel, and then began, demonlike, to compose a treatise that would be the definitive application of evolutionary theory. After eighteen-hour days over the period of a year, he delivered to his publisher a large, two-volume work of more than one thousand pages, with a concluding forest of plates depicting tree diagrams of systematic relationships. His *Generelle Morphologie der Organismen* (1866) sought to explain those relationships through the devices that Darwin had advanced: namely, natural selection and the inheritance of acquired characteristics. Depending on the traits and the situation of the organism, one of these devices might be emphasized more than the other.[18] Through the course of Haeckel's career, he tilted to the Lamarckian notion, but kept natural selection at the ready. Haeckel, though, emphasized another explanatory principle—namely, the principle of recapitulation.[19] He gave an extended definition of the principle in his *Generelle Morphologie*: "The organic individual . . . repeats during the quick and short course of its individual development the most important of those changes in form that its ancestors had gone through during the slow and long course of their paleontological development according to the laws of inheritance and adaptation."[20] As he more succinctly phrased it, "Ontogeny is nothing other than a short recapitulation of phylogeny."[21] For Haeckel, the principle was both evidence for evolutionary transmission as well as a causal explanation for the early developmental features of the embryo. He did not assume that recapitulation would be perfect. The more the embryo was subject to environmental forces—for example, when the larvae of insects or marine organisms were exposed to the impact of natural selection—the greater the differences between ancestor and embryo. Thus Haeckel modified the principle of recapitulation with the corollaries of *palingenesis*—when the recapitulation was very close—and *cenogenesis*—when the differences were more extreme.[22]

The final two chapters of Haeckel's magnum opus transformed the dream that Anna had realized in living form into something else—a conception of unity in the depths of nature. Chapter 29 was prefaced by Goethe's poem *Prometheus*, in which Prometheus renounced Zeus and chose to cast his lot with suffering humanity. The chapter declared that the only sure path to truth for human beings was through the rational and empirical methods of science, a science that had no room for an anthropomorphic God, an "imaginary, gaseous substance" (*Vorstellungen von gasförmigen Materien*).[23] The last chapter asserted the metaphysical perspective that lay behind the previous thousand pages of technical discussion of evolutionary systematics and morphology. Haeckel advanced a monistic conception of nature as understood by Goethe: "the unity of God with the whole of nature"—it was Spinoza's "Deus sive Natura." This kind of monism postulated mind and matter as properties of an underlying substance that was neither. In Haeckel's naturalistic theology, "God is the almighty; he is the sole creator, the cause of all things; in other words: God is the universal causal law."[24] Haeckel's version of monistic metaphysics is exactly of the sort that branded Spinoza with the insignia of atheism. But it also had a different kind of religious meaning. Since the conservation laws of physics held that force and matter could not be destroyed, Anna might be yet preserved in nature. She would not die forever.

The dream of a life with Anna had been transformed into one of a different kind: the pursuit of nature through the scientific instruments of evolutionary theory. Of course, Haeckel had already begun the pursuit while Anna was still a warm presence. But now he undertook it as a religious mission. He informed Darwin of this new urgency. In a letter of 7 July 1864, he wrote of the great tragedy that had "hardened me against the blame as well as the praise of men, so that I am completely untouched by external influence of any sort, and only have one goal in life, namely, to work for your descent theory, to support it, and perfect it."[25] In a subsequent letter in October, he made clear to Darwin that he sought to recover in his work the love that he had lost: "Now in my isolation, which since the death of my wife is so lonesome, this engrossing work is a great consolation, and I toil at it with so great an enthusiasm as if my Anna herself drove me to its completion and had left this task as a memorial."[26]

After the exhausting work on his *Generelle Morphologie*, Haeckel and his assistants escaped to the Canary Islands for research. But on the way, they passed through London, and then Haeckel took a train to the village of Downe to meet his master, Charles Darwin. It would be the first of three visits, interspersed with a flow of letters between the two naturalists. Upon his return to Jena in spring of 1867, he renewed an acquaintance with the daughter of a former professor at the university, Agnes Huschke (1842–1915). In desperate hope and daring haste, he asked Agnes to marry him. Almost from the beginning, it became clear to him that this young woman could not replace the unforgettable Anna. Agnes hated the polemics in which her husband engaged, and she became afflicted with the nineteenth-century disease of neurasthenia. Haeckel's marriage was successful only in the biological sense; they had three children.

HAECKEL'S SUBSEQUENT RESEARCH: THE DARWINIAN DREAM REALIZED

Through the last quarter of the nineteenth century and into the first decades of the twentieth, Haeckel pursued his dream not only of bringing the realms of nature under the aegis of Darwinian theory but also of exorcising the kind of superstitious belief that had haunted so much of natural science. Though Jena remained his home base for the rest of his career, almost every year he was away conducting research or lecturing, returning for several months to meet his teaching duties and settle in with his family. His first major research venture after Anna's death took him to the Canary Islands and Spain during the fall and winter of 1866–1867. This research resulted in a prize-winning work on siphonophores and three volumes on sponges.[27] In the spring of 1873, he traveled to Egypt and down to the Red Sea, which yielded a beautiful study of the region's corals, augmented by several of his paintings of desert scenes.[28] He sailed with the "golden brothers" Hertwig in the spring of 1875 to Sardinia and Corsica, where he collected numerous specimens of medusa; and in February 1877 he journeyed to Corfu for more research on jellyfish. During the late summer of 1878, he spent several weeks off the coasts of Normandy and the Isle of Jersey, adding to the materials on radiolarians and medusae. These several excursions resulted in two beautifully illustrated volumes on medusae, his *System der Medusen*.[29] In October of 1880, Haeckel left Jena with sixteen large trunks to begin his first journey to the far east (a second to Sumatra and Java would occur in 1900). In mid-November he reached Colombo, the capital of Ceylon (now Sri Lanka), and headed down the coast, where he discovered ever more species of radiolarians and medusae. When he returned to Jena in mid-April of 1881, he began work on a new and extraordinary project: the description of material from the *Challenger* expedition.

The *Challenger* was a British research ship that spent three and a half years (1873–1876) dredging the Atlantic and Pacific oceans, pulling up all manner of marine life and testing the chemical composition of the seas. The materials were sent to experts all over the world to be described and classified, and the subsequent results appeared in fifty large folio volumes. Because of his extraordinary reputation for research, Haeckel was asked to work on radiolarians, medusae, siphonophores, and sponges. His study of radiolarians, which included much of his own material, described more than four thousand species in two large folios of 1803 pages; a third volume of 140 plates completed the study. Those volumes appeared in 1887. His research on medusae produced a preliminary volume in 1881, and two *Challenger* tomes the next year. The years 1888 and 1889 saw the remaining *Challenger* volumes on siphonophores and sponges.[30] The commission Haeckel received to describe the *Challenger* materials testifies to his standing as a research scientist in the eyes of his colleagues, as does the continued support given by such stalwarts as Thomas Henry Huxley, August Weisman, Hermann von Helmholtz, and Darwin himself.

Through his many publications, Haeckel contended that the specific areas of marine biological systematics could only be understood in terms of evolutionary theory. In the *Origin of Species*, Darwin had composed "one long argument." Haeckel's research solidified the long argument, planting it firmly on empirical ground of an extent unmatched by any other biologist of the period. But Haeckel not only drew upon his systematic observations of marine life and his ability to conceive of that life as branches of a living evolutionary tree, but he accomplished what virtually no other evolutionary scientist of the nineteenth century was able to do—namely, he introduced experimental procedures into his discipline.

During his stay in the Canary Islands in fall and winter of 1866–1867, just after his visit with Darwin, Haeckel performed a series of experiments on siphonophores (jellyfish-like organisms in the phylum of Cnidaria) that aimed to demonstrate species transitions. He undertook three kinds of experiments. First, he followed the larval development of species from ten different genera of siphonophores, showing their virtual identity at early developmental stages. In light of the biogenetic law, this identity suggested a common ancestral form. Second, during larval development, he altered environmental conditions (e.g., water temperature, water movement, amount of light, salinity, etc.). The effects were surprising: the manipulations not only showed the susceptibility of embryos to changed conditions (thus supporting the inheritance of acquired characters), but the alterations revealed morphologically distant species forms hidden beneath those of the particular species on which he was experimenting. The most significant set of experiments anticipated the work of his two students, Wilhelm Roux and Hans Driesch, in developmental mechanics some twenty years later.

In this third set of experiments, Haeckel used a fine needle to separate the cells of two-day-old embryos into two, three, or four groups of cells, and then watched the further development of these groups into separate embryos. In six cases, development got to the sixth day; in three of those, development proceeded to the eighth day; two reached the tenth day; and one went to day fifteen. The embryos were morphologically complete, though smaller than normal embryos. Though he didn't explicitly draw this conclusion, his experiments nonetheless showed that cells of embryos at early cleavage stages were totipotent—they had the capacity to develop into all the parts of the organism. Only in the late 1880s, did Roux and Driesch produce comparable experiments, which led to the enterprise of developmental mechanics.[31]

Haeckel's empirical work in establishing evolutionary theory—and the fame it brought—served as a tactical advantage: it gave him the authority and leverage for launching bristling attacks against the ingressions of anthropomorphic religion into science. Haeckel's religion was that of Spinoza and Goethe, that is, *Deus sive Natura.* In October of 1892, he gave a lecture explicating his brand of religion: *Der Monismus als Band zwischen Religion und*

Wissenschaft (Monism as the bond between religion and science). The lecture and subsequent publication, which would reach a seventeenth edition just before Haeckel's death in 1919, became the foundation for his even more successful *Die Welträthsel* seven years later. Both argued for a universe of atoms swimming through waves of ether and governed by attractive and repulsive forces. From the inorganic up through levels of biological organization, no unbridgeable barriers arose blocking evolutionary transformations. In this monistic universe, the properties of matter and mind ran together as united in an underlying substance, even down to the simplest atom. One might thus speak of the human soul and the central nervous system in one breath. A reviewer of the English translation of *Die Welträthsel* for the *New York Times* summed up the book as follows: "One of the objects of Dr. Haeckel—it would not be unfair to say the chief object—is to prove that the immortality of the human soul and the existence of a Creator, designer, and ruler of the universe are simply impossible."[32] While Haeckel's monistic system provided a philosophical foundation for then-current physics and biology and a counter to anthropomorphic religion, it had a more personal support.

CONCLUSION: LIFE IN NATURE

In the *System der Medusen,* in which he pictured a yellow-tinted *Mitrocoma Anna*, the medusa that reminded him of his first wife, Haeckel included another newly discovered medusa sent to him from South Africa by his cousin Wilhelm Bleek, also a cousin of Anna Sethe. He named it *Desmonema Annasethe*. It came to him in a soldered tin preserved in spirits of wine. By the time it arrived, it was a mess, crumpled and denatured of most of its color. Haeckel represented it in monochromatic brown—it's still preserved in the natural history museum in Jena as a ghostly white, mostly a translucent tangle lying at the bottom of a glass container. That poor creature underwent a transformation during the succeeding years of Haeckel's career, just as Anna had in his memory, becoming ever more lovely over time. In 1899 through 1904, Haeckel began issuing fascicles of an art book he was composing, his *Kunstformen der Natur* (Art-forms of nature, 1904).[33] He employed many of the illustrations from his monographs, now represented in new settings and with vibrant, lithographic colors. The wan *Mitrocoma Anna* found no place in the art book, while the original illustration of *Desmonema Annasethe* had been dramatically and beautifully transformed for a new appearance (fig. 2): "The species name of this extraordinary Discomedusa—one of the loveliest and most interesting of all the medusa—immortalizes [*verewigt*] the memory of Anna Sethe, the highly gifted, extremely sensitive wife of the

Figure 2.2. *Discomedusa Desmonema Annasethe*. From Haeckel's *Kunstformen der Natur* (1904).

author of this work, to whom he owes the happiest years of his life."[34] In his artistic imagination, the original dream of a German girl had been magically altered into a creature still living in the seas. Love had fled and hid her face among sea-creatures. Nature had not proved completely foreign to human aspiration and hope.

FURTHER READING

Bowler, Peter. *Evolution: The History of an Idea*. Berkeley: University of California Press, 1989.

Hopwood, Nick. *Haeckel's Embryos: Images, Evolution, and Fraud*. Chicago: University of Chicago Press, 2015.

Nyhart, Lynn. *Biology Takes Form: Animal Morphology and the German Universities, 1899–1900*. Chicago: University of Chicago Press, 1995.

Richards, Robert J. *Darwin and the Emergence of Evolutionary Theories of Mind and Behavior*. Chicago: University of Chicago Press, 1987.

———. *The Tragic Sense of Life: Ernst Haeckel and the Struggle over Evolutionary Thought*. Chicago: University of Chicago Press, 2008.

Ruse, Michael, ed. *The Darwin Encyclopedia*. Cambridge: Cambridge University Press, 2013.

NOTES

1. This essay is based on my book *The Tragic Sense of Life: Ernst Haeckel and the Struggle over Evolutionary Thought* (Chicago: University of Chicago Press, 2008).

2. Erik Nordenskiöld, *The History of Biology: A Survey*, trans. Leonard Eyte, 2nd ed. (1920–24; repr., New York: Tudor, 1935), 515.

3. For example, of the 684 registered radiolarian species discovered from before Haeckel's time to the present, Haeckel identified more than 22%. From the nineteenth century to the present, Haeckel described more of the recognized genera in the subclass Calcaronea (calcinated sponges) than any other researcher. His discoveries ranged over many classes of organisms.

4. Haeckel to a friend (14 September 1858), in *Himmelhoch Jauchzend: Erinnerungen und Briefe der Liebe*, ed. Heinrich Schmidt (Dresden: Carl Reissner, 1927), 67.

5. Haeckel to Anna Sethe (26 September 1858), in Schmidt, *Himmelhoch Jauchzend*, 72–73.

6. Haeckel to Anna Sethe (25 June 1859), in Schmidt, *Himmelhoch Jauchzend*, 69.

7. Charles Darwin to Haeckel (3 March 1864), in *The Correspondence of Charles Darwin*, vol. 12, *1864*, ed. Frederick Burkhardt et al. (Cambridge: Cambridge University Press, 2001), p. 61.

8. Charles Darwin, *Über die Entstehung der Arten im Thier- und Pflanzen Reich durch natürliche Züchtung; oder Erhaltung der vervollkommnesten Rassen in Kampfe um's Daseyn*, trans. Georg Bronn (Stuttgart: Schweizerbart'sche, 1860).

9. Haeckel to Anna Sethe (17 June 1862), in Schmidt, *Himmelhoch Jauchzend*, 281.

10. Haeckel to Anna Sethe (7 June 1861), in Schmidt, *Himmelhoch Jauchzend*, 187.

11. Haeckel to Charles Darwin (10 August 1864), in *The Correspondence of Charles Darwin*, 24 vols. to date (Cambridge: Cambridge University Press, 1985–), 12: 298–300.

12. Ernst Haeckel, "Ueber die Entwickelungstheorie Darwins," in *Amtliche Bericht über die acht und dreißigste Versammlung Deutscher Naturforscher und Ärtze in Stettin* (Stettin: Hessenland's Buchdruckerei, 1864), 17–30.

13. *Stettiner Zeitung*, no. 439 (20 September 1863).

14. Haeckel to Allmers (15 December 1863), in *Ernst Haeckel: Sein Leben, Denken und Wirken*, ed. Victor Franz, 2 vols. (Jena: Wilhelm Gronau und W. Agricola, 1943–44), 2:36.

15. Haeckel to his parents (21 March 1864), in Schmidt, *Himmelhoch Jauchzend*, 318–19.

16. Haeckel to Frieda von Uslar-Gleichen (14 February 1899), in *Das ungelöste Welträtsel: Frida von Uslar-Gleichen und Ernst Haeckel. Briefe und Tagebücher 1898–1900*, ed. Norbert Elsner, 3 vols. (Berlin: Wallstein, 2000), 1:128.

17. Ernst Haeckel, *Das System der Medusen*, 2 vols. (Jena: Gustav Fischer, 1879), 1:526–27.

18. Darwin, it is well recognized, held a belief in the inheritance of acquired characteristics from the beginning of his career to the end. In *On the Origin of Species* (London: Murray, 1859), he affirms this quasi-Lamarckian theory (134): "I think there can be little doubt that use in our domestic animals strengthens and enlarges certain parts, and disuse diminishes them; and that such modifications are inherited."

19. Bowler states that "recapitulation theory thus illustrates the non-Darwinian character of Haeckel's evolutionism." See Peter Bowler, *The Non-Darwinian Revolution* (Baltimore: Johns Hopkins University Press, 1988), 84. Other historians who have thought Darwin did not endorse the recapitulation hypothesis are E. S. Russell, Stephen Jay Gould, Dov Ospovat, and Ernst Mayr. However, on page 1 of Darwin's first transmutation notebook, *Notebook B*, he enunciates the principle, and he restates it in *On the Origin of Species* (London: Murray, 1859): "As the embryonic state of each species and group of species partially shows us the structure of their less modified ancient progenitors, we can clearly see why ancient and extinct forms of life should resemble the embryos of their descendants,—our existing species" (449). By "their less modified ancient progenitors," Darwin meant the adult ancestors of the current species, something he is explicit about in the sixth edition of the *Origin*. See my discussion in *The Meaning of Evolution* (Chicago: University of Chicago Press, 1992), 152–66.

20. Ernst Haeckel, *Generelle Morphologie der Organismen*, 2 vols. (Berlin: Georg Reimer, 1866), 2:300.

21. Haeckel, *Generelle Morphologie*, 2:7.

22. Haeckel introduced this terminology in "Die Gastrula und die Eifurchung der Thiere," *Jenaische Zeitschrift für Naturwissenschaft* 9 (1875): 409.

23. Haeckel, *Generelle Morphologie*, 2:442n.

24. Haeckel, *Generelle Morphologie*, 2:451.

25. Haeckel to Darwin (10 August 1864), in *Correspondence of Charles Darwin*, 12:298–300.

26. Haeckel to Darwin (26 October 1864), in *Correspondence of Charles Darwin*.

27. Ernst Haeckel, *Zur Entwickelungsgeschichte der Siphonophoren* (Utrecht: C. Van der Post, 1869); and *Die Kalkschwämme*, 3 vols. (Berlin: Georg Reimer, 1872).

28. Ernst Haeckel, *Arabische Korallen* (Berlin: Georg Reimer, 1876).

29. Ernst Haeckel, *Das System der Medusen*, 2 vols. (Jena: Gustav Fischer, 1879).

30. Ernst Haeckel, *Die Tiefsee-Medusen der Challenger-Reise* (Jena: Gustav Fischer, 1881); and the following volumes from *Report on the Scientific Results of the Voyage of H.M.S. Challenger during the Years 1873–1867* (London: Her Majesty's Stationery Office): vol. 14 (2 parts), *Report on the Deep-Sea Medusae Dredged by H.M.S. Challenger* (1882); vol. 18 (3 parts), *Report on Radiolaria* (1887); vol. 28, *Report on the Siphonophorae Collected by H.M.S. Challenger* (1888); and vol. 32 *Report on the Deep-Sea Keratosa Collected by H. M. S. Challenger* (1889).

31. I have discussed Haeckel's experiments and those of Roux and Driesch in *The Tragic Sense of Life*, 185–95.

32. "A Little Riddle of the Universe," *New York Times*, 27 July 1901.

33. Ernst Haeckel, *Kunstformen der Natur* (Leipzig: Bibliographisches Institut, 1904).

34. Haeckel, *Kunstformen der Natur*, text to plate 8.

OREN HARMAN

PETER KROPOTKIN
ANARCHIST, REVOLUTIONARY, DREAMER

INTRODUCTION

Peter Kropotkin is known to most as a Russian anarchist, a prince-turned-philosopher who sought to combat czarist injustices and inspired millions. But Kropotkin was also a biologist—one, in fact, who offered a crisp challenge to nineteenth-century evolutionary orthodoxy. In an environment in which Darwinism had been co-opted by metaphors of conflict and rivalry, the anarchist dared to look at the flip side. Although the precise causal arrows linking his politics to his biology remain uncertain, the two became inextricably linked and inspired each other. Together, they offered a dramatic alternative both to the perceived natural order and political currents of the day.

It would take nearly a century for cooperation to assume its rightful place alongside variation and selection in the trivium of evolution, though some, including Darwin, had addressed it. By this time, for many, Kropotkin had been forgotten. Today especially, as the roles of symbiosis, empathy, altruism, and partnership in evolution are vigorously studied and theorized, it is worth remembering the thinker with whom it all started. As we shall see, his was an idiosyncratic approach, born of idiosyncratic circumstances, but it does perhaps carry salutary lessons. In the personage of Peter Kropotkin, human cooperation and animal mutual aid melded into two sides of a dream, one that adamantly refused to countenance any separation between morality and nature.

"THE DOCTRINE OF MALTHUS APPLIED TO THE WHOLE ANIMAL AND VEGETABLE KINGDOMS"

The story is well known: In October 1838, Darwin read *An Essay on the Principle of Population* by the clergyman and former professor of political economy Thomas Malthus and was alight. The idea that population increases geometrically, while food supply increases arithmetically, was meant by Malthus to prove that starvation, wars, death, and suffering were never the consequence of the defects of one political system or another, but the necessary results of a natural law. A Whig and a supporter of Poor Law action to ameliorate the condition of the destitute, Darwin was not sympathetic to Mal-

Figure 3.1. Pyotr Alexeyevich Kropotkin, c. 1900. New York Public Library; photo by
F. Nadar.

thus's reactionary politics, but applying the clergyman's law to nature was another matter. Immediately he realized that given the struggle for existence everywhere, "favorable variations would tend to be preserved, and unfavorable ones to be destroyed. The result of this would be the formation of new species. Here, then," he wrote, "I had at last got a theory by which to work." Evolution by natural selection was nothing more and nothing less than "the doctrine of Malthus, applied to the whole animal and vegetable kingdoms."[1]

Darwin loved to wax poetic over nature's pulchritude. On the massive vines of kelp off the coast of Tierra del Fuego, plummeting forty-five fathoms into the darkness, he found patelliform shells, troche, mollusks, bivalves and innumerable crustaceans, and when he shook, he wrote, out came "small fish, shells, cuttle-fish, crabs of all orders, sea-eggs, star-fish, beautiful holuthuriae, planariae, and crawling nereidous animals of a multitude of forms." The "great entangled roots" reminded Darwin of tropical forests, swarming with every imaginable species of ant and beetle rustling beneath the feet of giant capybaras and slit-eyed lizards under the watchful gaze of carrion hawks. The splendor and variation of Earth's offspring were endless.[2] And yet, in his heart of hearts he knew that the beauty was also a terrible deception; a glimmering sea concealing ferocious undercurrents. For nature's beauties could only be the result of an infinity of cacophonous battles—brutal, unyielding, and cruel. If populations in the wild have such high rates of fertility that their size would increase exponentially if not constrained; if it is known that, excepting seasonal fluctuations, the size of populations remains stable over time; if Malthus was right, as he surely was, that the resources available to a species are limited—then it follows that there must be intense competition, or a *struggle for existence*, among the members of a species.[3] And if no two members of a population are identical, and some of these differences render the life chances, or *fitness*, of some greater than others—and these are inherited—then it follows that the selection of the fitter over the less fit will lead, over time, to evolution. The consequences were unthinkable, yet Darwin's logic was spotless. From the "war of nature, from famine and death," the most exalted creatures had been created. Malthus had brought about in him a complete "conversion," one which, he famously wrote to his trusted friend Joseph Hooker in 1844, was "like confessing a murder."[4]

"AS NO QUARTER IS GIVEN"

Forty-three years later, having just lost a beloved daughter to pneumonia, Thomas Henry Huxley traveled to Manchester to give a talk to which he felt honor-bound. As the train sped north through the West Midlands—Coventry, Birmingham, Wolverhampton, Stock-on-Trent—he glanced at England passing by. For more than four years now, he had been president of the Royal Society, the winner of medals, and the very sun of the scientific *orbis terranum*.[5] But if Huxley had come a long way from above a butcher's shop in Ealing, so too had England from its affluent Age of Equipoise. Dissenters and Nonconformists had waged a battle for meritocracy against Church and Crown in the 1850s, 1860s, and 1870s, but this was long yesterday's triumph. Great boring machines were now miraculously digging the Channel Tunnel deep beneath the sea, yet millions in the cities and countryside took to bed at night hungry. The "interminable Depression" had coincided with "a specialist age"; at its finest hour, technology was failing the masses.[6]

And the masses were swelling. Britain's population had reached thirty-six million and was adding nearly 350 thousand hungry mouths every year. English Darwinians had by tradition little qualms with folding the social into the biological: for them, people and animals bowed just as humbly before Nature and its laws. But as political socialism took a bite at the Malthusian core of survival of the fittest, and as suffrage, labor unrest, and the "Woman Question" ushered in a new age of extremes, a currency was needed to remind civilization of its beastly beginnings. Were the teeming congestion and competitive strife not confirmation enough of Malthus's prediction?

Huxley had fashioned himself the very embodiment of "science as panacea," and, he thought he knew, nature really *was* brutal, like "a surface of ten thousand wedges," each representing a species being "driven inward by incessant blows."[7] Success always came at the expense of another's failure, but how then to wrest morality for the masses from the talons of nature "red in tooth and claw"?

These were his mind's torments as the train pulled into London Road Station, Manchester. At Town Hall, before his crowd, the darkness in his soul poured itself unto the natural world. Glossy-eyed and imagining his daughter, Huxley unmasked the vision of Nature's butchery: "You see a meadow rich in flower & foliage and your memory rests upon it as an image of peaceful beauty. It is a delusion . . . Not a bird that twitters but is either slayer or [slain and] . . . not a moment passes in that a holocaust, in every hedge & every copse battle murder & sudden death are the order of the day."[8]

As "melancholy as a pelican in the wilderness," Huxley was sinking into his own depression.[9] The Manchester Address was printed in February's *Nineteenth Century* and soon became a disputed *cause célèbre*. In "The Struggle for Existence in Human Society: A Programme," Huxley asked readers to imagine the chase of a deer by a wolf. Had a man intervened to aid the deer, we would call him "brave and compassionate" as we would judge an abetter of the wolf "base and cruel." But this was a hoax, the spoiled fruit of man's translation of his own world onto nature. Under the "dry light of science," none could be more admirable than the other: "the goodness of the right hand which helps the deer, and the wickedness of the left hand which eggs on the wolf" neutralize each other. Nature was "neither moral nor immoral, but non-moral"; the ghost of the deer was no more likely to reach a heaven of "perennial existence in clover" than the ghost of the wolf was to be consigned to a boneless kennel in hell. "From the point of view of the moralist the animal world is on about the same level as a gladiator's show," Huxley wrote, "the strongest, the swiftest, and the cunningest . . . living to fight another day." There was no need for the spectator to turn his thumbs down, "as no quarter is given," but "he must shut his eyes if he would not see that more or less enduring suffering is the meed of both vanquished and victor."[10]

Darwin and Spencer believed that the struggle for existence "tends to final good," the suffering of the ancestor paid for by the increased perfection of its future offspring. But this was nonsense unless, "in Chinese fashion, the present generation could pay its debts to its ancestors." Otherwise, it was unclear to Huxley "what compensation the *Eohippus* gets for his sorrows in the fact that, some millions of years afterwards, one of his descendants wins the Derby." Besides, life was constantly adapting to its environment. If a "universal winter" came upon the world, as the "physical philosophers" watching the cooling sun and Earth now warned, Arctic diatoms and Protococci of the red snow would be all that was left on the planet. Christians, perhaps, imagined God's fingerprint on nature, but it was Ishtar, the Babylonian goddess, whose meddling seemed to Huxley more true. A blend of Aphrodite and Ares, Ishtar knew neither good nor evil, nor, like the Beneficent Deity, did she promise any rewards. She only demanded that which came to her: the sacrifice of the weak. Nature-Ishtar was the heartless executioner of necessity.[11] Desperately seeking the cure for social ills, Huxley nevertheless would not search for it in nature.

"A PLANT ON THE EDGE OF A DESERT"

Enter our dreamer, Peter Alexeyevich Kropotkin, born in Moscow in the winter of 1842, the fourth child of an artistically inclined daughter of a Cossack army officer and, on his father Aleksei Petrovich's side, scion of the great Riurik dynasty, first rulers of Russia before the Romanovs.[12] At a time when family wealth was measured in numbers of serfs, the Kropotkins owned nearly twelve hundred souls in three different provinces. It was a world of birch trees, governesses, samovars, sailor suits and sleigh-rides that young

Peter was born into, "the taste of tea and jam sharpened and sweetened by the sense of the vast empty steppes beyond the garden and immanent end of it all."[13]

Not all was idyllic. Like other famous sons of Russian landed nobility—Herzen, Bakunin, Tolstoy—Peter would come to despise the particular flavor of oriental despotism baked in the juices of Prussian militarism and overlaid with a foreign veneer of French culture. The writer Ivan Turgenev's short story *Mumu*, which describes the misfortunes of the serfs, came as a startling revelation to an apathetic nation: "They love just as we do; is it possible?" was the reaction of sentimental urban ladies who "could not read a French novel without shedding tears over the troubles of the noble heroes and heroines."[14] Images etched themselves on young Peter's mind: the old man who had gone gray in his master's service and chose to hang himself under his master's window, the cruel laying to waste of entire villages when a loaf of bread went missing, the young girl who found her only salvation from a landlord-arranged marriage in drowning herself, the floggings. Increasingly, thinking and caring sons of the ruling elites of Imperial Russia witnessed up close the meanness and sterility of the feudal world into which they were born and fretted over the future of their beloved Russia. Many wondered, What is to be done?[15]

The Corps of Pages in St. Petersburg was the training ground for Russia's future military elite; only a hundred and fifty boys, mainly sons of the courtly nobility, were admitted to the privileged corps, and upon graduation they could join any regiment they chose. The top sixteen would be even more lucky: they became *pages de chambre* to members of the imperial family—*the* card of entry to a life of influence and prestige. When Peter was sent there by his father at fifteen, he already considered this a misfortune. But despite himself, he graduated at the top of his class and was made personal liege to Alexander II, Nicholas having died some years earlier. It was 1861; insurrections were growing more violent and expensive, and opposition was becoming more damagingly vocal. The new czar was under increasing pressure to grant freedom to serfs. When he finally signed the Edict of Liberty on the Fifth of March (old Russian calendar), Alexander seemed to Peter transcendent. The sentiment was fleeting. The glamour of richly decorated drawing rooms flanked by chamberlains in gold-embroidered uniforms at first took his breath away, but soon he saw that such trifles absorbed the court at the expense of matters of true importance. Power, he was learning, was corrupting.

Shadowing the czar at a distance, in the requisite combination of "presence with absence," the aureole he once imagined over the imperial ruler's person gradually, gloomily faded. Secretly, he began to read Herzen's London review, *The Polar Star*, and even to edit a revolutionary paper. When the time came to pick a commission, he determined to travel to the far expanses of eastern Siberia, to the recently annexed Afar region. Prisons to reform, schools to build, tribunals to assemble—the great administrative apparatus of the state was waiting to be marshaled. Wide-eyed, Kropotkin had joined the Cossack regiment, eager to bring justice to far away districts. Gradually, he saw his considered recommendations all dying a silent death on the gallows of bureaucracy and official corruption. When a Polish insurrection broke out in the summer of 1863, Alexander II unleashed a terrible reaction, all reforms and their spirit long forgotten. Disillusioned, Kropotkin gradually turned to nature. Fifty thousand miles he traveled—in carts, aboard steamers, in boats, but chiefly on horseback, with a few pounds of bread and a few ounces of tea in a leather bag, a kettle and a hatchet hanging at the side of his saddle. Trekking to Manchuria on a geographical survey, he slept under open skies, read Mill's *On Liberty*, and beheld with astonishment "man's oneness with nature."[16]

Kropotkin's primary concern now became working out a theory of mountain chains and high plateaus, but he was keen, too, to find evidence for Darwin's great theory. He had read *On the Origin of Species* at the Corps of Pages, and in a way this was his polar voyage of the *Beagle*. What he saw, then, came as a great surprise: Darwin, and especially his "bulldog" Huxley, spoke of a fierce struggle between members of the same species, but everywhere he looked Kropotkin found collaboration: Horses forming protective rings to guard against predators; wolves coming together to hunt in packs; birds helping each other at the nest; fallow deer marching in unison to cross a river. Mutual aid and cooperation were everywhere.

Like Darwin upon returning from his journey, Kropotkin had yet to develop a full-blown theory of nature after five years of adventure. But if Darwin's belief in the fixity of species had been shaken on the *Beagle*, Kropotkin's assurance regarding the struggle for existence was completely shattered on the Afar. By the time he arrived in St. Petersburg in April 1867, he wrote that "the poetry of nature" had become the philosophy of his life.[17] At the same time, he had lost all faith in the state: Once a constitutionalist who believed, as did Huxley, in the promise of benevolent administration, Kropotkin emerged from the great Russian expanse fully "prepared to become an anarchist."[18]

It was in Switzerland some years later that he became a full-fledged revolutionary. The death of his father finally setting him free, Kropotkin was drawn to Europe by news of the Paris Commune. In Zurich he joined the International, gaining a taste for revolutionary politics. But it was in Sonvilliers, a little valley in the Jura hills, that something really moved him. In the midst of a heavy snowstorm that "blinded us and froze the blood in our veins," fifty isolated watchmakers, most of them old men, braved the weather in order to discuss their no-government philosophy of living. This was not a mass being led by and made subservient to the political ends of a few apparatchiks. It was a union of independents, a federation of equals, setting standards by fraternal consensus. He was touched and deeply impressed by their wisdom. "When I came away from the mountains," Kropotkin wrote, "my views upon socialism were settled. I was an anarchist."[19]

Back in St. Petersburg, he joined the Chaikovsky Circle, an underground outfit working to spread revolutionary ideas. For two years, between learned debates at the Geographical Society and lavish imperial soirees, Kropotkin became "Borodin." Disguised as this peasant, he ducked into shady apartments to lecture on everything from Proudhon to reading and arithmetic, slipping away like a phantom. Communalism and fraternity were the anarchist response to the state: order without Order. Here was the creed: Left to his devices, man would cooperate in egalitarian communes, property and coercion replaced by liberty and consent.

In truth, Malthus was already dead in Russia when *Russian Nights* became a best seller in the 1840s. The novel's author, Prince Vladimir Odoesky, had created an economist antihero, driven to suicide by his gloomy prophecies of reproduction ran amok. The suicide was cheered on by the Russian reading masses: after all, in a land as vast and under-populated as theirs, Malthusianism was a joke. England was a cramped furnace on the verge of explosion; Russia, an expanse of bounty almost entirely unfilled. But it was more than that. "The country that wallowed in the moral bookkeeping of the past century," Odoesky explained, "was destined to create a man who focused in himself the crimes, all the fallacies of his epoch, and squeezed strict and mathematically formulated laws of society out of them." Malthus was no hero in Russia.[20]

And so when the *Origin of Species* was translated in 1864, Russian evolutionists found themselves in something of a quandary. Darwin was the champion of science, the father of a great theory, but also an adherent of Malthus, that "malicious mediocrity" according to Tolstoy.[21] How to divorce the kind and portly naturalist Whig from Downe from the cleft-palated, fire-spitting, reactionary reverend from Surrey?[22] Both ends of the political spectrum had good grounds for annulment. Radicals, such as Herzen, reviled Malthus for his morals: unlike bourgeois political economy, the cherished peasant commune allowed "everyone without exception to take his place at the table." Monarchists and conservatives, on the other hand, such as the Slavophile biologist

Nikolai Danilevsky, contrasted czarist Russia's nobility to a nation of shop-keepers pettily counting their coins. Danilevsky saw Darwin's dependence on Malthus as proof of the inseparability of science from cultural values. "The English national type," he wrote, "accepts [struggle] with all its conse-quences, demands it as his right, tolerates no limits upon it. . . . He boxes one on one, not is a group as we Russians like to spar." Darwinism for Danilevsky was "a purely English doctrine," its pedigree still unfolding: "On usefulness and utilitarianism is founded Benthamite ethics, and essentially Spencer's also; on the war of all against all, now termed the struggle for existence—Hobbes's theory of politics; on competition—the economic theory of Adam Smith . . . Malthus applied the very same principle to the problem of popu-lation. . . . Darwin extended both Malthus's partial theory and the general theory of the political economists to the organic world." Russian values were of a different timber.[23]

But so, according to Kropotkin, was Russian nature. Darwin and Wal-lace had eavesdropped on life in the shrieking hullabaloo of the tropics. But the winds of the arctic tundra whistled an altogether different tune. And so, wanting to stay loyal to Darwin, the Russian evolutionist now turned to his sage, training a torch on those expressions Huxley and the Malthusians had swept aside. "I use this term in a large and metaphorical sense" Darwin wrote of the struggle for existence in the *Origin*. "Two canine animals, in a time of dearth, may be truly said to struggle with each other which shall get food and live. But a plant on the edge of a desert is said to struggle for life against the drought, though more properly it should be said to be dependent on the moisture."[24] Here was the merciful getaway from *bellum omnium contra omnes*, even if Darwin had not underscored it.[25] For if the struggle could mean both competition with other members of the same species *and* a battle against the elements, it was a matter of evidence which of the two was more important in nature. And if harsh surroundings were the enemy rather than rivals from one's own species, animals might seek other ways than conflict to manage such struggle. Here, in Russia, the fight against the elements could actually lead to cooperation.

"THEREFORE COMBINE—PRACTICE MUTUAL AID!"

In March 1881 Alexander II was assassinated. Once his trusted liege, Kropotkin welcomed news of his death as a harbinger of the coming revo-lution. Increasingly, he turned to science to find backing for his revolutionary activities: the science of anarchy and the science of nature. They had evolved apart from one another, but the two sciences were now converging, even be-coming uncannily interchangeable. When Darwin died in the spring of 1882,

Kropotkin penned an obituary in *Le Révolté*. Celebrating, in true Russian fashion, the sage of evolution entirely divorced from Malthus, the prince judged Darwin's ideas "an excellent argument that animal societies are best organized in the community-anarchist manner."[26] In "The Scientific Basis of Anarchy," some years later, he made clear that the river ran in both directions. "The anarchist thinker," Kropotkin wrote, "does not resort to metaphysical conceptions (like the 'natural rights', the 'duties of the state' and so on) for establishing what are, in his opinion, the best conditions for realising the greatest happiness for humanity. He follows, on the contrary, the course traced by the modern philosophy of evolution."[27] Finding the answers to society's woes "was no longer a matter of faith; it [is] a matter for scientific discussion."[28]

Meanwhile, back in Britain, Huxley had found an uneasy path to allay his heart's torments. If instincts were bloody, morality would be bought by casting away their yoke. This was the task of civilization—its very *raison d'etre*: to combat, with full force, man's evolutionary heritage. It might seem "an audacious proposal" to create thus "an artificial world within the cosmos," but of course this was man's "nature within nature," sanctioned by his evolution, a "strange microcosm spinning counter-clockwise." Nature's injustice had "burned itself deeply" into his soul, but Huxley remained hopeful. It was an optimism born of necessity: for a believing Darwinist, any other course would mean utter bleakness and despair.[29]

Years in the Amur, in prisons, and in revolutionary politics had coalesced Kropotkin's thoughts, too, into a single, unerring philosophy. Quite the opposite of Huxley's tortured plea to wrest civilized man away from his savage beginnings, it was rather the *return* to animal origins that promised to save morality for mankind. And so, when in a dank library in Harrow, perusing the *Nineteenth Century*, Kropotkin's eyes fell on Huxley's "The Struggle for Existence," anger swelled within him. He would need to rescue Darwin from the "infidels," men like Huxley, who had "raised the 'pitiless' struggle for personal advantage to the height of a biological principle."[30] Moved to action, the "shepherd from the Delectable Mountains" wrote to James Knowles, the *Nineteenth*'s editor, asking that he extend his hospitality for "an elaborate reply." Knowles complied willingly, writing to Huxley that the result was "one of the most refreshing & reviving aspects of Nature that ever I came across."[31]

"Mutual Aid Among Animals" was the first of a series of five articles, written between 1890 and 1896, that would become famously known in 1902 as the book *Mutual Aid*. Here Kropotkin finally sunk his talons into "nature red in tooth and claw." For if the bees and ants and termites had "renounced the Hobbesian war" and were "the better for it," so had shoaling fish, burying beetles, herding deer, lizards, birds, and squirrels. Remembering his years in the great expanses of the Afar, Kropotkin now wrote, "Wherever I saw animal life in abundance, I saw Mutual Aid and Mutual Support."[32]

This was a general principle, not a Siberian exception. There was the common crab, as Darwin's own grandfather Erasmus had noticed, stationing sentinels when its friends were moulting. There were the pelicans forming a wide half-circle and paddling toward the shore to entrap fish. There was the house sparrow, who "shares any food," and the white-tailed eagles spreading apart high in the sky to get a full view before crying to each other when a meal is spotted. There were the little tee-tees, whose childish faces had so struck Alexander von Humboldt, embracing and protecting one another when it rains, "rolling their tails over the necks of their shivering comrades." And, of course, there were the great hordes of mammals: deer, antelope, elephants, wild donkeys, camels, sheep, jackals, wolves, wild boar—for all of whom "mutual aid [is] the rule." Despite the prevalent picture of "lions and hyenas plunging their bleeding [sic] teeth into the flesh of their victims," the large hordes were of astonishingly greater numbers than the carnivores. If the altruism of the hymenoptera (ants, bees and wasps) was imposed by their physiological structure, in these "higher" animals it was cultivated for the benefits of mutual aid. There was no greater weapon in the struggle for existence. Life *was* a struggle, and in that struggle the fittest *did* survive. But the answers to the questions, "By which arms is this struggle chiefly carried on?" and "Who are the fittest in the struggle?" made abundantly clear that "natural selection continually seeks out the ways precisely for avoiding competition." Putting limits on physical struggle, sociability left room "for the development of better moral feelings." Intelligence, compassion and "higher moral sentiments" were where progressive evolution was heading, not bloody competition between the fiercest and the strong.[33] It was a revolutionary view of the march of life.

But where had mutual aid come from? Some thought from "love" that had grown within the family, but Kropotkin was at once more hardened and expansive.[34] To reduce animal sociability to familial love and sympathy meant to reduce its generality and importance. Communities in the wild were not predicated on family ties, nor was mutualism a result of mere "friendship." Despite Huxley's belief in the family as the only refuge from nature's battles, for Kropotkin the savage tribe, the barbarian village, the primitive community, the guilds, the medieval city—all taught the very same lesson: for mankind, too, mutualism beyond the family had been the natural state of existence.[35] "It is not love to my neighbor—whom I often do not know at all," Kropotkin wrote, "which induces me to seize a pail of water and to rush towards his house when I see it on fire; it is a far wider, even though more vague feeling or instinct of human solidarity and sociability which moves me. So it is also with animals."[36]

The message was clear: "Don't compete! Competition is always injurious to the species, and you have plenty of resources to avoid it." Kropotkin had a powerful ally on his side. "That is the watchword," he wrote, "which comes to us from the bush, the forest, the river, the ocean." Nature herself would be man's guide. "Therefore combine—practice mutual aid! That is the surest means of giving to each other and to all the greatest safety, the best guarantee of existence and progress, bodily, intellectual, and moral."[37]

If capitalism had allowed the industrial "war" to corrupt man's natural beginnings; if overpopulation and starvation were the necessary evils of progress—Kropotkin was having none of it. Darwin's Malthusian "bulldog" and his Victorian "X Club" minions had gotten it precisely the wrong way around. Far from having to combat his natural instincts in order to gain a modicum of morality, all man needed to find goodness was to train his gaze within. This was the anarchist's creed and dream.

"I CONSIDER IT A DUTY TO TESTIFY": A DREAMER'S CODA

When the revolution in Russia finally broke out in February 1917, Kropotkin was already old and famous. On May 30, after forty-one years in exile, thousands flocked to the Petrograd train station to welcome him home.[38] Czarless and reborn, Russia had revived his optimism in the future. But then came October and the Bolsheviks, and, as it had years earlier in the Afar, the spirit of promise soon wasted into disappointment. Trying his best to remain loyal to himself, Kropotkin moved from Moscow to the small village of Dimitrov, where a cooperative was being constructed. Increasingly frail and working against the clock on his magnum opus, *Ethics*, he still found time to help the workers.[39] "I consider it a duty to testify," he wrote to Lenin on March 4, 1920, "that the situation of these employees is truly desperate. The majority are literally starving. . . . At present, it is the party committees, not the soviets, who rule in Russia. . . . If the present situation continues, the very word 'socialism' will turn into a curse."[40] Lenin never replied. But he did give his personal consent when Peter Kropotkin died on February 8, 1921, that the anarchists should arrange his funeral. It would be their last mass gathering in Russia.

For a time—a very long time—it was Huxley's legacy that won the day: evolution was a game of survival of the fittest, glory belonging to the fiercest individual combatants in the bloody battle for survival. Gradually that began to change, sporadically at first, with people like the University of Chicago ecologists Warder Clyde Allee and Alfred Emerson, taken in by Morton Wheeler's notion of the "superorganism," writing in the 1930s and 1940s about the importance of cooperative forces and of the subordination of the individual to the population, respectively. They were followed by V. C. Wynne-Edwards, with his controversial work on Arctic fulmars. Change proceeded in a more sustained fashion after George Price and Bill Hamilton attempted to make group selection respectable again in the 1970s—a torch picked up by such thinkers as the theoretician David Sloan Wilson and the experimentalist Michael J. Wade. At the same time, beginning in the 1950s, Cold War preoccupations saw game theory applied to the problem of cooperation. Whether due to the logic of groups or the imperatives of self-interest as outlined in the Tit-for-Tat game, altruism and reciprocity were making a comeback.[41]

Nowadays cooperation is everywhere. Wherever you look, an evolutionist will explain to you how most of the great innovations on the planet, from genes to genomes, cells, and societies, have been due to the creative powers of cooperation. However much we've gotten used to "nature red in tooth and claw," evolution is just as much about cooperation as competition, and everything, from the origin of life to chromosomes, ant colonies, language, and morals, is implicated.[42] Even as group selection remains controversial, mutual aid, or "the snuggle for existence," has firmly ensconced itself alongside mutation and selection as a third, and vital, pillar of evolution. The extent to which this newfound interest in cooperation is related to political and social developments in our times remains to be carefully studied by historians.[43]

But precisely because this is now beginning to be taken for granted, it is worth remembering the late nineteenth- and early twentieth-century dreamer who latched on to a lesser-noticed metaphor of Darwin's—and put forth an alternative view of how nature works. Kropotkin was not the only person to stress the role of mutualism in nature,[44] but his beliefs stemmed from a unique entanglement between up-close encounters with nature and the political world he inhabited. This is what makes him such a fascinating dreamer, for his dreams of social justice and natural order grew in tandem, reinforcing and buttressing each other rather than deterministically producing each other.[45] It is a comment, one might argue, on the relationship between seeing and believing, and—there being plenty of competition to countenance in nature—on the power of framing. Perhaps he was looking for it, but Prince Peter Kropotkin truly did glimpse mutual aid both among people and animals, and it was this apparition that bolstered his challenge to the political, as well as the scientific, consensus of his day. Ultimately, Kropotkin's embrace of mutual aid would meld with his revolutionary anarchism, transforming his life dramatically. It would also alter our ever-changing view of evolution, moving as it does with the times.

FURTHER READING

Desmond, Adrian. *Huxley: From Devil's Disciple to Evolution's High Priest.*
Harmondsworth: Penguin, 1997.

Graham, Loren R. *What Have We Learned About Science and Technology from the
Russian Experience.* Stanford, CA: Stanford University Press, 1998.

Hale, Piers. *Political Descent: Malthus, Mutualism, and the Politics of Evolution in
Victorian England.* Chicago: University of Chicago Press, 2014.

Huxley, Thomas Henry. "The Struggle for Existence in Human Society: A
Programme." *Nineteenth Century* 23 (February 1888): 161–80.

Kropotkin, Peter. *Memoirs of a Revolutionist.* London: Folio Society, 1978.

———. *Mutual Aid: A Factor in Evolution.* 1902. Reprint, Boston: Extending Horizons
Books, 1955.

Todes, Daniel P. *Darwin without Malthus: The Struggle for Existence in Russian
Evolutionary Thought.* Oxford: Oxford University Press, 1989.

NOTES

1. Charles Darwin, *The Autobiography of Charles Darwin* (New York: W. W. Norton, 1993), 120; Charles Darwin, *The Origin of Species,* 2nd ed. (Oxford: Oxford University Press, 1996), 6. The current essay draws from materials in Oren Harman, *The Price of Altruism: George Price and the Search for the Origins of Kindness* (New York: W. W. Norton, 2010).

2. Charles Darwin, *The Voyage of the Beagle* (Hertfordshire: Wordsworth, 1997), 228–29.

3. Darwin's concept of struggle was, nonetheless, more complex. Alongside competition, there were also ecological dependence and chance, as well as sacrifice for the greater good of the community. Nature was surely "red in tooth and claw" but, as the ants and bees and termites made clear, not exclusively.

4. Darwin, *Origin of Species,* 396; Charles Darwin letter to J. D. Hooker (January 11, 1844), in *The Correspondence of Charles Darwin,* ed. Frederick Burkhardt, vol. 3, *1844–46* (Cambridge: Cambridge University Press, 1987), 2.

5. He had won the Royal, the Wollaston and the Clarcke; the Copley, the Linnaean, and the Darwin still awaited him.

6. Adrian Desmond, *Huxley: From Devil's Disciple to Evolution's High Priest* (Harmondsworth: Penguin, 1997), 572–73.

7. Darwin had used this image already in his essay from 1844, but it was made public in C. R. Darwin and A. R. Wallace, "On the Tendency of Species to Form Varieties; and on the Perpetuation of Varieties and Species By Natural Means of Selection" (Read 1 July, 1858), *Journal of the Proceedings of the Linnaean Society of London. Zoology* 3:46–50.

8. Huxley's notes for his Manchester Address, quoted in Desmond, *Huxley,* 558.

9. Huxley to Foster, 8 January 1888, in *The Life and Letters of Thomas Henry Huxley,* by Leonard Huxley (London: Macmillan, 1900), 198. Quoted in Lee Dugatkin, *The Altruism*

Equation: Seven Scientists Search for the Origins of Goodness (Princeton, NJ: Princeton University Press, 2006), 19.

10. T. H. Huxley, "The Struggle," in *Collected Essays* (London: Macmillan, 1883–84), 197, 198, 199, 200.

11. Huxley, "The Struggle," 198, 199, 200.

12. Besides Kropotkin's own *Memoirs of a Revolutionist* (London: The Folio Society, 1978), see George Woodcock and Ivan Avakumovich, *The Anarchist Prince: A Biographical Study of Peter Kropotkin* (London: T. V. Boardman, 1950); and the more scholarly Martin A. Miller, *Kropotkin* (Chicago: University of Chicago Press, 1976).

13. Colin Ward, introduction to Kropotkin, *Memoirs*, 8.

14. Kropotkin, *Memoirs*, 56; Ivan Sergeevich Turgenev, *Mumu* (Moskva: Detgiz, 1959).

15. This was the title of an influential treatise by Chernyshevsky and was later used by both Tolstoy and Lenin.

16. Kropotkin, *Memoirs*, 94.

17. Kropotkin, *Memoirs*.

18. Kropotkin, *Memoirs*, 157.

19. Kropotkin, *Memoirs*, 201, 202. The best book on Russian anarchism remains Paul Avrich's *The Russian Anarchists* (Princeton, NJ: Princeton University Press, 1967).

20. Daniel P. Todes, "Darwin's Malthusian Metaphor and Russian Evolutionary Thought, 1859–1917," *Isis* 87 (1987): 537–51; quotations on 539–40. See his broader treatment in *Darwin without Malthus: The Struggle for Existence in Russian Evolutionary Thought* (Oxford: Oxford University Press, 1989). See also Stephen Jay Gould, "Kropotkin Was No Crackpot," in *Bully for Brontosaurus* (Harmondsworth: Penguin, 1991), 325–39.

21. Quoted in Todes, "Darwin's Malthusian Metaphor," 542.

22. On Malthus, see Patricia James, *Population Malthus: His Life and Times* (London: Routledge and Kegan Paul, 1979); Samuel Hollander, *The Economics of Thomas Robert Malthus* (Toronto: University of Toronto Press, 1997); William Peterson, *Malthus, Founder of Modern Demography*, 2nd ed. (New Brunswick, NJ: Transaction, 1999); J. Dupâquier, "Malthus, Thomas Robert (1766–1834)," in *International Encyclopedia of the Social and Behavioral Sciences* (Amsterdam: Elsevier, 2001), 9151–56. For an account of English dissent from Malthus in the second half of the nineteenth century, see Piers Hale, *Political Descent: Malthus, Mutualism, and the Politics of Evolution in Victorian England* (Chicago: University of Chicago Press, 2014).

23. Todes, "Darwin's Malthusian Metaphor," 542, 540, 541–42. See also Thomas F. Glick, ed., *The Comparative Reception of Darwinism* (Chicago: University of Chicago Press, 1988), 227–68. See also, Engels's letters to Lavrov, 12–17 November 1875, available at the Marx/Engels Internet Archive, http://marxists.org.

24. Darwin, *Origin of Species*, 53.

25. Darwin did, however, write about forms of cooperation and what would be called "altruism" that were due to natural selection working, in his words, "for the good of the community"; he focused in particular on the social insects.

26. Peter Kropotkin, "Charles Darwin," *Le Révolté*, 29 April 1882.

27. "Without entering," he added, "the slippery route of mere analogies so often resorted to by Herbert Spencer." See Peter Kropotkin, "The Scientific Basis of Anarchy," *Nineteenth Century* 22, no. 126 (1887): 238–52; quotations on 238.

28. Kropotkin, "Scientific Basis of Anarchy," 239.

29. Desmond, *Huxley*, 599. The oft-quoted sentence is "The ethical progress of society depends, not on imitating the cosmic process [evolution], still less in running away from it, but in combating it"; see T. H. Huxley, "Evolution and Ethics," in *Evolution and Ethics and Other Essays* (New York: D. Appleton, 1898), 83; Desmond, *Huxley*, 598.

30. Peter Kropotkin, *Mutual Aid: A Factor in Evolution* (1902; repr., Boston: Extending Horizons Book, Porter Sargent Publishers, 1955), 4. Kropotkin thought Darwin, especially in his *Descent of Man*, had emphasized the role of cooperation, whereas his followers took to the narrower definition of the struggle for existence. "Those communities," Darwin wrote in the *Descent*, "which included the greatest number of the most sympathetic members would flourish best, and rear the greatest number of offspring." (2nd ed. [New York: Wallachia, 2015], 163). "The term," Kropotkin added, "thus lost its narrowness in the mind of one who knew Nature".

31. This was George Bernard Shaw's description of Kropotkin; Kropotkin, *Mutual Aid*, xiv; Desmond, *Huxley*, 564.

32. Kropotkin, *Mutual Aid*, 14, ix.

33. Kropotkin, *Mutual Aid*, 51, 40, 60–61; Kropotkin freely admitted that there was much competition in nature, and that this was important. But intra-species conflict had been exaggerated by the likes of Huxley; it also often left all combatants bruised and reeling. True progressive evolution was due to the law of mutual aid.

34. Louis Buchner, *Liebe und Liebes-Leven in der Thierwelt* (Leipzig: Theodor Thomas, 1885); Henry Drummond, *The Ascent of Man* (New York: J. Pott, 1894); Alexander Sutherland, *The Origin and Growth of the Moral Instinct* (London: Longmans Green, 1898). Kessler, too, thought that mutual aid was predicated on "parental feeling," a position from which Kropotkin was careful to detach himself. See Kropotkin, *Mutual Aid*, x.

35. Kropotkin marshaled evidence from varied sources, especially liberally interpreted archaeological evidence, to argue that man's "natural" state was in small, self-sustaining, communal groups. For English thought on mutualism beyond Huxley, including such leftist figures as George Bernard Shaw, Marx's daughter Eleanor, and William Morris (who was a regular dining partner of the exiled Kropotkin), see Hale, *Political Descent*.

36. Kropotkin, *Mutual Aid*, xiii.

37. Kropotkin, *Mutual Aid*, 75.

38. Kerensky was there, and offered him a ministry in the new government, which Kropotkin declined. Still, he did become active in party politics from the outside. See Miller, *Kropotkin*, 232–237.

39. An attempt to lay the foundations of a morality free of religion and based on nature, *Ethics* was published posthumously in 1922.

40. P. A. Kropotkin, *Selected Writings on Anarchism and Revolution*, ed. Martin A. Miller (Cambridge, MA: MIT Press, 1970), 336.

41. See Harman, *Price of Altruism*, for a description of these developments, as well as Mark Borrello, *Evolutionary Restraints: The Contentious History of Group Selection* (Chicago: University of Chicago Press, 2010).

42. See works by James Attwater and Philipp Hollinger, Robert Axelrod, Frans de Waal, Martin Nowak, Christopher Boehm, Samuel Bowles, Sarah Blaffer Hrdy, E. O. Wilson, Michael Tomasello, among many others.

43. For an incisive comment on the importance of time and place, culture and ideology, when it comes to science and technology in the Russian case, see Loren R. Graham, *What Have We Learned About Science and Technology from the Russian Experience* (Stanford, CA: Stanford University Press, 1998).

44. Hale, in *Political Descent*, has written eloquently about contemporary Victorians who took up Kropotkin with relish. In this debate Darwin and Malthus were often pitted against Lamarck and socialism. But see also Gregory Radick, "Dissent of Man," *Times Literary Supplement*, 1 July 2015, in which he discusses, among other things, eugenic aftereffects.

45. In this sense, like Patricia Churchland and others today, Kropotkin offers a challenge to our complacent acceptance of what has become known as the "naturalistic fallacy." See Oren Harman, "Is the Naturalistic Fallacy Dead? (And If So, Ought It Be?)," *Journal of the History of Biology* 45, no. 3 (2012): 557–72.

Part II: The Medicalists

Figure 4.1. Mrs. Mary Lasker, half-length portrait, presenting award. *World Telegram & Sun* photo by Fred Palumbo, 1957. Retrieved from the Library of Congress, https://www.loc.gov/item/95513859/.

KIRSTEN E. GARDNER

MARY LASKER
CITIZEN LOBBYIST FOR
MEDICAL RESEARCH

INTRODUCTION

Mary Woodard Reinhardt Lasker, described by one US senator as "an American hero and a legend bigger than life," spent much of her life dreaming about health policy that might produce cancer cures, new vaccines, mental health relief, and more.[1] She believed passionately that federally funded medical research would decrease disease incidence, inspire cures, and yield better health outcomes for all Americans. She introduced the public and politicians to popular health discourse that detailed disease incidence, outcomes, and research programs as a means to advocate for better health in the United States.

Throughout her fifty-year career as a health advocate, Mary Lasker effectively lobbied politicians, championed medical philanthropy, and gained broad support for her ideas. She built a framework for medical fund-raising that included a clear and accessible discussion of disease incidence, a promising proposal for scientific research that would improve disease outcome, and a celebration of health improvements. Throughout her life (1900–1994), Lasker realized many of her dreams, and her health advocacy transformed medical research protocol in the United States. She promoted scientific research, increasing federal support and spending for medical research. She advocated for health institutes that could engage in long-term research projects. She tapped into a powerful network of politicians who shared her dreams, built reputations based on their support of medical research funding, and ensured that federal medical research funding would continue to exist long after Lasker's personal work came to an end.

Born Mary Woodard in Watertown, Wisconsin, in 1900, Lasker's earliest memories speak to her precocious childhood personality. Raised by a quiet father and influential Irish Protestant mother, Lasker's worldview seemed framed by her mother, Sarah Johnson. Johnson emigrated from Northern Ireland, moved to Canada, and then settled in the United States, where she became part of a small but increasing cohort of independent career women. At a time when women had limited independence and career options, Johnson achieved economic and business success running retail dress departments in Chicago. Like her mother, Lasker acquired business skills and more, imagining ways that women could expand their influence to better achieve their goals.[2]

Lasker's young adulthood was framed by a well-rounded education and a clear memory of the profound burdens of health and disease. In numerous recollections of her life history, Lasker shared vivid memories of the burdens of disease. For example, early in life she learned that the family laundress, Mrs. Belter, had had a double mastectomy, and she witnessed her younger sister's near brush with death due to pneumonia. Her childhood memories were framed by her repeated cases of dysentery and recurring mastoid pain. She also recalled her time in the health ward during the great influenza epidemic. Additionally, when she was a young adult, the mother of one of her closest friends died from breast cancer. Her first husband, Paul Reinhardt, experienced a severe eye infection. Finally, her father experienced multiple strokes, and her mother's death was instigated by stroke.

Such life experiences evoked frustration at the limits of medicine and, more significantly, inspired an intense desire to improve health in the United States. As a case in point, after her parents' death from strokes, she wrote renowned author and microbiologist Paul de Kruif and philanthropist John D. Rockefeller inquiring about research on strokes. She learned that very little research funding existed for cardiac health. Gradually, via marriage, she gained access to incredible power and privilege. She used such power to propel political projects designed to promote medical research—most notably through her influential support for the expansion of the National Institutes of Health and Cancer Research.[3]

Educated at the Milwaukee-Downer Seminary, University of Wisconsin, Radcliffe College, and Oxford, Lasker studied fine arts and cultivated a network of important friendships—many that would last a lifetime. As a young adult, she traveled and studied throughout Europe, expanding her rich network of friends. After graduation, she maintained an adventuresome and confident spirit. As Lasker explained, "Well, when I got out of college it was an era when no respectable girl would ever think of going home to a small town in the Middlewest. It was always absolutely required that they work in New York for a certain time at least, and I considered it absolutely too dull for words to return to Watertown, Wisconsin, absolutely impossible."[4] In 1923, after graduating from Radcliffe College, Lasker moved to New York City, finally overcoming the health issues (often defined as frailty and fatigue) of her childhood.

Once in New York City, she moved into a "girls' home" on Fifty-seventh Avenue and began working in the art world. She started work at Eric Galleries, and soon thereafter moved to Reinhardt Gallery to organize remarkable exhibits. Her most noteworthy assignments included hosting an early Marc Chagall exhibit and an Ignacio Zuloaga show. Lasker's success in the art world synthesized her skills at organizing, advertising, and building a community that shared her interest in a specific topic. Approximately two years after beginning work at Reinhardt Gallery, Lasker married gallery owner Paul Reinhardt. This marriage proved short-lived however, due to Reinhardt's alcoholism.

A divorcee in 1934, Lasker launched several business ventures. Her most successful, "Hollywood Patterns," tapped into the expanding department store model and offered consumers affordable Hollywood dress designs. As a successful businesswoman, Lasker appreciated the potential of turning ideas into realities. She also continued to nurture her vast array of powerful friendships, living in the heart of Manhattan and hosting regular parties. As one indication of her fame, in the 1940s she and her dear friend Kay Swift hosted annual soirees that included the likes of Wendell Wilke, Margaret Sanger, Walter Mack, Albert Lasker, David Sarnoff, and Karl Menninger.[5] This guest list reflects Mary Lasker's access to influential and powerful people with whom she associated throughout her life. She would ultimately rely on such influential networks to lobby on behalf of medicine and science with remarkable success.

Far more momentous than her first marriage, Mary Lasker's second marriage to public relations guru Albert Davis Lasker proved transformative. On June 21, 1940, the Laskers wed and soon thereafter formally merged their shared interest in improving health. Like many before her, when Mary first noticed Albert Lasker, she recognized that he had a sense of "business genius" about him. But as they began courting in 1939, she appreciated his keen interest in and knowledge about public affairs. Mary and Albert Lasker also shared a mutual respect for one another's independence and business acumen. Perhaps as important, they also shared philanthropic visions.[6] As one example, Mary Lasker believed that birth control was the most important project in early twentieth-century public health efforts. She served as the secretary of the Birth Control Federation of America's (BCFA) and would become vice-president of Planned Parenthood. Soon after meeting Albert Lasker, she shared her enthusiasm for the cause. Almost immediately, he invested money in the organization, encouraged his sisters to join the cause as well, and offered his public relations skills to the promotion of reproductive control.[7] As Mary shared more of her dreams—supporting universal health insurance, sponsoring cancer research, ending tuberculosis—Albert continued to share her enthusiasm. In particular, Albert and Mary strategized to reframe biomedical research funding. In short, the Laskers realized that effective lobbying within the federal government could provide far more funding than any amount of private philanthropy. Therefore, the Laskers tapped into political networks, allies, and candidates who would support increasing federal spending on health and research. This strategy of transforming private concerns into public causes would define Mary Lasker's work for the rest of her life.[8]

The Laskers shared a belief in scientific research and a passion for advancing medical research funding. Mary Lasker assumed a visible position in their health advocacy partnership, while Albert Lasker (known as a "boy wonder" when he first started with his public relations company in 1898) provided useful advertising strategies for health lobbying efforts. The Laskers inaugurated a political discourse about health policy that centered on data and evidence—from enumerating disease incidence to documenting dollars invested in medical research. Moreover, they introduced the idea that health spending was an investment in the future.

Mary Lasker's marriage to one of the most important businessmen of the twentieth century certainly provided access to influential audiences. In the early 1940s, Mary Lasker befriended First Lady Eleanor Roosevelt. Roosevelt and Lasker shared a faith in family planning and a commitment to reproductive health. The first lady valued Lasker's ideas and introduced her to Surgeon General Dr. Thomas Parran of the Public Health Services. In October of 1941, Lasker asked the US Public Health Service to incorporate child spacing into their regular health programs. While the Public Health Services refused to take a public stand on the issue (due to the political volatility of the subject), they agreed to consider funding a child-spacing initiative. Lasker celebrated such victories and appreciated the long-term impact of family planning, "It seems to me that the whole course of mankind depends on intelligent ability to limit the size of people's families, from every possible point of view: economic, emotional and every other point of view."[9]

Though keeping her faith in the value and purpose of family planning, Lasker soon began dreaming of more systemic change for the US health-care system. In short, dismayed by the limited federal funds available to support health research, she began to envision a federally funded research program. She learned about existent health agencies, such as the Office of Scientific Research and Development (OSRD). This wartime agency had helped usher in one of the greatest medical discoveries of the century by collaborating to facilitate penicillin production, yet the OSRD faced shutdown at the end of the war. Lasker supported continued spending on such research efforts and easily imagined the federal government as a permanent funding source for medical research. She strategized with others by discussing health policy that could be productive and popular. On a very practical level, she supported politicians who shared her vision and supported expanded health policy.

In a parallel vein, by 1942, Mary and Albert Lasker crystallized their shared support of medical research with the creation of the Albert and Mary Lasker Foundation. This foundation would promote biomedical research and create a home for their causes. As Mary Lasker described it, the Lasker Foundation provided "the fact books, the facts on all the major killing and crippling diseases: what was being spent, what the incidence of the diseases were, how many deaths, what pay-offs there had been due to research. Our factual material provided the background for witnesses to testify with."[10] Within two years, in 1944, the Laskers established the Albert Lasker Award to recognize outstanding clinical and basic medical research. This award is frequently described as "the most prestigious American award for biomedical research."[11] Often defined as one of the most coveted awards in medical research, the award signified an emerging respect and public awareness of the value of medical and scientific research. (For many reasons, including the parallel award criteria, the Lasker Award has often been recognized as a precursor to the Nobel Prize. To date, more than eighty-five Lasker prize winners have gone on to win the Nobel Prize.)[12]

While the Laskers formalized their foundation, award structure, and personal dedication to reframing the funding structure for medical research, they also became increasingly invested in the cancer problem. In short, in the mid-1940s the Laskers successfully ushered in an era of increased personal philanthropy for cancer; supported renaming of the American Society for the Control of Cancer to the American Cancer Society, with a promise to give 25% of its budget to cancer research; included lay persons on a former medical board; and demonstrated how a business model that put research at its core could be quickly adopted by a philanthropic organization. In 1945, the American Cancer Society (ACS) sought public support via the radio and popular publications such as *Reader's Digest* and raised an unprecedented $4 million, directing $1 million to research as promised.

In the post–World War II era, Mary Lasker dreamed bigger. She began meeting with politicians in earnest, including President Truman and his wife. Through a combination of political connections, strategic friendships, and firm political pressure, Lasker started to build a broad base of support for federal funding of medical research. Her support base was broad and effective, and she maintained connections to the executive office, counsel to the president, and congressional leaders for five decades.

Throughout her life, Mary Lasker championed a new way of thinking about medical research and slowly enfolded powerful Democrats into her way of thinking. By the mid-1940s, Mary Lasker and Florence Mahoney became friends and partners in an incredible political lobbying effort. They shared a commitment to health research and funding, and Mahoney introduced another network of political allies to Lasker, including Senator Claude Pepper.[13] (Pepper had introduced the 1937 National Cancer Act.) Over time, Pepper would be a key ally for Lasker and Mahoney, sponsoring many legislative and appropriation bills in support of biomedical research. Lasker and Mahoney gradually persuaded more and more powerful Democrats to their way of thinking. Another ally, Dr. Claude Lenfant, former Director of the National Heart, Blood, and Lung Institute, highlighted Lasker's attention to the intersections of the public and political: "Mary was also a keen politician. She knew how to garner support for biomedical research. She could marshal all of the players who turn policy into reality—the public, government officials, and the biomedical community itself. She knew powerful persons. . . . She was able to form private-sector alliances to advocate greater federal support."[14]

In many ways, Lasker's careful networking strategy reflected her life's magnum opus. Lasker promoted federally funded medical research far and wide. By selling her belief to politicians and those who voted for them, she ensured a broad base of support. By recruiting well-spoken medical authorities, she could organize impromptu educational meetings and seminars. By collecting facts and figures, she and her allies could offer a persuasive argument for federal funding whenever possible. In an oral interview conducted in the 1960s, Lasker recalled the early years when she, Florence Mahoney, Senator Pepper, President Truman, and others dreamed of creating enormous funding structures for medical research. She recalled with passion, "We were the grassroots rising!"

Albert Lasker died of abdominal cancer in 1952, and from that point onward Mary Lasker became the voice and public persona of Lasker fame—a position she occupied until her death in 1994. As she explained, "After Albert's death on May 30, 1952, I was more determined than ever to go ahead and obtain more funds for research against cancer and heart [disease] and other major killers and cripplers of our time. I realized even more pointedly how little was known."[15] While she often described herself as a "citizen health lobbyist," that title belies the power she accumulated during and after Albert's life. As reporters for *US News and World Reports* summarized at the end of Mary Lasker's life, "For nearly half a century she was the single most influential figure behind the buildup of federal funds for biomedical research."[16] She lobbied Congress and the executive office; served two terms on the Cancer Institute Council and the Heart Institute Council; supported organizations, including the American Cancer Society, Research to Prevent Blindness,

Cancer Research Institute, National Committee for Mental Hygiene, and more; and tirelessly made telephone calls, sent telegrams, organized dinners and seminars, appealed to decision makers, and accumulated a database of facts and figures to educate the public about why medical research should be funded.[17]

Mary Lasker's dedication to cancer research strengthened in the 1950s with her support of appropriations for chemotherapy research in 1954.[18] Perhaps most notably of all, Mary Lasker is often credited with the political acumen that pushed Nixon into signing the National Cancer Act in 1971. Clearly, her commitment to cancer research remained strong over time, but so too did her commitment to expanding the role of the federal government in research funding. As *U.S. News and World Report* noted, Lasker witnessed the expansion of the NIH budget from $2.5 million in 1945 to $11 billion in 1994, and "when Mary Lasker died, biomedical research lost the most effective friend it ever had."[19] As early as 1971, when Congress approved of the National Cancer Institute, Lasker could take credit for a political culture that valued medical research. She earned a reputation that included eliciting support from Democratic administrations, Washington legislators, and others.

Lasker created a discourse about health policy and funding that still frames contemporary political discourse. Medical research became an investment in the future. Her constant innovation and planning allowed for congressional discussion of disease incidence, medical research funding and distribution, and an evaluation of the federal government's commitment to improved health outcomes. She recruited politicians to her cause by bestowing financial support, using cronyism, cultivating political networks, and more. Perhaps most significantly, however, she tapped into the concerns and expectations of American citizens, defining an emerging popular belief that the federal government should fund medical research. To that end, she furnished politicians with language that the public could easily support. "Fact sheets" quantified persuasive data, "research payoffs" quantified the value of the investment, and larger "fact books" that the Lasker Foundation updated every two to three years introduced new language for new causes. As Lasker explained about political lobbying, "We furnished them with explanations of how the funds were used and with how much money was being lost by illness and premature death."[20]

Lasker's political efforts to expand federal funding for medical research inspired the rapid expansion of the National Institutes of Health (NIH), between 1945 and 1960. Within this fifteen-year window, six new institutes formed, and federal appropriations increased to more than $90 million. Lasker directed much attention to research on cancer and viruses in the 1960s. At the same time, impressed by the US Space program that had successfully dreamed big and realized its goal, Lasker zeroed in on the idea of a cancer cure. By 1969, convinced that a cancer cure could only be discovered with a similar groundswell of support, Lasker created and led a Citizens Committee for the Conquest of Cancer (CCCC). The organization adopted many of the successful publicity tactics that Lasker had initiated since the postwar years. It inspired public engagement, urged citizens to write to their political representatives, and claimed tangible and material goals. In perhaps the most illustrative example of its activism, the CCCC purchased full-page advertisements in popular newspapers, including the *New York Times* and the *Washington Post.* On December 9, 1969, readers opened their newspapers to find a huge and bold statement: "Mr. Nixon: You Can Cure Cancer." The word "Cancer" was emphasized by visual representations of cancer cells infiltrating the word. The script beneath the claim used both emotional appeal and expert testimony to emphasize the need for vast funding for cancer immediately. If one read the much smaller script, one learned that the appeal for funding suggested that science could discover a cure within the next five years.[21] By all accounts, this was an outrageous claim—however, it spoke to the emotions that had now become part of the discourse about health, funding research, and federal obligations to advancing science.

Strategic advertisements, political lobbying efforts, and a well-placed plea from columnist Ann Landers in 1971 urging her readers to write to their representatives all contributed to Nixon's support for the National Cancer Act. On December 23, 1971, President Nixon signed the act, which allowed for the creation of the National Cancer Program. Although Lasker had drawn some criticism over the years, in the early 1970s it became more pronounced. In short, critics argued that Lasker was not evaluating science through a rigorous lens; rather, she had learned how to make emotional pleas for funding.

In a 1972 photograph of the first meeting of the National Cancer Advisory Board, twenty-four individuals stand shoulder to shoulder in front of a table covered with academic papers. Mary Lasker is positioned in the front, a bit off to the left side. Twenty-three of the twenty-four people are wearing suits and ties. All appear white, and all but four have earned the title "Dr." This board membership reflects the dominant faces of medicine in the mid-1960s. Although she had never earned any formal science degrees and was often the sole female in a predominantly male environment, Lasker appears completely at ease in this photo—confident, poised, and standing tall. The photo captures her powerful presence. Remarkably, Lasker spent much of her career lobbying in a male-dominated field—both within medicine and politics. She straddled the gender divide with great grace, claiming friends frequently and bridging her personal relationships and political associates so carefully that it was hard to distinguish the two.[22]

In 1983, the Cold Spring Harbor Laboratory, long recognized as a premier research institute for cancer, held a conference entitled "The Cancer Cell" and dedicated the meeting to Mary Lasker. This ceremony signified the gratitude that academic and research facilities felt for Lasker's work. The evening celebration included a dinner; a program that included such medical leaders as Cold Spring Harbor Director Dr. James Watson, Director of the National Cancer Institute Dr. Vincent DeVita, and Secretary of Health and Human Services Margaret Heckler. A tribute booklet provided to attendees detailed how Lasker prioritized hope in an era when many felt desperate about disease. She possessed "a single-minded denial of hopelessness [that] became the cornerstone of a lifetime commitment to increasing public and governmental support of biomedical research."[23]

Lasker gained significant recognition for her work during her lifetime. In 1969, President Nixon awarded her the Presidential Medal of Freedom, created to recognize, "citizens of the United States of America who have performed exemplary deeds of service for their country or their fellow citizens."[24] Pediatric physician and early oncologist Dr. Sidney Farber used the occasion to articulate what Lasker's work meant to him. As he wrote in 1969, "I hope that I will have the privilege of working with you at least until the problems of cancer are no more. As for you, my heartfelt wish is that you continue to inspire, lead and even push and pull until all the dread diseases are no more. The health of the world owes more to you than any one person."[25] President Lyndon B. Johnson added to the praise: "Humanist, philanthropist, activist— Mary Lasker has inspired understanding and productive legislation which has improved the lot of mankind. In medical research, in adding grace and beauty to the environment, and in exhorting her fellow citizens to rally to the cause of progress, she has made a lasting imprint on the quality of life in this country. She has led her President and the Congress to greater heights for justice for her people and beauty for her land."[26] President Johnson's reference to Lasker's love of "beauty" refers to her support for parks and gardens. Throughout her life, Lasker also funded projects that planted bulbs and trees, especially in urban areas. As one testament to her purchasing power, in 1985 a pink tulip was named in her honor. Additionally, she worked to create and cultivate more green space in urban settings (especially New York City and Washington, DC), while supporting the fine arts.

In 1984, the NIH recognized the enormous role Lasker had played in assuring that the NIH would be created, expanded, and funded over time by naming a building in her honor—the Mary Woodard Lasker Center for Health

Research and Education. A few years later, in 1989, Congress bestowed the Congressional Gold Medal on Mary Lasker. When she died in her sleep from heart failure in 1994, at the age of 93, a statement on the Senate floor named her "an American hero and a legend bigger than life."[27] This sentiment was echoed in numerous obituaries from the *New York Times* to local newspapers throughout the country.

In short, the innovative funding structure for biomedical research that emerged in the post–World War II era owes much to the novel strategies, political connectedness, and tireless advocacy of Mary Lasker. She believed that investments yielded dividends—in business, politics, and philanthropy. She surrounded herself with allies who offered clever suggestions and necessary votes, sympathetic friends who championed her cause, and supporters who often held influential political seats or board seats. Although she held no official job title or office, Lasker could gain an audience with the US president, US senators and congressional representatives, influential CEOs, and the leaders of key philanthropic organizations. As a smart, bold, informed, and well-networked philanthropist, Lasker merged her political, social, and economic worlds and ushered in an era of government-supported research projects based on her inspiring dreams of medical cures and pathbreaking treatments.[28] She integrated a faith in medical research into the public consciousness, fostering government spending on issues near and dear to all. Lasker strongly believed that a better would emerge as science improved outcomes for diseases that ranged from cancer, to tuberculosis, mental health, and beyond. Throughout her lifetime, she pioneered innovative, imaginative, and significant funding structures that brought these dreams a bit closer to reality.

FURTHER READING

Gardner, Kirsten E. *Early Detection: Women, Cancer, and Awareness Campaigns in Twentieth-Century United States*. Chapel Hill: University of North Carolina Press, 2006.

Katz, Esther. "Mary Lasker." In *Notable American Women: A Biographical Dictionary Completing the Twentieth Century*, ed. Susan Ware, vol. 5. Boston: Harvard University Press, 2004.

Lerner, Barron. *Breast Cancer Wars: Hope, Fear and the Pursuit of a Cure in Twentieth-Century America*. New York: Oxford University Press, 2003.

National Library of Medicine. History of Medicine Division. Finding Aid to the Albert and Mary Lasker Foundation. Albert Lasker Awards Archives, 1944–1987. Online at http://oculus.nlm.nih.gov/lasker.

Oliver, Thomas R., ed. *Guide to U.S. Health and Health Policy*. Washington, DC: CQ
  Press, 2014.
Patterson, James T. *The Dread Disease: Cancer and the Modern American Culture*.
  Cambridge, MA: Harvard University Press, 1987.
Reminiscences of Mary Lasker. Oral History Research Office Collection of the
  Columbia University Libraries (OHRO/CUL).

NOTES

1. "Congressional Document," accessed 11 May 2016, at http://congressional.pro
quest.com/congressional/result/congressional/pqpdocumentview?accountid=7122&
groupid=114734&pgId=0458e42e-84a4-4487-9aeb-1a9ad9bb327c&rsId=154067A7B19.

2. Reminiscences of Mary Lasker (1962), part 1, session 1, in the Oral History Research Office Collection of the Columbia University Libraries (OHRO/CUL). Beginning
in 1962, and continuing until 1982, John T. Mason Jr. completed fifty-eight oral history sessions with Mary Lasker. The Columbia University Oral History Research Office
has transcribed the entire 1143-page record of these sessions. This essay is framed
by the content of this collection, as well as dozens of published accounts of Mary
Lasker's life.

3. Reminiscences of Mary Lasker (1962), part 1, session 4, 105. Mrs. Belter worked as
a laundress for the Woodards; Alice Woodard Fordyce (1906–1992) was Mary's younger
sister; Mrs. Dorr was the mother of one of Mary's closest friends; Paul Reinhardt was
Mary's first husband; Paul de Kruif published *Microbe Hunters* in 1926; and John D.
Rockefeller was one of the most significant private benefactors of early twentieth-
century medical research.

4. Reminiscences of Mary Lasker (1962), part 1, session 1, 25.

5. Reminiscences of Mary Lasker (1962), "Notable New Yorkers," 60–61, online at
http://www.columbia.edu/cu/lweb/digital/collections/nny/laskerm/transcripts/laskerm
_1_2_74.html. Wendell Wilke would become a presidential nominee in 1940; Margaret Sanger ran the American Birth Control League, soon to become Planned Parenthood; Walter Mack served as Republican national committeeman; Albert Lasker ran
the PR firm Lord & Thomas; David Sarnoff ran American Radio Broadcasting; and Karl
Menninger was a renowned psychiatrist. Her friend Kay Swift, aka Katharine Faulkner
Swift, was an American composer, of both popular and classical music.

6. In the 1930s, Albert Lasker donated $1 million to the University of Chicago for
disease research, but was disappointed with the impact. He renewed his interest in
securing medical funding after meeting Mary. See Cold Spring Harbor Laboratory,
"Program for 'A Tribute to Mrs. Albert D. Lasker'" (Cold Spring Harbor Laboratory Archives, 13 September 1983), online at http://libgallery.cshl.edu/items/show/72608.

7. Albert Lasker led the advertising firm Lord & Thomas until he sold it in 1940.
He was recognized as a founder of advertising, and certainly one of its most popular
practitioners. He often shared ideas and strategies learned in advertising in support of
Mary's health goals. For example, in 1950, Albert suggested renaming the Birth Control

Federation of America as Planned Parenthood in order to direct attention away from the "birth control" debate.

8. Reminiscences of Mary Lasker (1962), part 1, session 5, 128–30.

9. Reminiscences of Mary Lasker (1962), part 1, session 4, 126.

10. Reminiscences of Mary Lasker (1962), part 1, session 10, 278.

11. Joan Arehart-Treichel, "Lasker Award: Passport to a Nobel Prize?," *Science News* 102, no. 23 (1972): 365, doi:10.2307/3957395.

12. Arehart-Treichel, "Lasker Award," 365.

13. Carla Baranauckas, "Florence S. Mahoney, 103, Health Advocate," *New York Times*, 16 December 2002, sec. US, online at http://www.nytimes.com/2002/12/16/us/florence-s-mahoney-103-health-advocate.html. Mahoney and Lasker formed an effective political lobbying team, both securing political appointments and audiences critical to advancing medical research funding.

14. E. Bagg, "A Little Heart Trouble: Mary Lasker and the Founding of the National Heart Institute," *Texas Heart Institute Journal* 25, no. 2 (1998): 97.

15. Reminiscences of Mary Lasker (1962), part 1, session 10, 276.

16. Gary Cohen and Shannon Brownlee, "Mary and Her 'Little Lambs' Launch a War," *U.S. News & World Report* 120, no. 5 (1996): 76.

17. Cohen and Brownlee, "Mary and Her 'Little Lambs,'" 76; "Mary W. Lasker, Philanthropist or Medical Research, Dies at 93," *New York Times*, February 23, 1994, sec. Obituaries.

18. Vincent T. DeVita and Edward Chu, "A History of Cancer Chemotherapy," *Cancer Research* 68, no. 21 (2008): 8643–53.

19. Cohen and Brownlee, "Mary and Her 'Little Lambs,'" 76. See also "The NIH Almanac," online at https://www.nih.gov/about-nih/what-we-do/nih-almanac/appropriations-section-2.

20. Reminiscences of Mary Lasker (1962), part 1, session 11, 312.

21. "Display Ad 109—No Title," *Washington Post, Times Herald (1959–1973)*, 9 December 1969, sec. City Life.

22. "National Cancer Advisory Board First Meeting," Cold Spring Harbor Laboratory Archives Repository, Reference JDW/1/11/10, online at http://libgallery.cshl.edu/items/show/51481.

23. Cold Spring Harbor Laboratory, "Program for 'A Tribute.'"

24. "The Presidential Citizens Medal Criteria," online at https://www.whitehouse.gov/node/7913.

25. Cold Spring Harbor Laboratory, "Program for 'A Tribute.'"

26. Cold Spring Harbor Laboratory, "Program for 'A Tribute.'"

27. "Congressional Document" (see n. 1).

28. "Mary W. Lasker, Philanthropist for Medical Research, Dies at 93."

CHARLOTTE DECROES JACOBS

JONAS SALK

AMERICAN HERO,

SCIENTIFIC OUTCAST

Hope lies in dreams, in imagination and in the
courage of those who dare to make dreams into reality.
—Jonas Salk, 1977

INTRODUCTION

Jonas Salk all but eradicated a crippling disease, and the scientific community seemingly never forgave him. When the public learned on April 12, 1955, that his vaccine had prevented poliomyelitis, Salk became a hero overnight. Born in a New York tenement, he had dreamed since childhood of aiding mankind. The press portrayed this modest, congenial man as a lionhearted physician who had overpowered a deadly disease. The public showered him with gifts; international leaders bestowed their highest honors. Yet the scientific community all but snubbed him. This somewhat unconventional scientist had challenged one of their firmly held tenets, and they accused him of grabbing the limelight and debasing the field of science with the fanfare surrounding the vaccine's announcement.

Salk's role in preventing polio overshadowed his collaboration in developing the first influenza vaccine, his efforts to meld the humanities and sciences in the Salk Institute, and his pioneering work on AIDS. Although they represent further examples of the visionary, romantic Salk, I concentrate on his polio work. Within the framework of his life, I consider some broader issues of scientific discovery: the inherent tribulations, the politics of investigation, and the price of fame. Salk's journey exemplifies the ordeals and gratification of the visionary in science by relating how one man took on the medical establishment, at significant personal cost, to attain his dream—and the consequences thereof.

BORN WITH A CAUL

Jonas Salk was born into a Russian Jewish immigrant family on October 28, 1914, in East Harlem. Later in life he said he was shaped by his ancestors and by world events that occurred during his childhood. He revered his forefathers, who had survived the pogroms through their tenacity—a trait that would come to define him. Several memories haunted him: At a 1918

Figure 5.1. Jonas Salk in his laboratory. Photograph courtesy of University of Pittsburgh Archive Services.

Armistice Day parade, Jonas was saddened to see soldiers missing an arm or leg. That same year, influenza slaughtered thousands in New York, and he recalled standing on the sidewalk watching wagons pass by, filled with coffins.

According to his mother, Jonas was born with a caul, a thin amniotic membrane covering his face. She told him it was an omen, that he had special powers and was destined to perform noble deeds—quite unlikely for this slight, reticent boy who was most comfortable with his books. Yet he believed her. For Jews, a person is defined by his good works, and Jonas prayed that when he grew up, he would help relieve man's suffering. His hero, Louis Pasteur, had conquered devastating diseases with his scientific creativity, brilliant insights, perseverance, and daring, driven by his concern for the human condition. Young Jonas strived to emulate him.

BECOMING A SCIENTIST

Initially Jonas told his mother he wanted to be a lawyer and support the oppressed, but she pointed out that she prevailed over him in every debate. Besides, like his father, he disliked altercations. At City College of New York, Jonas became intrigued by the sciences and decided to become a doctor. His less than stellar grades, plus the unstated Jewish quotas, did not bode well for entrance into medical school. One aspect of his application did distinguish him—his ambition. He didn't plan to be a practitioner as did most applicants; he wanted to conduct medical research.

A matriculate at New York University College of Medicine, Salk began to question accepted dogma. "We were told that it was possible to immunize against diphtheria and tetanus by the use of a chemically treated toxin," he recalled in later years. "In the very next lecture we were told that in order to immunize against a viral disease, it was necessary to go through the experience of infection."[1] When Salk asked how both could be true, he found the answer unsatisfactory and determined to resolve the dichotomy one day.

Although a skilled clinician, Salk later said, "I didn't see myself practicing medicine. I saw myself trying to bring science into medicine."[2] During his senior year, he sought the mentorship of Thomas Francis Jr., the first US scientist to isolate a human influenza virus. Impressed by Salk's penchant for research, Francis assigned him the task of vaccinating mice against influenza, igniting his desire to prevent one of the world's worst contagions.

TACKLING INFLUENZA

Salk was just completing his internship at Mt. Sinai Hospital when Pearl Harbor was bombed. As the United States entered the war, influenza threatened the troops. During World War I, the 1918 influenza pandemic had killed almost as many young men as had died in battle. They desperately needed a vaccine, and Francis, now at the University of Michigan, was making some headway. Twenty-four-year-old Salk rushed to Francis's side, initially uninvited, to assist him. With a determination that exceeded his expertise, Salk worked tirelessly to help constitute and test an influenza vaccine. While Francis was traveling to suspected sites of outbreaks, Salk conducted clinical trials at Eloise Hospital for the Insane to prove its effectiveness. Although it was common practice to use inmates as research subjects prior to the adoption of ethical principles in 1947, Salk had a highly developed moral sense and likely rationalized his action by considering the potential catastrophe from another influenza pandemic. He went on to inoculate more than twelve thousand soldiers, reducing the rate of infection by 92%.

Salk reported that those inoculated with the influenza vaccine had anti-influenza antibody levels equivalent to those who had experienced an infection, and those levels remained elevated a year later. Furthermore, he and Francis observed what they called the "herd effect": In an immunized population, there were fewer infected individuals to pass disease on to others, thus limiting its spread. This would have major implications for future immunization programs.

In just five years, Salk matured from apprentice to independent researcher as he mastered the techniques for growing a virus and making a vaccine. Driven by the exigencies of war, he leaped from the laboratory to clinical application. Under Francis's guidance, one historian observed, "Salk acquired the administrative polish, the technical virtuosity and, above all, the philosophic grasp of viral disease that later enabled him to cope with polio."[3]

With time, Salk's relationship with Francis became tense. "I was not functioning in the expected way," Salk conceded years later. "I engaged in extrapolation because I had always felt that it was a legitimate means of provoking scientific thought and discussion. I engaged in prediction because I felt it was the essence of scientific thought . . . [but] neither extrapolation nor prediction were popular in virological circles."[4] Added to that, he discussed his ideas with reporters and accepted consultant fees from pharmaceutical companies—both considered improper among academicians at that time. As Salk's ambitions became more apparent, Francis became more critical. Although Salk could have benefited from further mentorship, he longed for independence and took a position at the University of Pittsburgh.

There Salk set out to tackle a major issue in preventing influenza. Since epidemiologists could not accurately predict which of many viral strains would emerge each year, they needed a vaccine with wide coverage. Yet the amount of virus one could add to each vaccine was limited. Immunologist

Jules Freund had shown in animals that mineral oil, added to a vaccine, concentrated virus at the inoculation site, enhancing the antibody response. Based on this, Salk proposed using mineral oil as an adjuvant to reduce the amount of virus per inoculate, allowing several influenza strains to be added to one vaccine. He tested this in monkeys, then in medical students, concluding in the *Bulletin of the New York Academy of Medicine* that "there is considerable latitude for the inclusion of many more strains . . . to cover the entire antigenic spectrum of both the type A and type B viruses."[5]

He was working with two drug companies, preparing to manufacture his new vaccine, when senior influenza investigators, thinking young Salk too rash in his conclusions, insisted his claims be substantiated by more seasoned researchers. Concerned that adjuvants might cause kidney damage and cancer, they blocked its approval. (Years later, Anthony Fauci at the National Institutes of Health wrote that adjuvants might be critically important in future influenza vaccines.) Salk was frustrated by his thwarted attempts to make a universal flu vaccine. So in 1947, when Harry Weaver, research director for the National Foundation for Infantile Paralysis (NFIP), invited him to join the effort to prevent polio, he agreed.

POLIO

By 1947, clinicians had described the clinical course of poliomyelitis, starting with a fever and sore throat. Scientists knew the disease was caused by a virus that attacked the anterior horn cells of the spinal cord, impairing motor control of one or a group of muscles, even causing respiratory failure. The virus was transmitted via the oral-fecal route and had a propensity for children. Prior to the 1900s, polio was endemic in the United States; most children were exposed in infancy when protected by their mothers' antibodies, transmitted in utero, imparting lifelong immunity. With the introduction of sanitation, those in the middle to higher socioeconomic groups, who practiced good hygiene, had no early exposure and were left defenseless. The first major epidemic in the US occurred in 1916 with 27,000 afflicted. Thereafter the numbers increased almost every summer, and no one could predict which community would be affected.

Franklin Roosevelt was polio's most famous victim. His law partner and friend, Basil O'Connor, helped him found the NFIP. With O'Connor as president, the March of Dimes was launched, and millions of dollars collected to aid victims and initiate a directed research program to solve the polio problem. Weaver invited thirty-three-year-old Salk to participate in an NFIP project to determine the number of different types of polioviruses, a prerequisite to making a vaccine. The typing project provided Salk entry into the elite ranks of polio researchers, among them Albert Sabin, a brilliant and contentious virologist who had first grown poliovirus in nervous tissue. Although Salk enjoyed their collegiality, he realized his junior status when he proposed two novel ways to speed the typing process. The group dismissed them outright. "It was like being kicked in the teeth," Salk said.[6]

In 1951, Salk made his polio debut at the International Poliomyelitis Conference in Copenhagen, where he reported identification of three distinct types of poliovirus. On his trip home, Salk developed a coveted friendship with O'Connor. Both aspired to improve the health of the public—O'Connor with his magnanimity, Salk with his optimistic vision. O'Connor appreciated Salk's cooperative spirit, lacking in many scientists, and how he looked "beyond the microscope," calling him a "human scientist."[7]

That same year, O'Connor appointed the Committee on Immunization, an advisory group to help him devise the optimal approach to polio. Salk was invited to join this distinguished group, which included Sabin, Francis, and leading microbiologists David Bodian, John Paul, John Enders, Thomas Rivers, Joseph Smadel, and Thomas Turner, with Weaver as chair. There was hardly a diffident scientist among the group; they argued vehemently over almost every detail of the polio prevention effort.

At that time, the standard approach to making an antiviral vaccine was to isolate the offending microbe, grow it on the proper media, then weaken its virulence by repeated passage though laboratory animals (which could take years), using this attenuated virus for immunization. Most virologists believed that only a live-virus vaccine could provide lifetime immunity by inducing a low-grade infection, as was the case with smallpox and yellow fever.

Salk deemed infection unnecessary in achieving immunity. He thought a virus could be killed and still stimulate an immune response while being much safer. He considered diphtheria, a deadly disease caused by a bacterium that released a potent toxin. If that toxin was made harmless by adding formaldehyde, it still engendered an immune response. The influenza vaccine had been made from a killed virus, although that was considered a special situation. The responsible virus mutated regularly, and epidemiologists had to predict which strain would emerge each year, then quickly make a vaccine against that strain—almost impossible with an attenuated virus. A killed-virus vaccine could be made within months. Finally, immunity was expected to last only for that flu season. Furthermore, Salk did not believe the detailed structure of poliovirus had to be discerned before developing a vaccine. Averse to discord, however, Salk kept these thoughts to himself.

Members of the Committee on Immunization railed against the NFIP for directing their research, although it provided funding; yet they expected their junior colleagues to follow their direction. In 1952, 57,800 Americans contracted polio. Salk didn't want to stand in line behind committee members, working on assigned tasks to make a live-virus vaccine on their timeline. He had a different plan, supported by Weaver and O'Connor, who wanted a vaccine before the next polio season. "There was nobody like him in those days," Weaver recalled. "His approach was entirely different from that which had dominated the field. . . . He wanted to leap, not crawl. His willingness to shoot the works was made to order for us."[8]

With methods similar to those he had used for influenza and using a modification of John Enders's new cell-culture technique, Salk and his small laboratory team made a killed-poliovirus vaccine, containing all three types, in just over three months. The rapidity with which he made the vaccine lead some scientists to consider his work slapdash. Salk, however, was a meticulous, assiduous scientist. He conducted the first human trials at the D. T. Watson Home for Crippled Children in secret, fearing obstruction by committee members. Years later, Salk said the crowning point for him in the entire polio saga was the first time he measured antipolio antibodies in a child. At that moment, he believed he could prevent poliomyelitis.

When Salk disclosed his work to the committee, they were shocked. In the ensuing interrogation, some expressed doubt that antibody levels from inactivated virus could equal those from natural infection and expressed concern about using mineral oil adjuvant. Others worried that material derived from monkey kidney cell cultures could cause organ damage. Sabin depicted the results as implausible, humiliating Salk. Committee members took sides: those who favored a prompt nationwide trial of his vaccine and those who didn't. Among the most vocal in the latter group were Sabin and Enders. Weaver and several others felt they had a good vaccine at hand, and the next polio season would soon be upon them. It was time for a large, randomized trial. Salk didn't agree with either. He wanted to test his vaccine in several thousand children, and if it reduced the number of predicted polio cases, deem it a success. He felt it unethical to conduct a randomized trial where half the children would get placebo. "I would feel that every child who is injected with a placebo and becomes paralyzed," he told O'Connor, "will do so at my hands."[9]

The committee at an impasse, Weaver appointed a Vaccine Advisory Committee, composed of health-care leaders who favored an immediate, large, randomized trial. Furthermore, he told Salk it would be considered suspect if he tested his own product and turned over its design and analysis to Thomas Francis Jr. When pharmaceutical companies began large-scale production, some took shortcuts, which Salk worried might reduce its safety. The Public Health Service began dictating changes in his formulation, which Salk feared might impair its efficacy. Every week brought some new setback: vaccine tainted with live virus, rumors of possible kidney damage in children, the major needle manufacturer on strike. Despite these issues, on April 26, 1954, the biggest clinical trial in the history of medicine commenced, conducted in almost a million children, financed and carried out by March of Dimes volunteers. Across the country, first-, second-, and third-graders were randomized to receive inoculation with vaccine or placebo. Salk stood on the sidelines and received no preliminary results from Francis until the day of the announcement.

On April 12, 1955, Francis presented the highly-anticipated results of the trial in a public forum at the University of Michigan: The vaccine had been 80 to 90% effective against paralytic polio and harmed not one child. "POLIO IS CONQUERED,"[10] "TRIUMPH OVER POLIO"[11] announced newspapers around the world. Many compared the celebration to the end of a war. The vaccine was licensed that afternoon, and thousands of pounds of polio vaccine were shipped across the country to vaccinate millions of children.

Two weeks into the national immunization program, the surgeon general received reports that seven children had become paralyzed following inoculation, most with vaccine from Cutter Laboratories in Berkeley. He suspended all vaccinations. When government regulators reported no problems with Cutter's production, many scientists blamed Salk, calling his viral inactivation technique inadequate. Salk did his own investigation and found Cutter had deviated from his filtering procedures, leaving live virus in the vaccine. More than two hundred people contracted polio directly or indirectly from contaminated vaccine; eleven died. Salk was devastated. With stricter manufacturing guidelines and oversight, the vaccination program resumed.

BACKLASH

Salk received thousands of letters, gifts, and awards from the grateful public. His picture appeared on the front page of every newspaper, on the cover of every major magazine. Heads of state gave him their highest honors, among them the Presidential Citation and French Legion of Honor. Yet he received scant praise from the scientific community. Their rebuke cast a shadow over his achievement.

What lay beneath these aspersions? This forty-year-old physician-scientist had made and initially tested his polio vaccine in secret while challenging one of their firmly held principles. Despite concerns that several scientists could not reproduce his inactivation technique and that most senior virologists favored a live-virus vaccine, Basil O'Connor promoted Salk's vaccine.

Several resented Salk's favored position, referring to him as "the chosen."[12] Academic leaders criticized the Ann Arbor Symposium, where the results of the trial had been presented, calling the event and associated media frenzy an "extravaganza," a "circus" that had violated the usual peer review process. Although Salk had no part in its planning, anti-Salk sentiments reverberated, and he was chided for showmanship and greed. "I knew right away that I was through," he later reflected, "cast out."[13]

Some complained that Salk had grabbed the spotlight, neglecting to recognize other polio researchers. Salk had tried to share the credit, but the press rarely mentioned other investigators. Following the adage, "names make the news," they crowned Salk the icon for polio. With the vaccine's success, Salk became one of the most celebrated physician-scientists in medical history, so one cannot underestimate the role of jealousy, a repercussion experienced by many successful visionaries.

Salk's lack of recognition by the scientific community (save the Lasker Award) underscores their view of his achievement. Several nominations for the Nobel Prize were rejected by the Nobel Committee. Years after his death, the permanent secretary of the Royal Swedish Academy of Sciences revealed the details.[14] Sven Gard, a Karolinska Institutet virologist with expertise in polio, was on the five-member Nobel Committee that reviewed Salk's work. He said that Salk's demonstration that a killed virus could achieve immunity had already been demonstrated and that he had capitalized on the work of others. Gard went on to say he could not verify Salk's viral inactivation methods and held Salk responsible for the Cutter incident. This omission by the Nobel Committee has generated debate over the years. "The fact that a fundamental advance in human health [rather than a basic biologic breakthrough] could not be recognized as a scientific contribution," said Nobel laureate Renato Dulbecco, "raises the question of the role of science in our society."[15]

Surely his accomplishments merited admission into the National Academy of Sciences, but several members contended he had made no original scientific discovery. He was compared to a director of product development at a pharmaceutical company. Some suspected that Sabin had blackballed him. Calling his rival's work "kitchen chemistry," Sabin berated Salk. "He never had an original idea in his life," he said. "You could go into the kitchen and do what he did."[16]

Salk was not blameless. He reached out to the public in ways few physician-scientists had. No serious scientist gave interviews to *Good Housekeeping* and *Parent* magazine or showed a television audience how to make a vaccine using a Waring blender. Salk told his secretary he felt obliged to the public; they had supported his work with their dimes. Nevertheless, scientists accused him of crossing the line of acceptable academic behavior by soliciting media attention. To some extent they were correct; he had learned early on how to cooperate with the press to get public support.

FATE OF THE SALK VACCINE

Five years following the NFIP trial, Salk's killed-polio vaccine had reduced the incidence of paralytic polio in the US by 90%. Even so, most senior virologists considered the killed-virus vaccine a stopgap. They maintained that only a live-virus vaccine could eradicate polio by providing lifelong immunity. Additionally, it generated a herd effect, as the excretion of live, weakened virus by vaccinees could immunize the unvaccinated. Sabin prepared such a vaccine, which could be delivered in a sugar cube, and tested it extensively in Russia, reporting its safety and efficacy. In 1961, the US Public Health Service replaced Salk's vaccine with Sabin's oral vaccine, citing cost and convenience.

Salk warned that a live poliovirus, although weakened, could revert to its original form and cause polio. In his efforts to have Sabin's vaccine delicensed, he was overruled by the American Medical Association (AMA), American Academy of Pediatrics (AAP), the surgeon general, and Center for Disease Control (CDC). They argued that Salk's vaccine had not been 100% effective; in 1957, six thousand people in the United States had developed polio. Salk countered that the reason was socioeconomic not biologic, reporting low inoculation rates in some areas. Furthermore, he could still measure antibody six years after vaccination. Yet by 1968, US pharmaceutical companies had stopped manufacturing Salk's vaccine.

With time, most cases of paralytic polio in the United States could be traced to Sabin's vaccine, but the numbers were few, and Sabin denied the charge. For years, Salk tried to reverse what he called a dangerous, politically driven decision. He argued that after being vaccinated, individuals excreted live virus for up to eight weeks and could spread it to others. If the virus reverted to its more virulent form, an epidemic could ensue. Salk sent articles and editorials to the leading medical journals, but most were rejected. He repeatedly appealed to the major health organizations, testified before two congressional committees, and even approached President Jimmy Carter. The press began to depict the debate between Salk and Sabin as a medical feud.

When policy did not change, Salk reached out to the public. "In the absence of any other voice," he commented in the *New York Times*, "I feel a responsibility to inform the public that they can justifiably demand . . . a vaccine which is not only effective but completely safe." He disclosed that authorities knew the live-virus vaccine caused most of the current polio cases in the United States. He called the decision to switch vaccines "unnecessary and ill-advised."[17] Furthermore, Salk prompted newsmen to report individual tragic stories.

Salk's tenacity became legendary. Late in his career, he teamed up with pharmaceutical tycoon Charles Merieux and Dutch scientist Toon van Wezel to develop an enhanced killed-virus vaccine which could be mixed with other childhood vaccines. In 1999, the US government recalled the Sabin vaccine, replacing it with this newer version of Salk's vaccine. He never could rejoice in the decision however; he had died suddenly from complications of heart failure on June 28, 1995, at the age of eighty.

CONCLUSIONS

With his polio vaccine, Jonas Salk changed the health of the public. He faced enormous criticism when he challenged a basic biologic principle, but he persisted. He cast himself as a bench-to-bedside physician-researcher, a rare species at the time he began his career. While researchers were debating how to prevent polio, Salk, having never conducted any work on the poliovirus, leaped ahead to his vision of a preventive vaccine. He was enabled in his quest by the financial and moral support of the NFIP, specifically Basil O'Connor. In challenging an established doctrine, Salk broadened thinking about how to achieve a lasting immune response. The future would see vaccines not only made from killed viruses but also from parts of the microbe (subunit vaccines) or from its DNA.

Salk was the people's scientist, glorified by the public not only because he had made a vaccine, but also because he had relieved them of an overwhelming fear of one of the most frightening diseases of the twentieth century. Salk influenced how the public viewed science, stimulating its participation in health issues. With development of his vaccine came one of the first widespread vaccination programs. Salk called the immunization of all the world's children a moral commitment, and thanks to Rotary International's Polio Plus Program and the Global Polio Eradication Initiative, we are approaching worldwide extinction of polio.

Two aspects of Salk's personality annoyed public health officials and the scientific community. Confident about his approach to problems, he rarely expressed self-doubt. And he had a fierce tenacity, repeating his views over and over. Many found this doggedness exhausting. Sabin's status among scientists proved a strong advantage. They accepted his views over those of Salk, whose status as a hero hampered his acceptance by the academic world.

The medical world needs big thinkers and visionaries. Pasteur was a beacon for Salk, and Salk has been and, we may hope, will be for future physician-scientists. But academic medicine changed drastically during and since Salk's lifetime. The bench-to-bedside physician is becoming obsolete. The solo clinical investigator has been replaced with large clinical trial groups, which require collaborative agreement, stifling the visionary; and pharmaceutical-sponsored trials, with specific endpoints, provide little leeway for the revolutionary.

Besides that, government rules and regulations, funding sources, peer review, and public scrutiny limit the dreamer in his or her quest. Only three and a half years elapsed from Salk's initial work on the vaccine until its public release. Today, with no NFIP equivalent, the physician-scientist competes for National Institute of Health grants, of which less than 15% are funded. Then the dreamer must negotiate the FDA's Investigational New Drug application, Institutional Review Boards, and a Biologics License Application. Trials are conducted by research professionals, not volunteers, adding substantially to the expense. Today, development of a new vaccine could take ten years and cost more than a billion dollars.

Salk's life serves as an exemplar of how a visionary can succeed. Dreamers need brilliance and/or creativity, fortitude, self-confidence, clear underlying motivation, as fame and fortune may elude them, and acceptance of the possible price—minimal recognition by peers and lack of academic promotion. The visionary must be prepared to face failure. When in later life, a journalist asked Salk what he considered his failures, he replied: "Failure is not a term that I would use. My whole life has been made up of challenges."[18]

Central to Salk's character was a spiritual belief that he had been born with a mission. Despite attacks by the scientific community, he never struck back; he personified equanimity. Only in his private notes did he reveal himself: "My intention and purpose has been to do good in the world through the advancement of knowledge; that of others is only the advancement and dissemination of knowledge. That's the difference. . . . That's what they can't forgive me for. . . . Is that beneath our dignity as scientists?"[19] At the core of Salk's being lay a passion—a passion to solve medical problems and aid humanity, which helped him weather the storms of criticism. In the end, his romantic vision, combined with his idealism and tenacity, did improve the health of mankind.

FURTHER READING

Carter, Richard. *Breakthrough: The Saga of Jonas Salk.* New York: Trident, 1966.
Hellman, Hal. *Great Feuds in Medicine.* New York: John Wiley & Sons, 2001.
Jacobs, Charlotte DeCroes. *Jonas Salk: A Life.* New York: Oxford University Press, 2015.

Norrby, Erling. *Nobel Prizes and Life Sciences*. Singapore: World Scientific Publishing, 2010.

Oshinsky, David M. *Polio: An American Story*. New York: Oxford University Press, 2005.

Paul, John R. *A History of Poliomyelitis*. New Haven, CT: Yale University Press, 1971.

NOTES

1. Judith Bronowski, producer/writer, *Jonas Salk: Personally Speaking* (San Diego: KPBS, 1999), VHS videotape.

2. Jonas Salk interview, 14 November 1990, March of Dimes Archives, March of Dimes Foundation, White Plains, NY.

3. Richard Carter, *Breakthrough: The Saga of Jonas Salk* (New York: Trident Press, 1966), 46.

4. Carter, *Breakthrough*, 51.

5. Jonas Salk, "An Interpretation of the Significance of Influenza Virus Variation for the Development of an Effective Vaccine," *Bulletin of the New York Academy of Medicine* 28 (1952): 761.

6. Carter, *Breakthrough*, 81.

7. Carter, *Breakthrough*, 121.

8. Carter, *Breakthrough*, 68–69.

9. Salk to O'Connor, 16 October 1953, Jonas Salk Papers, Mandeville Special Collections Library, University of California, San Diego, La Jolla, CA.

10. *Pittsburgh Press*, 12 April 1955.

11. *South China Morning Post*, 13 April 1955.

12. Interview with Lorraine Friedman, conducted by Charlotte Jacobs, 18 August 2004.

13. Carter, *Breakthrough*, 3.

14. Erling Norrby, *Nobel Prizes and Life Sciences* (Singapore: World Scientific Publishing, 2010).

15. R. Dulbecco, "Obituary: Jonas Salk (1914–95)," *Nature*, no. 376 (1995): 216.

16. Hal Hellman, *Great Feuds in Medicine* (New York: John Wiley & Sons, 2001), 126.

17. "Polio: The Cure for the New Controversy," *New York Times*, 26 May 1973.

18. "Twentieth Century Miracle Worker," *Modern Maturity*, December 1984.

19. Personal notes, Jonas Salk Papers.

6

THE ORIGINS OF
"DYNAMIC RECIPROCITY"

MINA BISSELL'S EXPANSIVE PICTURE

OF CANCER CAUSATION

INTRODUCTION

Open any textbook in cancer biology, and the story will be more or less the same: Cancer is a disease of disorderly cell growth, caused primarily by mutations to cancer cells. These mutations specify the aberrant phenotypes of cancer cells: decreased dependence on external stimuli to continue dividing, the ability to attract a blood supply, evade growth suppressors, resist cell death, enable replicative immortality, and activate invasion and metastasis. These "hallmarks" of cancer are all taken to be due (primarily) to changes in gene expression: "cancer [is] a disease involving dynamic changes in the genome. The foundation has been set in the discovery of mutations that produce oncogenes with dominant gain of function and tumor suppressor genes with recessive loss of function."[1]

This picture of cancer emerged gradually over the course of the twentieth century. At the height of this period, in 1989, the Nobel Prize in Physiology or Medicine was awarded to J. Michael Bishop and Harold E. Varmus "for their discovery of the cellular origin of retroviral oncogenes." Bishop and Varmus found that "c-src" was a gene shared by all vertebrates. Initially Howard Temin had proposed that it was a "viral" gene imported into chicken cells via Rous sarcoma virus, or RSV, a retrovirus associated with sarcoma in chickens.[2] In fact, RSV was the carrier of a mutated version of c-src. Bishop and Varmus demonstrated that all vertebrates share a vulnerability to cancer, due to the acquisition of somatic mutations to genes like src. Of course, no single mutation was sufficient for a cell to behave like a cancer cell; so, the discovery of src launched a search for more "oncogenes" and "tumor suppressor" genes, mutations to which play essential roles in the regulation of cell birth and death, and in DNA repair. This research program is sometimes called the "oncogene paradigm."[3]

Mina Bissell had the great fortune (or misfortune) to begin her career in cell biology at Berkeley in the 1970s, just when this research program was beginning to take off. She came to work with Harry Rubin at Berkeley in 1970. Rubin had worked with Temin at the California Institute of Technology

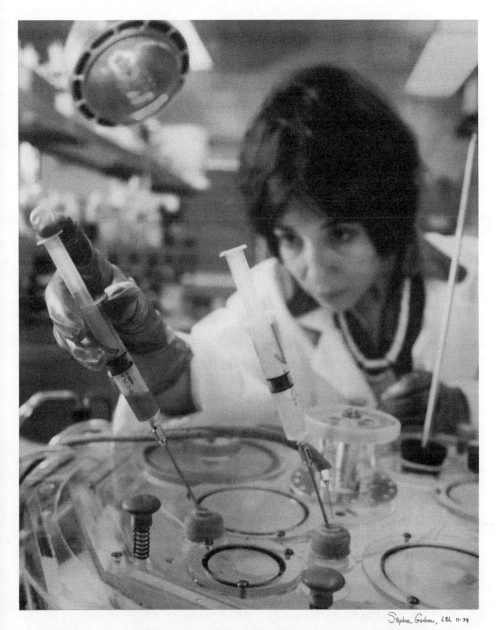

Stephen Garber, LBL 11-74

Figure 6.1. Mina J. Bissell performing a quantitative tracer study of the metabolic dynamics of animal cells growing in tissue culture.

(Caltech) in the 1950s and 1960s on RSV. Bissell entered Rubin's lab as a relative "outsider"; her graduate research was in bacteriology and biochemistry. As an outsider, she was less likely to take for granted what others regarded as obvious. Her persistent curiosity and habit of asking questions about matters that others took for granted served Bissell well. Indeed, they eventually led her to challenge mainstream views of both cell biology and cancer causation. Bissell argued that cancer cells and their microenvironment are in constant interaction over the course of cancer initiation and progression. What at the time was a radical hypothesis about "dynamic reciprocity" between cells and their environment has now become mainstream. It is now widely accepted that the tumor microenvironment plays a significant role in cancer progression.[4]

Over the course of twenty years of painstaking research, Bissell, along with her colleagues and students, showed how the causal pathways regulating cell behavior were a two-way street. Healthy cells and cancer cells are both highly context-dependent in their behavior. The pathway to this insight was not direct. Bissell's work began with research into cellular metabolism. As a result of this early research, she found herself challenging the idea that intercellular signaling molecules that had previously been regarded as serving "housekeeping" functions were not simply permissive conditions, but could be altered in ways that changed cell phenotype. Cells, it turns out, can "change their fate"—once a cell becomes differentiated into a liver, kidney, or skin cell, its fate is not necessarily sealed. This insight had significant implications: it was not simply intrinsic properties—such as mutations to "oncogenes" and "tumor suppressor" genes—that determine the typical behaviors of cells, but also the extracellular matrix, the signaling molecules and cell and molecular milieu. Even cancer cells, she found, were profoundly influenced by signals in their microenvironment. "Context," in Bissell's view, "is everything."[5]

BECOMING A BIOLOGIST

Bissell was not only a disciplinary "outsider." She was also an immigrant to the United States, and a woman in a male-dominated field. Over the course of her career, Bissell became an immigrant into many different fields, drawing upon the expertise of developmental biologists, mammologists, geneticists, and engineers. Bissell deliberately chose research that was very far from the mainstream. By working (in her words), "away from the rat race," she could indulge in novel experimental and theoretical approaches to questions that interested her. Rather than jump on the "bandwagon of oncogenes" in the 1980s, she chose projects that allowed her to have a family alongside a career in science.

Mina Bissell was born in Iran. Her grandfather was an ayatollah, a member of the religious elite in Iran. "I was lucky to be in an educated family. My dad came from a family that were intellectually very different—all passionately debating each other."[6] Her father, the eldest of ten children, was expected to follow his father and become an ayatollah. However, Bissell's father broke this tradition, receiving a law degree and later a PhD in law. "[My father] was a rebel and I think I took it from him. I was very curious—that was not a problem—I was an atheist—and that was not a problem. The family did not try to muzzle me. We would debate at the table. . . . My father said, 'You can do anything, but don't become a lawyer. The religious stuff is a joke. You can do anything you want'" (MB, 8/15). Bissell grew up in an environment where tolerance for competing views was permitted and indeed, encouraged. Nonetheless, respect for authority was also a core value: "One of the amazing characteristics of Iranians—they are extremely loyal. You always revere your teacher" (MB, 8/15). Bissell had great admiration for her teachers and colleagues; nonetheless, her earliest insights came from a willingness to take notice of anomalies, resist authority, and ask questions.

Bissell came to the United States at eighteen to study at Bryn Mawr and transferred to Radcliff College. She entered Harvard Medical School in pursuit of a PhD in biochemistry and bacteriology, recruited by Dr. Luigi Gorini, a bacteriologist and biochemist. Bissell's project when first arriving at the lab was testing the hypothesis that calcium induced secretion of a proteinase in a bacterium. Gorini was already convinced that this was true; he was, essentially, asking her to demonstrate what he had already become convinced must be true. However, despite years of effort, she could not prove what Gorini was convinced he already knew. She became very discouraged: "after four years, I concluded that my data did not support the hypothesis." She went to her advisor and suggested an alternative explanation: "Looking over the data we had accumulated over the years, I realized that all could be explained if we assumed that this proteinase was threading its way out of the cell 'unfolded,' possibly from membrane-bound ribosomes. It was not detected in the absence of Ca^{2+} because bacterial growth required constant shaking, and the protein, which had no disulfide bridges, was destroyed as it was secreted owing to surface tension and, possibly, transient proteolytic activity. Ca^{2+} would stabilize the protein by substituting for disulfide bridges" (MB, 8/15). Gorini was dismissive. Bissell remembers him saying, "'What do you think this protein is? Spaghetti?! Perhaps you should go back to ballet dancing, because you will not succeed as a scientist.' I was devastated" (MB, 8/15). She was not, however, deterred. She scheduled a meeting with a professor in the same department (Elmer Pfefferkorn), who explained both how radical her theory was and why the burden of proof was on her, and required her to do further experiments. She was able to demonstrate using radioactive tagging of the enzyme that "the enzyme was indeed secreted at all times, but because it was unfolded, it degraded in the absence of Ca^{2+}." That is, contrary to Gorini's initial hypothesis, calcium did not induce proteinase synthesis or excretion, but merely arrested degradation of the enzyme after its release, thus permitting accumulation of the active enzyme. Gorini conceded that she was right about the spaghetti hypothesis after all, and her results were published in 1971:[7] "Luigi and I went over all the experiments and agreed that only the new hypothesis could explain the data" (MB, 8/15).

This experience was a turning point for Bissell. The work was a lesson on the importance of persistence in the face of recalcitrant data. Moreover, the experiment demonstrated that the "obvious" explanation—the one that even respected authorities accepted—was not true. Bissell learned from this case that what may at first appear to be a direct cause may be an indirect one, with a far more complex causal pathway, involving causes that persisted in the local environment. The presence of the proteinase was due to a complex set of interactions between several factors, not a linear pathway from "initiator" to the outcome, as her advisor had assumed.

Bissell was very busy as a young mother and graduate student at Harvard; moreover, her life was rather chaotic: "I left my first husband in the fifth year of graduate school. He was a ladies' man. I was so busy with a baby; his eyes would go here and here, and I thought, 'Enough.' I was about to leave him, and then I ran into a guy over a centrifuge. Six-foot-five and very handsome. A year or two later he and I got married" (MB, 8/15). Her second husband wished to return home to California, and so she looked into postdoctoral positions at Berkeley. Though she knew "almost nothing in particular" about Harry Rubin, other than that he had produced important experimental work on protein products in bacteria, she applied to join his lab.

It turned out that Rubin's lab was a very exciting place to be. The Bay Area was a hotbed of research into cancer causation. The focus of Rubin's lab was RSV. Among the postdocs who passed through during the time were Peter Vogt, Peter Dosberg, and G. Steve Martin, several of the scientists who were among the first to show "the nature of the oncogene." According to Bissell, "Harry's lab was exciting" (MB, 8/15). Apart from her excitement, Bissell often found herself questioning the methods and assumptions of her colleagues; for instance, she noted that when cultures were removed from the incubator, the change in pH affected the cells they were studying. This observation was what prompted her to develop several novel tools for studying cells in culture.

Cancer genetics became the focus of a good deal of these researchers' work, especially after Bishop and Varmus showed the common vertebrate origin of *src*. But despite the excitement around cancer genetics in the lab, Bissell was unwilling to take up the "hot" new topic "Everybody was doing oncogenes, competing with each other . . . I wanted to have another kid" (MB, 8/15). So, she took up a project that was considered (at least at first) relatively marginal: "I found the Warburg effect. Tumors become aerobic, and they have glycolysis. No one was paying attention to Warburg."[8] Bissell thought Warburg's work on glycolytic metabolism in cancer was promising, contrary to popular sentiment:

> People thought he was out to lunch in the '70s. Furthermore, there was a huge amount of controversy. People would say, "Well, the normal cells are glycolytic, and he doesn't know what he is talking about. . . ." There were a lot of very dogmatic things in the literature. I did not know if they were right or wrong . . . I was reading all these things . . . people were saying: "housekeeping functions" (meaning they're not important) . . . and, "Luxury molecules"—meaning that they are tissue specific . . . everyone was calling glycolysis a housekeeping function. And, in the '70s, they discovered the gene for the RSV that causes cancer. So, everybody was doing oncogenes. And, glucose was a housekeeping function—it wasn't important. I said to Harry (Rubin)—I think that there is a real problem here. (MB, 11/15)

Bissell's work on the Warburg effect was the first stage in a gradual process leading to her rethinking of how the tumor microenvironment shaped cell behavior.

HOUSEKEEPING FUNCTIONS AND "GO-GO GIRLS"

Bissell in part stumbled upon the Warburg effect. She was initially inspired to work on metabolism because of how struck she was by the way cell culturists treated some by-products of glycolysis as proxies for enzymatic activity that was not well understood. She was curious about how the process of glycolysis went forward in the actual intercellular environment of the cell, rather than in artificial cell culture. She suspected that cells' behavior was affected by features of their environment—temperature, pH, intercellular signals—in ways not well understood, partly because these things were ignored in standard cell-culture studies. Her early work in Rubin's lab on metabolism in normal and RSV-transformed chick embryos tested exactly this assumption. It required, however, that she work with local engineers to develop a "steady state" apparatus to keep temperature and pH constant. She explained,

. . . in that paper, I did very careful work of measuring. What I realized was that when people will say, "anaerobic glycolysis," they were just measuring lactate production. . . . Or they would measure an enzyme activity in the beginning of the glucose pathway. And nobody was saying, "What goes in, and what comes out?" So I measured glucose uptake in relation to lactate production. And I did kinetic studies; . . . we developed this "steady state apparatus"; . . . the temperature and everything was constant. . . . We did fibroblast, we did muscle, we did liver, we did mammary gland (that came a little bit later)—and I could show the pattern of glucose metabolites, . . . that the pattern was completely different in fibroblasts and liver, and all the other places, . . . so I said, "Now wait a second? What is this housekeeping stuff? Everything about these tissues is different—even in glucose metabolism!" So, that was the beginning of that bright awakening. (MB, 11/15)

The steady state apparatus maintained constant pH and temperature for cells in culture, which was essential to Bissell's careful measurement of metabolic processes.

This early success inspired Bissell to question whether indeed it was appropriate to think of the behavior of cells as falling into two mutually exclusive categories: "constitutive" versus "inducible." Inducible behaviors were tissue specific, whereas constitutive behaviors were things that all cells did, and therefore (it was assumed) all cells did the same way. One such thing that all cells did was the so-called "housekeeping" function of precursors to glycolysis. Bissell became convinced that "housekeeping" functions were not "merely" housekeeping; in fact, she suspected this way of speaking about intercellular signaling assimilated the role of such molecules to the role of women and thus blinded researchers to their potential significance. Or, in her words, "Pff! Housekeeping functions and Go Go girls!"

Bissell found that the literature on cell metabolism and related work in cell culture was flawed, in part because the environment was not suitably controlled. In order to track cellular metabolism, she used a radioactive tracer to follow the kinetics. Again and again, she found that there were significant differences in rates of transfer of glucose in different cell types.

In one important experiment in 1976, she was able to induce normal rates of glucose transfer in cancer cells by reducing the amount of glucose added to culture, and to achieve the reverse effect in normal cells: "I brought tumor cells to the level of normal, and I brought normal to the level of tumor. And, normal cell pattern became tumorigenic, and the cancer cells became normal. I was jumping up and down—excited about this!" (MB, 11/15).[9]

What Bissell had shown was that what were merely considered "housekeeping" functions of the cell (performed the same everywhere), were in fact different in different cells in the body. The merely "constitutive" functions of the cell were highly variable, not "mere" housekeeping at all, but highly context sensitive. More radically, Bissell had demonstrated that the constitutive features of cell were inducible. That is, the distinctive glucose metabolism of cancer cells could be shifted by shifting the *environment* of the cell. This was the first in a series of experiments that suggested to Bissell that a cancer cell's signaling environment could have a significant influence on its phenotype.

Bissell invited many young women to work as graduate students and post-docs in her lab at Berkeley, and some of these women played an important role in her research on the role of context in shaping cell behavior. For instance, work by her postdoc Joanne Emerman demonstrated that one could change the shape of a mammary cell by changing the extracellular matrix.[10] Emerman showed that growing the cells in a different type of culture (3D culture) caused them to maintain milk protein (called beta casein), whereas growing them in a standard culture led to a different phenotype: "We showed that the pattern of metabolism was very different in 2D and 3D [cell culture]. That led to the question of why was it that this happened? Is it because the cells maintain the milk and don't degrade it, or is it because when you float them and change the shape, the whole rate of synthesis and transcription and translation changes?" (MB, 11/15). Bissell urged another graduate student, Eva Lee, to address this question. Namely, "Is this endogenous production of albumin, which then says that the shape signals? Or is it maintenance and inability to degrade as much as you do under this condition?" Lee was able to demonstrate that the mammary cells in this novel environment were taking up radioactivity, incorporating it into the milk protein, and making milk endogenously, something that had never been shown before. Environmental conditions permitted this effect to come about: "[The] floating collagen gel is both permissive and inductive for casein mRNA levels—i.e., the gel allows a preferential induction of casein mRNA in the presence of lactogenic hormones."[11] In other words, environmental conditions could affect the phenotypic expression of cells in culture, and regulation of cellular behavior was influenced at several points.

Bissell was struck by these experiments; they caused her to question the idea of "terminal differentiation." In other words, cells simply did not have a fixed phenotype, but could change in response to different environmental conditions: "You can change the behavior of the cells depending on the context" (MB, 11/15)

In a single-authored critical review paper published in 1981, "The Differentiated State of Normal and Malignant Cells in Culture; or, How to Define 'Normal' Cells in Culture," Bissell summarized these results and reviewed a vast body of literature on developmental and cell biology that likewise challenged many assumptions about "terminal differentiation." She opens her paper with a bold claim: "If there is one generalization that can be made from all the tissue and cell culture studies with regard to the differentiated state, it is this: since most, if not all, functions are changed in culture, quantitatively or qualitatively, there is little or no 'constitutive' regulation in higher organisms; i.e., the differentiated state of normal cells is unstable and the environment regulates gene expression."[12]

Bissell opens the paper with a puzzle: What are the defining conditions of cancer cells in culture? She argued that any attempt to identify a defining feature of cancer cells was absurd, in part because cancer was a heterogeneous and complex disease, and in part because cell biologists had been unable so far to characterize *normal* cells in culture. This is because tissue-specific traits were often lost in culture. Regulation of function *in vivo* depended on complex reciprocal interactions between cells and their microenvironment. Though developmental biologists had already demonstrated this in many cases, it had failed to come to the attention of cell biologists. One of her central arguments in this paper (and in several subsequent papers in the next ten years) was that cell biologists had paid insufficient attention to advances in developmental biology, which seemed to show that the very idea of "terminal differentiation" should be abandoned.[13] In her words,

> To my mind, this distinction between the "differentiated" and "nondifferentiated" traits is both arbitrary and unnecessary. The distinction may have served a useful purpose in the past. However, at present it causes confusion, and worse, implies a scientific categorization and truth, which simply is not well taken. . . . The initial definitions and the connotations of these words may have kept us from appreciating and understanding the eukaryotic cell in its totality. This confusion may also have contributed to our disregard for culture conditions and our inability to keep cells differentiated in culture.[14]

In sum, Bissell was challenging a received view in cell biology about terminal differentiation and suggesting that the relationship between cells and their environment was far more dynamic than had been imagined. Moreover, many of the same signals and interactions that shaped development might also play a role in cancer. This was the launching pad for her subsequent work on the possibility of reversing the cancer phenotype by altering the tissue microenvironment.

The next two decades were the most productive in Bissell's career. At least four innovations are worth mentioning. First, Bissell's lab was instrumental in demonstrating how and why wounding can induce tumors' growth, via the promotion of molecular mediators such as TGF-I (transforming growth factor).[15] Second, Valerie Weaver, a postdoc of Bissell's, and the Bissell lab, developed and modified a three-dimensional extracellular matrix assay to study the progression of human breast cancer in a functionally relevant cell-culture model. Essentially, the goal was to develop a tumor environment assay that would mirror the tumor microenvironment histologically.[16]

Third, this same group demonstrated that human breast cancer cells treated with an inhibitory antibody (Beta-integrin) could revert the cancer cell to the normal phenotype. That is, signaling molecules associated with tissue structure, the cellular integrins, could promote the assembly of adherens junctions and influence the cytostructure of these cells, suggesting that, "despite a number of prominent mutations, amplifications, and deletions, signaling events which are linked to the maintenance of normal tissue architecture are sufficient to abrogate malignancy and to repress the tumor phenotype. It is thus fair to state that cellular and tissue architecture act as the most dominant tumor-suppressor of all, and that the phenotype can— and does—override the genotype as long as the tissue architecture is maintained." Tumor cells treated with the same antibody and injected into nude mice resulted in significantly reduced number and size of tumors.[17]

Fourth, in a series of papers, Bissell's lab investigated how the ECM (extracellular matrix) regulates apoptosis and, in particular, which molecules in the extracellular environment promote invasion and metastasis. In particular, one MMP (matrix metalloproteinase), MMP3/Str-1 (Stromelysin 1) is active in development and during the process of tissue remodeling in the breast, during involution, when ECM remodeling and alveolar regression take place (the process that is initiated after breast feeding ceases). Such molecules facilitate angiogenesis by breaking down the extracellular matrix, but they also are associated with cell proliferation. Bissell's group showed that they have a tumor-promoting effect and drew a parallel between development and tumor formation that was relatively poorly understood. The molecules are synthesized not by cancer cells but by stromal cells in the tissue microenvironment. In other words, "an altered stromal environment can promote neoplastic formation."[18]

Throughout these decades, there was a central theme in Bissell's work and that of her students: namely, that the study of cells required attention to their environment; that intercellular structures and signaling were not merely passive "conditions" for the maintenance of cells ("housekeepers"), but played an active role in cell phenotype and could indeed lead cells to reverse conditions thought to be fixed. In other words, cellular behavior was not intrinsically determined, but context-dependent. They studied the regulatory roles of many extracellular factors. Bissell's close study of development led her to realize that it was inappropriate to speak of "cell fate," because interactions among signaling molecules and cells could modify the phenotype of a cell over time and even reverse the course of incipient cancers. This insight was in part a product of Bissell's willingness to traverse disciplinary boundaries. Bissell credits her study of developmental biologists' work with the insight that contextual interactions are fundamental in creating diverse differentiated tissues (MB, 11/15).

THE RESPONSE AND AFTERMATH

The response to Bissell's work was—at least initially—quite dismissive. The rise of the oncogene paradigm in the 1980s overshadowed much of her research into cell metabolism. When she attempted to bring cancer biologists' attention to her work on the Warburg effect, she explains, "No matter what I would try to do, to tell people why these [metabolic] curves were important, nobody would listen. Chicken cells!? (MB, 11/15)." Nonetheless, Bissell continued to produce a steady stream of publications that have critically transformed our understanding of the role of the microenvironment in cancer initiation and progression.

Her research strategies seemed to have been working with women or non-US citizens, as well as cross-disciplinary collaboration. Students were made welcome who might not have found a home in a (then) more prestigious lab, because they were pregnant, had young children, or were non-native English speakers. Bissell mentioned specifically Joanne Emerman ("She had three children, and amazing: she did all this stuff after she had her children") and Eva Lee ("No one would accept her because she had a baby, and then she was pregnant again, . . . and she was a very smart woman") (MB, 11/15). Bissell cultivated relationships with women from similar backgrounds, women who were similarly brilliant, ambitious, and mothers, as well as scientists.

CONCLUSIONS

In a paper published in the *Journal of Theoretical Biology* in 1982, "How Does the Extracellular Matrix Direct Gene Expression?" Bissell and coauthors invert the direction of causality in development, previously often described as driven by gene expression.[19] The paper made use of the expression "dynamic reciprocity." While the expression was not new to Bissell and colleagues (it was first used by Bornstein, et. al. in 1981), it marked a transition from a long-standing view of cell differentiation.[20] Rather than view cells' phenotypes as fixed by the time embryonic development was completed, they argued that the cell is in constant dynamic interaction with its intercellular environment, and that this interaction affects everything from gene expression to cancer.[21] This picture eventually became the accepted view among most biologists today. The paradigm according to which genes (alone) direct the behavior of cells, has gone the way of "terminal differentiation."

For instance, in 2011, Hanahan and Weinberg revisited their famous 2001 "Hallmarks of Cancer," in a paper entitled "Hallmarks of Cancer: The Next Generation" to include not only "cell-intrinsic" determinants of the cancer phenotype (mutations to oncogenes and tumor suppressor genes), but also cell-*extrinsic* factors. Hanahan and Weinberg explain that "conceptual progress in the last decade has added two emerging hallmarks: . . . reprogramming of energy metabolism and evading immune destruction." They suggest that taking into account the active role of the "tumor microenvironment." "will increasingly affect the development of new means to treat human cancer."[22]

Bissell's story illustrates several features of the "dreamer's" or visionary's experience: working outside of the mainstream, drawing upon interdisciplinary work, and using innovative tools. Despite the fact that Bissell chose to work outside of the "rat race," she also promoted big and controversial ideas, taking decades to build her case for a dynamic and reciprocal relationship between cells and their environment. She also invited postdocs and graduate students to join her lab, many of whom were viewed as marginal—either because of training, language, ethnicity, nationality, or sex. In this way, Bissell was instrumental in a profound shift in thinking about cell biology, and cancer in particular. She was also instrumental in a shift in the culture of science, creating a more inclusive environment.

FURTHER READING

Bertolaso, Marta. "Towards an Integrated View of the Neoplastic Phenomena in Cancer Research." *History and Philosophy of the Life Sciences* 31, no. 1 (2009): 79–97.

Bissell, Mina J., and William C. Hines. "Why Don't We Get More Cancer? A Proposed Role of the Microenvironment in Restraining Cancer Progression." *Nature Medicine* 17, no. 3 (2011): 320–29.

Fujimura, Joan H. *Crafting Science: A Sociohistory of the Quest for the Genetics of Cancer.* Cambridge, MA: Harvard University Press, 1996.

Malaterre, Christophe. "Organicism and Reductionism in Cancer Research: Towards a Systemic Approach." *International Studies in the Philosophy of Science* 21, no. 1 (2007): 57–73.

Marcum, James A. "Cancer: Complexity, Causation, and Systems Biology." *Medicina & Storia* 9, no. 17–18 (2009): 267–87.

Morange, Michel. "From the Regulatory Vision of Cancer to the Oncogene Paradigm, 1975–1985." *Journal of the History of Biology* 30, no. 1 (1997): 1–29.

Moss, Lenny. *What Genes Can't Do*. Cambridge, MA: MIT press, 2004.

Plutynski, Anya. "Cancer and the Goals of Integration." *Studies in History and Philosophy of Science, Part C: Studies in History and Philosophy of Biological and Biomedical Sciences* 44, no. 4 (2013): 466–76.

NOTES

1. Douglas Hanahan and Robert A. Weinberg, "Hallmarks of Cancer: The Next Generation," *Cell* 144, no. 5 (2011): 657.

2. H. M. Temin, "Homology between RNA from Rous Sarcoma Virus and DNA from Rous Sarcoma Virus-Infected Cells," *Proceedings of the National Academy of Sciences of the United States of America* 52 (1964): 323–29.

3. See, for example, Joan H. Fujimura, *Crafting Science: A Sociohistory of the Quest for the Genetics of Cancer* (Cambridge, MA: Harvard University Press, 1996).

4. Hanahan and Weinberg, "Hallmarks of Cancer."

5. Mina Bissell, "Mina Bissell: Context Is Everything [an interview by Ben Short]," *Journal of Cell Biology* 185, no. 3 (2009): 374–75.

6. Mina Bissell, interview by Anya Plutynski, August 2015, Berkeley, CA. Hereafter cited parenthetically in the text as (MB, 8/15).

7. Mina J.Bissell, Roberto Tosi, and Luigi Gorini, "Mechanism of Excretion of a Bacterial Proteinase: Factors Controlling Accumulation of the Extracellular Proteinase of a Sarcina Strain (Coccus P)," *Journal of Bacteriology* 105, no. 3 (1971): 1099–1109.

8. Mina Bissell, interview by Anya Plutynski, November 2015. Hereafter cited parenthetically in the text as (MB, 11/15).

9. MB, 11/15. See also Mina J. Bissell, "Transport as a Rate-Limiting Step in Glucose Metabolism in Virus-Transformed Cells: Studies with Cytochalasin B," *Journal of Cellular Physiology* 89, no. 4 (1976): 701–9.

10. Joanne T. Emerman and Mina J. Bissell, "A Simple Technique for Detection and Quantitation of Lactose Synthesis and Secretion," *Analytical Biochemistry* 94, no. 2 (1979): 340–45.

11. E. Y. Lee, Wen-Hwa Lee, Charlotte S. Kaetzel, et al., "Interaction of Mouse Mammary Epithelial Cells with Collagen Substrata: Regulation of Casein Gene Expression and Secretion," *Proceedings of the National Academy of Sciences* 82, no. 5 (1985): 1419–23.

12. Mina J. Bissell, "The Differentiated State of Normal and Malignant Cells; or, How to Define a 'Normal' Cell in Culture," *International Review of Cytology* 70 (1981): 27–100.

13. Mina J. Bissell, H. G. Hall, and G. Parry, "How Does the Extracellular Matrix Direct Gene Expression?," *Journal of Theoretical Biology* 99, no. 1 (1982): 31–68; Mina J. Bissell,

EY–H Lee, M–L Li, et al., "Role of Extracellular Matrix and Hormones in Modulation in Tissue-Specific Functions in Culture: Mammary Gland as a Model for Endocrine Sensitive Tissues," in *Benign Prostatic Hyperplasia*, ed. H. Rogers, D. C. Coffey, G. R. Cunha, et al., vol. 2, NIH Publ. No. 87-2881 (Washington, DC: U.S. Department of Health and Human Services, 1985); David S. Dolberg, Robert Hollingsworth, Mark Hertle, and Mina J. Bissell, "Wounding and Its Role in RSV-Mediated Tumor Formation," *Science* 230, no. 4726 (1985): 676–78; Mina J. Bissell and J. Aggeler, "Dynamic Reciprocity: How Do Extracellular Matrix and Hormones Direct Gene Expression?," in *Mechanisms of Signal Transduction by Hormones and Growth Factors*, ed. M. C. Cabot and W. L. McKeehan (New York: Alan Liss, 1987), 251–62; Mina J. Bissell and M. H. Barcellos-Hoff, "The Influence of Extracellular Matrix on Gene Expression: Is Structure the Message?" [review], *Journal of Cell Science*, suppl., 8 (1987): 327–43.

14. Bissell, "Differentiated State," 33.

15. Dolberg, Hollingsworth, Hertle, and Bissell, "Wounding and Its Role"; Michael H. Sieweke, Nancy L. Thompson, Michael B. Sporn, and Mina J. Bissell, "Mediation of Wound-Related Rous Sarcoma Virus Tumorigenesis by TGF-beta," *Science* 248, no. 4963 (1990): 1656–60; M. Martins-Green, C. Tilley, R. Schwarz, et al., "Wound-Factor-Induced and Cell Cycle Phase-Dependent Expression of 9E3/CEF4, the Avian Gro Gene," *Cell Regulation* 2, no. 9 (1991): 739–52.

16. V. M. Weaver, A. R. Howlett, B. Langton-Webster, et al., "The Development of a Functionally Relevant Cell Culture Model of Progressive Human Breast Cancer" [review], *Seminars in Cancer Biology* 6, no. 3 (1995): 175–84.

17. Valerie M. Weaver, Ole William Petersen, F. Wang, et al., "Reversion of the Malignant Phenotype of Human Breast Cells in Three-Dimensional Culture and in Vivo by Integrin Blocking Antibodies," *Journal of Cell Biology* 137, no. 1 (1997): 243.

18. Mark D. Sternlicht, Andre Lochter, Carolyn J. Sympson, et al., "The Stromal Proteinase MMP3/stromelysin-1 Promotes Mammary Carcinogenesis," *Cell* 98, no. 2 (1999): 137–46; Nancy Boudreau, Carolyn J. Sympson, Zena Werb, and Mina J. Bissell, "Suppression of ICE and Apoptosis in Mammary Epithelial Cells by Extracellular Matrix," *Science* 267, no. 5199 (1995): 891.

19. Bissell, Hall, and Parry. "How Does the Extracellular Matrix Direct Gene Expression?"

20. P. Bornstein, J. McPherson, and H. Sage, "Synthesis and Secretion of Structural Macromolecules by Endothelial Cells in Culture," in *Pathobiology of the Endothelial Cell*, ed. H. L. Nossel and H. J. Vogel (New York: Academic Press, 1982) , 215–28.

21. Bissell, Hall, and Parry, "How Does the Extracellular Matrix Direct Gene Expression?"

22. Hanahan and Weinberg, "Hallmarks of Cancer," 646.

Part III: The Molecularists

Figure 7.1. W. Ford Doolittle. Courtesy of W. Ford Doolittle.

MAUREEN A. O'MALLEY

W. FORD DOOLITTLE

EVOLUTIONARY PROVOCATIONS

AND A PLURALISTIC VISION

INTRODUCTION

Although many scientists see their work as distinct from academic endeavors that are not classified as natural science, others draw resources from more distant disciplines. Ford Doolittle has maintained throughout his career a deep interest in the humanities and arts, which might be explained to some extent by his father's academic interests. His father was a professor of art (painting) at the University of Illinois in Champaign-Urbana, the city in which Ford was born in 1942. As an undergraduate at Harvard, Doolittle was torn between science and literature. Science won, at least in part because his applications to literature programs were rejected.[1] When he was interviewed in 1970 for a position at Dalhousie University (Halifax, Nova Scotia, Canada), the department head asked him what he would do if he didn't get the job. "Write science fiction," said Doolittle, suggesting an alternative career that was not utterly unimaginable to people who knew him at the time.[2] Non-counterfactually, Doolittle pursued part-time for several years a degree in Fine Arts at the Nova Scotia College of Art and Design (completed in 2012). He worked in several media, but his photography in particular has engaged with his scientific worldview and been exhibited in its own right. Some of the more postmodern theory that informs the study of art these days may also have crept into his philosophical research. As he says, "In both science and art I struggle to be deconstructive, in a polite way."[3]

As part of this "deconstructive" approach, philosophy has been an implicit and explicit spur to Doolittle's work for at least two decades. The explicit commitments began with the formation of Dalhousie's Evolution Studies Group in the 1990s. Its members include philosophers of science and mathematics, moral philosophers, psychologists, cognitive scientists, economists, anthropologists, sociologists, biophysicists, and a huge variety of biologists. The group appointed a postdoctoral fellow with Doolittle's funding in 2001 (the author of this essay), and now, with a Herzberg award (often called Canada's highest science honor), Doolittle has chosen to invest a large portion of it in more philosophy research, including the appointment of two more postdoctoral fellows.

As well as advancing discussion on core topics in philosophy of biology, such as function, phylogeny, adaptation, and evolutionary theory gener-ally, Doolittle has used the philosophy of science to reflect on his own meta-methodology and the nature of the scientific debates in which he has been involved. Increasingly of late, he has been working out his commitments to pluralism, and—ultimately—the relationship of artistic creativity to scien-tific sensibility.

Among the several major contributions Doolittle has made to the scien-tific literature, perhaps the best known, for a general audience, is what he has had to say about organisms promiscuously passing around genetic mate-rial and the implications of that for the very possibility of phylogenetic re-construction (the graphic representation of evolutionary lineages and their divergence points). But just as pioneering are numerous other debates that Doolittle has either initiated or transformed during his engagement with them. I will show how his broad research agenda is driven by a commitment to questioning orthodoxy, by examining its assumptions and implicit limi-tations. I suggest that it is philosophical because it proposes a view of how science should proceed (sometimes by going backwards, and often by mov-ing sideways); moreover, it is a practical philosophy that serves as a meta-methodology, allowing innovative thinking to flourish. Its aim is explicitly to strengthen scientific inquiry, not weaken it, and several themes in Doolittle's career attest to how such an approach can produce novel insight.

EARLY CAREER

After an undergraduate degree in biochemistry at Harvard, Doolit-tle did his PhD at Stanford with Charles Yanofsky, of co-linearity and tryp-tophan biosynthesis fame. Doolittle's PhD topic involved an analysis of the transcriptional repression of tryptophan in *Escherichia coli*.[4] Yanofsky was a product of the Beadle-Tatum "biochemical genetics" lineage,[5] and it is in this illustrious tradition that Doolittle forged his own trajectory in molecular mechanisms and their evolution. Just as Yanofsky saw his career as one that attested to the value and creativity of basic science,[6] so too have these aspira-tions guided Doolittle throughout his subsequent career.[7]

From Stanford, Doolittle returned to Illinois in 1968 for a postdoctoral fel-lowship with Sol Spiegelman (1914–1983). Spiegelman, whom Doolittle has described as "a major figure who should have gotten the Nobel Prize"[8]—something it seems Spiegelman himself felt rather bitter about[9]—was re-nowned for his groundbreaking work on RNA and particularly DNA-RNA hy-bridization, where complementary strands of DNA and RNA bond to form a hybrid helix.[10] Doolittle had already gained some undergraduate experience in Spiegelman's lab, primarily as a dishwasher and laboratory drudge, but he had also been exposed to the excitement and sheer dedication of laboratory work.[11] As a postdoctoral fellow, Doolittle shared more equally in these sentiments and made strong connections with Spiegelman's colleagues. Notable among them were Carl Woese (1928–2012), running his own "revolutionary" research program down the hallway,[12] and Norman Pace, Spiegelman's postdoc, who also collaborated with Woese. Doolittle took up a subsequent research fellowship with Pace, and they worked together on ribosomal RNA[13]— a molecule that Woese was revealing as central to understanding evolution. In 1971, Doolittle carried these interests with him to Dalhousie University, where he developed the research that has made him famous.

Doolittle became interested in joining Dalhousie after making a connection with Stanley Wainwright (1927–2003), a biochemist who had come to Dalhousie via Yale, Columbia, and the Pasteur Institute. Wainwright's wife, Lillian Schneider Wainwright had worked and published with molecular biology illuminati Francis Ryan (1916–1963) and Joshua Lederberg (1925–2008). Despite Dalhousie being a somewhat out-of-the-way place from an American perspective, the omens were therefore good for settling into a strong scientific community. Not only that, but as Doolittle has observed himself, there was none of the heavy pressure to be a star that existed at more prestigious institutions across the border.[14] Be that as it may, Doolittle rapidly became one of Dalhousie's stars, but he did so by working with and shining alongside local scientific talent. A major collaboration developed with Michael Gray. He and Doolittle did the formative molecular work that established once and for all the endosymbiont hypothesis of chloroplast origins plus much of the groundwork that did the same for the origin of the mitochondrion. Both of these origin scenarios propose that these organelles originated as separate prokaryotic organisms and were then incorporated into eukaryotic cells.[15] Gray and Doolittle's solid evidential anchoring of the then slightly dodgy endosymbiont hypothesis may have looked like the positive reinforcement of an idea whose time had finally come. Eventually, however, this work connected with another radical agenda about lateral acquisitions (see "Upsetting the Tree of Life" below). Doolittle's proclivity for doubting orthodoxy began much earlier with questions about the power of adaptationism—the idea that the task of evolutionary biology was the explanation of ubiquitous adaptations.

QUESTIONING ADAPTATIONISM

Although as a PhD student Doolittle encountered "panadaptationist" thinking that claimed that every significant trait was an adaptation, the very nature of working with molecules meant that he would have to grapple with non-adaptive or neutral evolution at work in nucleotide and amino acid sequences.[16] In fact, Doolittle has claimed that Gould and Lewontin's "Spandrels" paper advocating a reconsideration of panadaptationism was his greatest influence.[17] Doolittle's advocacy for non-adaptationist explanations was honed during his involvement in a major debate about the evolution of introns. In 1977, which was a "big year"[18] because it also saw the discovery of Archaea (the nonbacterial prokaryotes), eukaryote protein-coding genes were found to have introns. These genes are "interrupted by 'silent' DNA," said Walter Gilbert of subsequent sequencing technique fame. Gilbert then tried to explain why these "intergenic regions"—for which he coined the word *intron*, in contrast to *exon* for the coding regions—were selected.[19] "What are the benefits of this intronic/exonic structure for genes?" Gilbert mused, taking for granted there would be benefit. His answer was that evolution puts aside some "scattered" genes in the form of introns so that the potential for variation and speedy evolution is preserved. Introns are thus "both frozen remnants of history and . . . the sites of future evolution."[20]

Doolittle, who was then on sabbatical in Gilbert's lab, elaborated on this historical scenario of the ancient emergence of introns.[21] According to Doolittle, genes may have been interrupted from their very origin, and prokaryotes (which don't have the sorts of introns being discussed) were then streamlined from this primordial architecture. Rather than thinking of the tidy prokaryote genome as "primitive" and the messy eukaryote genome as an evolutionary refinement, this view of "introns early" (i.e., the basic state of genome organization) undercut a very traditional simple-to-complex view of the evolution of life. Moreover, argued Doolittle, Gilbert's adaptationist explanation was teleological, because it suggested that introns were retained by selection for the future benefit of the lineages carrying them.[22]

The "introns early" perspective was eventually overwhelmed by the "introns late" view (the position that introns originated in eukaryotes, and prokaryotes never had them). Doolittle himself produced some of the evidence undermining his original position, which he then announced as "untenable."[23] He was ultimately able to explain the evolution of these introns as an example of "constructive neutral evolution" (see below).[24] It may be the case that "introns early" is not completely dead, though it would have to be considerably different from the original simple view. In Eugene Koonin's recent revisionary efforts, the introns-early hypothesis, which tried to explain important differences between prokaryotes and eukaryotes while acknowledging their shared features, "incorporated too many good ideas to just go out with a whimper."[25] So although Doolittle's early perspective was wrong, it was nevertheless productive and helped steer the future development of research in this area.

Several other lines of Doolittle's research have questioned adaptationist thinking, including his groundbreaking work on selfish DNA and some early collaboration with Arlin Stoltzfus, who put these discussions into print as the theory of constructive neutral evolution.[26] The "selfish gene" paper that Doolittle wrote with graduate student Carmen Sapienza is his second-most cited article, with more than sixteen hundred citations.[27] It argues that there is no theoretical or empirical reason to expect most of the DNA in a genome to be phenotypically functional—the very opposite, in fact, is more likely to be the case. Transposable elements were the focus of the paper's discussion of this "non-phenotypic selection." Published at the same time in *Nature* was Leslie Orgel and Francis Crick's 1980 outline of the same phenomenon and explanation.[28] They too had converged on the importance of this idea, in an era of molecular biology that was not very strongly based in population genetics, and when assumptions of adaptive function tended to be the default. Both papers argued that this was a dangerous default.

Doolittle's proclivity to question adaptationist claims, especially but not only at the molecular level, found its strongest voice in response to the EN-CODE debate. ENCODE (the Encyclopedia of DNA Elements) was and still is a huge sequencing project that aims to discern the function of the DNA in the human genome. By conflating an evolutionary (selected) definition of function with a minimalist biochemical definition (i.e., "some sort of reaction will occur"), ENCODE's publications triggered scientific outrage, especially when associated media releases suggested almost the entirety of the human genome was functional.[29] Although many of these objecting papers attacked the underlying definition, Doolittle and Graur et al. were the only ones to dwell on classic philosophical discussions of function.[30] They distinguished "selected effects" from "causal roles" to devastate ENCODE's overweening claims.[31] But even though it could appear to many readers as if there were clearly a wrong and a right side to the debate, Doolittle was doing something more nuanced. His subsequent musings on the role of provocation and debate in science make this clear.

CONJECTURES, REFUTATIONS, AND FRUITFUL LINES OF INQUIRY

A more abstract change of direction in Doolittle's views occurred in relation to discussions of Gaia. In 1974 Lynn Margulis (1938–2011)[32] joined forces with James Lovelock to advance a cybernetic view of Earth as a single, organism-like biosphere (see chapter 17).[33] Using both theory and evidence (particularly biogeochemical), Margulis and Lovelock pushed the notion of a feedback-controlled system that was regulated from a planetary level to maintain Earth in the state it currently and optimally has. Not everyone was convinced. Although much of the criticism was directed at the idea of Earth as a homeostatic system, Doolittle went on the attack via an evolutionary argument, as did Richard Dawkins.[34] If Gaia were like an organism and had evolved by natural selection to be that way, as Margulis and Lovelock had argued,[35] then how would selection have worked, asked Doolittle.[36] There is only one putative Gaia, and no competition to be had, and therefore selection—the operation of which is often used to identify living systems—could not be at work. In addition, just as for introns, the subsystems (organisms) could not structure their efforts to benefit a future entity.

Doolittle used an analogy with another famous Dolittle, notably the fictional one who talked to animals and, in one volume of a popular series,[37] went to the moon. On the species-diverse moon, Dr. John Dolittle found that instead of competition, the animals and plants (all of them sentient and enormous) had formed a council to regulate their interactions (fig. 7.2). The council's aim and accomplishment was to eliminate warfare at every level and bring about a "balance." No such top-down mechanisms exist on Earth, nor indeed balance, observed Doolittle,[38] and there was no choice but to understand all evolutionary hypotheses in light of individual Darwinian fitness.

However, even though these criticisms were widely acclaimed (despite being published in an obscure journal, the *CoEvolution Quarterly*), Doolittle has gone on to rethink some of his argument.[39] In particular, he has reflected on the notion that not all evolving systems need to reproduce to have fitness, and that outcompeting other entities might not be as crucial to evolutionary fitness as he had supposed. Gaia might be a persisting entity ("immortal"), and thus a living system evolving in a way that is not captured by classical Darwinian tenets. What might be the point of making this quite abstract argument, which in the end does not vindicate the Gaia concept? A major reason for Doolittle is that it fills in some Darwinian gaps by adding "selection by survival" to everyday natural selection.[40] He believes that this kind of persistence can be understood as minimally Darwinian (rather than paradigmatically Darwinian), and is thus in accord with Peter Godfrey-Smith's philosophical efforts to describe the continuum of Darwinian individuality created by varying combinations of different processes.[41] More generally, I suggest, this sort of philosophical turn is part of Doolittle's effort to understand science itself as a dynamic evolving process that rewards serious consideration of nonorthodox ideas—even when they are found ultimately to be wrong.

Figure 7.2. Dr. Dolittle experimenting with sentient plants on the moon to determine their cognitive capacities and distance-communication abilities. Hugh Lofting, *Doctor Dolittle in the Moon* (1928; repr., London: Jonathan Cape, 1929). From the free Gutenberg version of this book: http://gutenberg.net.au/ebooks06/0607691h .html#032.

UPSETTING THE TREE OF LIFE

Doolittle's anti-orthodoxy is most recognized in his revision of how the tree of life is understood. Molecular phylogeny had become by the late 1980s the new gold standard of evolutionary tree reconstruction.[42] However, while molecules made phylogeny possible for organisms without much morphology (i.e., prokaryotes), they also revealed wayward genetic movements behind minor and major organismal differences. Lateral gene transfer (LGT) occurs between organisms (often unicells but not exclusively) and involves the "gift" of genetic material that may have a very different evolutionary history from that of the recipient. Rather than orthodox bifurcating lineages with strict genetic continuity, LGT requires messier, weblike representations. The impact of LGT on phylogeny is the theme that brought Doolittle's work to a broader audience. His paper in *Science* in 1999, which has also been cited more than eighteen hundred times, spelled out what the conceptual implications for phylogeny would be if LGT were as common as recent data indicated it was.[43] The hand-drawn cartoon with which Doolittle illustrated the alternatives played a central role in bringing home the conclusions of his argument (fig. 7.3).

What is crucial to understand here is that Doolittle was not a lone voice crying out in the wilderness of orthodoxy. The message was indeed revolutionary, but several other members of the evolutionary community were simultaneously reflecting broadly on the theoretical and phylogenetic implications of LGT.[44] The distinctiveness of Doolittle's contribution was how he framed it. This framing has implications for how he is interpreted as a "visionary" scientist. My point is not that he and his colleagues were right, and radically so (such that their position now is the new orthodoxy), but that Doolittle's special contribution was and is meta-methodological in a philosophical way.

Any understanding of Doolittle as a visionary "dreamer" needs to recognize that he couched his argument about LGT conditionally. As he noted subsequently, "I only wanted to point out how interesting the consequences for phylogeny would be *if* lateral gene transfer were a major force."[45] The way in which the conditionality of this statement was lost, as reams of data accumulated in its favor, has been covered by numerous review articles— scientific, philosophical, and historical.[46] How Doolittle sees this transformation, however, is not exclusively as an episode of careful data gathering and evaluation of competing hypotheses, but as an instance of giving entrenched frameworks a good kick to see what gets dislodged. In other words, "the debate . . . shows how strongly unstated philosophical commitments can influence how we collect and interpret hard facts."[47] Much of his career, including many of its challenges to well-accepted ways of thinking, can be understood as a disciplined inquiry into the philosophies guiding and even driving the science.

PLURALISM

One of Doolittle's high-profile recent papers (at more than three hundred and fifty citations, it is "only" near the bottom of his top-twenty most-cited publications) is one he wrote with Eric Bapteste (then a postdoctoral fellow at Dalhousie) for Doolittle's inauguration as a member of the National Academy of Sciences (USA) in 2009. This paper explores what it means to be a pluralist in phylogenetic practice.[48] Doolittle and Bapteste outline "pattern pluralism" as "the recognition that different evolutionary models and representations of relationships will be appropriate, and true, for different taxa or at different scales, or for different purposes."[49] In other words, just as the

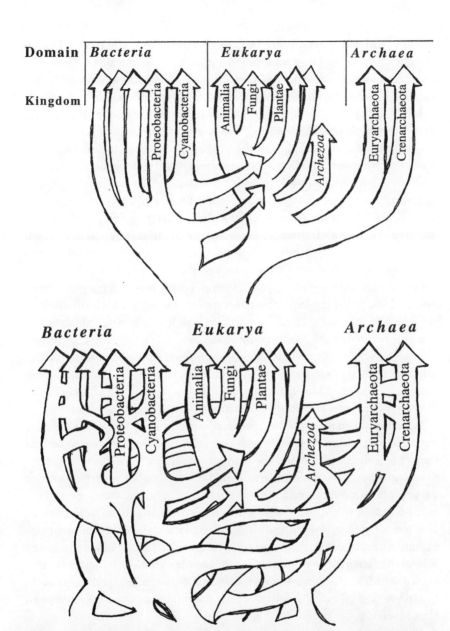

Figure 7.3. Doolittle's 1999 cartoons of the standard Tree of Life (*A*) and the alternative "net" of life (*B*). From W. Ford Doolittle, "Phylogenetic Classification and the Universal Tree," *Science* 284 (1999): 2124–28. Reprinted with permission from American Association for the Advancement of Science.

modern synthesis of evolution embraces mutation, drift, and recombination in addition to natural selection, so must phylogeny be pluralist.[50] Phylogeny needs "a versatile and well-stocked *explanatory toolkit*," argues Doolittle, not only for the sake of science but also for its public reception.[51]

Methodologically, Doolittle calls himself an "instinctive" reductionist.[52] By this he simply means a tendency to ask molecules for biological answers. His self-label does not imply that this is the only level at which to work, but it is where some very good data can be obtained and successful explanations generated. For philosophers, pluralism can be divided into a metaphysics of plurality (i.e., the nature of things is not fixed) and an epistemic stance (sometimes but not always associated with the metaphysical position): that multiple explanatory frameworks can capture the phenomenon of interest. Certainly, with regard to species and the tree of life, Doolittle could be read as an ontological pluralist.[53] But it is unlikely that he regards all or even many phenomena in this way (e.g., "junk" DNA is simply not on a par with "functional" DNA, given the explanatory frameworks in which such discussions are rightfully held). Explanation is what it's about then. And explanatorily, in a number of discussions throughout his career, Doolittle has come down hard on "wrong" accounts of biological phenomena: the ENCODE account of function, the Gaia view of evolution ("unquestionably false"),[54] and the autogenous (non-endosymbiotic) account of the chloroplast and mitochondrion.[55] How do we interpret Doolittle's pluralism in light of such assessments?

What pluralism seems to be associated with for Doolittle, when trying to understand his career-wide research trajectory, is his reluctance to restrict inquiry to a single "metanarrative" or "monistic view,"[56] and his openness to novel hypotheses and innovative explanations. Yes indeed, emerging and existing explanations all need to be evaluated rigorously, but shutting down the creative side of hypothesis generation seems to be anathema to his view of how science works. "I try to find alternatives to the standard evolutionary explanations for biological patterns and processes," he says, "in part because I believe that scientists are as liable as other humans to cling to dominant theories or attitudes, even when obvious alternatives become equally tenable."[57] In articulating this view, Doolittle draws some important parallels between art and science.

Both the arts and the natural sciences, from Doolittle's point of view, are very similar in that they possess "potential purity of motive" and "internal enemies, which are laziness and self-deceit."[58] There is, however, a big difference: "Scientists are not supposed to be too deliberately playful with their audience, nor supposed to speculate wildly in order to evaluate the response [to their work]. That's what artists must do to survive, and I think we [scientists] could learn from them," argues Doolittle.[59] He believes that self-regulated constraint in interpretation often occurs out of fear of appearing unconventional and "un-objective." But such a belief is built on a limited view of what science is: it is not an aggregation of rational, self-policing individuals (though that comes into it, albeit via the back door), but is instead a social institution that assesses and adjudicates at the institutional level. It has room for more wayward but occasionally productive interpretations, and may even need them. That is the reason behind Doolittle's opposition to increasingly commercial requirements of scientists and his advocacy for basic science driven by curiosity.

On the other hand, he observes that "science is way too uncritical of itself" and needs more people like Rosie Redfield, who is well known for debunking the notorious arsenic life hypothesis, which proposed that arsenic could be substituted for phosphorus in living organisms.[60] Redfield, herself an iconoclastic evolutionary microbiologist, fond of deflating a variety of allegedly puffed-up hypotheses,[61] sometimes hints that Doolittle's "philosophical" musings are mere indulgences and not the stuff of which science is made. As Doolittle explains it, many scientists fear that too much philosophy is not only at too great a remove from the science it purports to analyze, but it can also lead to "analysis paralysis" and an inability to make any further headway.[62] My sketch of how he has interrogated evolutionary and other biological explanations shows how in many respects, the opposite has occurred. For Doolittle and sometimes the entire area in which he is working, philosophical inquiry has brought about liberation from constraining frameworks and a capacity to move into new phases of inquiry. In his ongoing practical engagement with philosophy, Doolittle may be at his most visionary, transgressing some very traditional boundaries and attempting a synthesis of a fairly risky sort.

CONCLUDING THOUGHTS

Ford Doolittle has led a double life: as a scientist and as an "embedded" philosopher who has challenged assumptions ("deconstructively") and offered novel and sometimes left-field explanations ("playful speculation"). Although many substantive elements of Doolittle's research program are visionary in their own right, my suggestion is that underlying it is a normative view of how science works that offers advice about the responsibilities that any individual scientist has to his or her community and the broader institution of science. Part of this responsibility, he suggests, is to examine every argument and interpretation, and to be willing to doubt even the most established scientific framework. This is not done for destructive purposes, however, but to be constructively open to the expansion of explanatory options. Each of the topics in Doolittle's oeuvre above shows how valuable a methodology this is, not just for him personally but for the scientific community more generally.

FURTHER READING

Archibald, John. *One Plus One Equals One: Symbiosis and the Evolution of Complex Life*. Oxford: Oxford University Press, 2014.

Gitschier, Jane. "The Philosophical Approach: An Interview with Ford Doolittle." *PLoS Genetics* 11 (2015): e1005173.

NOTES

1. W. Ford Doolittle, "Q&A: W. Ford Doolittle," *Current Biology* 14 (2004): R178–R179.

2. John Archibald, *One Plus One Equals One: Symbiosis and the Evolution of Complex Life* (Oxford: Oxford University Press, 2014).

3. W. Ford Doolittle, "ViewPoint Gallery: Ford Doolittle Philosophy" (2016), online at http://viewpointgallery.weebly.com/ford_doolittle.html.

4. W. Ford Doolittle and Charles Yanofsky, "Mutants of *Escherichia coli* with an Altered Tryptophanyl-Transfer Ribonucleic Acid Synthetase," *Journal of Bacteriology* 95 (1968): 1283–94; Charles Yanofsky, "Transcription Attenuation: Once Viewed as a Novel Regulatory Strategy," *Journal of Bacteriology* 182 (2000): 1–8.

5. Charles Yanofsky, "The Favourable Features of Tryptophan Synthase for Proving Beadle and Tatum's One Gene-One Enzyme Hypothesis," *Genetics* 169 (2005): 511–16.

6. Charles Yanofsky, "Advancing Our Knowledge in Biochemistry, Genetics and Microbiology through Studies of Tryptophan Metabolism," *Annual Review of Biochemistry* 70 (2001): 1–37.

7. Doolittle, "Q&A."

8. Jane Gitschier, "The Philosophical Approach: An Interview with Ford Doolittle," *PLoS Genetics* 11 (2015): e1005173.

9. Benno Müller-Hill, *The lac Operon: A Short History of a Genetic Paradigm* (Berlin: Walter de Gruyter, 1996); David L. Nanney, *Candide in Academe Meets Tracy Agonistes: A Memoir of the Morning of Molecular Biology: Coming of Age in Bloomington, 1946–1951* (Draft of 23 March 2004), online at http://www.life.illinois.edu/nanney/autobiography/candide.html.

10. Masayasu Nomura, Benjamin D. Hall, and Sol Spiegelman, "Characterization of RNA Synthesized in *Escherichia coli* after Bacteriophage T2 Infection," *Journal of Molecular Biology* 2 (1960): 306–26; David Gillespie and Sol Spiegelman, "A Quantitative Assay for DNA-RNA Hybrids with DNA Immobilized on a Membrane," *Journal of Molecular Biology* 12 (1965): 829–42; Susie Fisher, "Not Just 'a Clever Way to Detect Whether DNA Really Made RNA': The Invention of DNA-RNA Hybridization and Its Outcome," *Studies in History and Philosophy of Biological and Biomedical Sciences* 53 (2015): 40–52.

11. Doolittle, "Q&A."

12. Jan Sapp, "The Iconoclastic Research Programme of Carl Woese," in *Rebels, Mav-*

ericks, and Heretics in Biology, ed. Oren Harman and Michael R. Dietrich (Princeton, NJ: Yale University Press, 2008), 302–20.

13. W. Ford Doolittle and Norman Pace, "Transcriptional Organization of the Ribosomal RNA Cistrons in *Escherichia coli*," *Proceedings of the National Academy of Sciences USA* 68 (1971): 1786–90.

14. Gitschier, "Philosophical Approach."

15. Michael W. Gray and W. Ford Doolittle, "Has the Endosymbiont Hypothesis Been Proven?," *Microbiological Reviews* 46 (1982): 1–42; Archibald, *One Plus One*.

16. Michael R. Dietrich, "The Origins of the Neutralist-Selectionist Debates,'" Dibner Workshop, May (2002), transcripts, online athttp://authors.library.caltech .edu/5456/1/hrst.mit.edu/hrs/evolution/public/transcripts/origins_transcript.html.

17. Doolittle, "Q&A."

18. Gitschier, "Philosophical Approach."

19. Walter Gilbert, "Why Genes in Pieces?," *Nature* 271 (1978): 501.

20. Gilbert, "Why Genes in Pieces?," 501.

21. W. Ford Doolittle, "Genes in Pieces: Were They Ever Together?," *Nature* 272 (1978): 581–82.

22. Doolittle, "Genes in Pieces"; W. Ford Doolittle, "The Origin and Function of Intervening Sequences in DNA: A Review," *American Naturalist* 130 (1987): 915–28.

23. Arlin Stoltzfus, David F. Spencer, Michael Zuker, et al., "Testing the Exon Theory of Genes: The Evidence from Protein Structure," *Science* 265 (1994): 206.

24. Olga Zhaxybayeva and J. Peter Gogarten, "Spliceosomal Introns: New Insights into Their Evolution," *Current Biology* 13 (2003): R764–R766; Gitschier, "Philosophical Approach."

25. Eugene V. Koonin, "The Origin of Introns and Their Role in Eukaryogenesis: A Compromise Solution to the Introns-Early Versus Introns-Late Debate?," *Biology Direct* 1 (2006): 22, doi:10.1186/1745-6150-1-22. (See Doolittle's referee comments for a very clear history of the debate.)

26. Arlin Stoltzfus, "On the Possibility of Constructive Neutral Evolution," *Journal of Molecular Evolution* 49 (1999): 169–81; Julius Lukeš, John M. Archibald, Patrick J. Keeling, et al., "How a Neutral Evolutionary Ratchet Can Build Cellular Complexity," *IUBMB Life* 63 (2011): 528–37.

27. W. Ford Doolittle and Carmen Sapienza, "Selfish Genes: The Phenotype Paradigm and Genome Evolution," *Nature* 284 (1980): 601–3.

28. Leslie E. Orgel and Francis H. C. Crick, "Selfish DNA: The Ultimate Parasite," *Nature* 284 (1980): 604–7.

29. Sean R. Eddy, "The ENCODE Project: Missteps Overshadowing a Success," *Current Biology* 23 (2013): R 259–R261.

30. W. Ford Doolittle, "Is Junk DNA Bunk? A Critique of ENCODE," *Proceedings of the National Academy of Sciences USA* 110 (2013): 5294–5300; Dan Graur, Yichen Zheng, Nicholas Price, et al., "On the Immortality of Television Sets: 'Function' in the Human Genome According to the Evolution-Free Gospel of ENCODE," *Genome Biology and Evolution* 5 (2013): 578–90.

31. W. Ford Doolittle, Tyler D. P. Brunet, Stefan Linquist, et al., "Distinguishing between 'Function' and 'Effect' in Genome Biology," *Genome Biology and Evolution* 6 (2014): 1234–37.

32. For a discussion of how Margulis's basic position was fleshed out molecularly by Doolittle and others, see Archibald, *One Plus One.*

33. Lynn Margulis and James E. Lovelock, "Biological Modulation of the Earth's Atmosphere," *Icarus* 21 (1974): 471–89.

34. Richard Dawkins, *The Extended Phenotype* (1982; rev. ed., Oxford: Oxford University Press, 1999).

35. Margulis and Lovelock, "Biological Modulation," 486.

36. W. Ford Doolittle, "Is Nature Really Motherly?," *CoEvolution Quarterly*, Spring 1981, 58–63.

37. Hugh Lofting, *Doctor Dolittle in the Moon* (1928; repr., London: Jonathan Cape, 1929).

38. Doolittle, "Is Nature Motherly?"

39. W. Ford Doolittle, "Natural Selection through Survival Alone, and the Possibility of Gaia," *Biology and Philosophy* 29 (2014): 415–23.

40. Doolittle, "Natural Selection," 421.

41. Peter Godfrey-Smith, *Darwinian Populations* (Oxford: Oxford University Press, 2009).

42. Edna Suárez-Diaz and Victor H. Anaya-Muñoz, "History, Objectivity, and the Construction of Molecular Phylogenies," *Studies in History and Philosophy of Biological and Biomedical Sciences* 39 (2008): 451–68.

43. W. Ford Doolittle, "Phylogenetic Classification and the Universal Tree," *Science* 284 (1999): 2124–28.

44. Elena Hilario and J. Peter Gogarten, "Horizontal Transfer of ATPase Genes—The Tree of Life Becomes a Net of Life," *Biosystems* 31 (1993): 111–19; William Martin, "Mosaic Bacterial Chromosomes: A Challenge en Route to a Tree of Genomes," *Bioessays* 21 (1999): 99–104; Michael Syvanen, "Horizontal Gene Transfer: Evidence and Possible Consequences," *Annual Review of Genetics* 28 (1994): 237–61.

45. Doolittle, "Q&A," R176; emphasis in original.

46. James O. McInerney, James A. Cotton, and Davide Pisani, "The Prokaryotic Tree of Life: Past, Present . . . and Future?," *Trends in Ecology and Evolution* 23 (2008): 276–81; Maureen A. O'Malley and Yan Boucher, "Paradigm Change in Evolutionary Microbiology," *Studies in History and Philosophy of Biological and Biomedical Sciences* 36 (2005): 183–208; Jan Sapp, *The New Foundations of Evolution: On the Tree of Life* (New York: Oxford University Press, 2009).

47. Doolittle, "Q&A," R176.

48. W. Ford Doolittle and Eric Bapteste, "Pattern Pluralism and the Tree of Life Hypothesis," *Proceedings of the National Academy of Sciences USA* 104 (2007): 2043–49.

49. Doolittle and Bapteste, "Pattern Pluralism," 2043; W. Ford Doolittle, "The Attempt on the Tree of Life: Science, Philosophy and Politics," *Biology and Philosophy* 25 (2010): 455–73; and "The Practice of Classification and the Theory of Evolution,

and What the Demise of Charles Darwin's Tree of Life Hypothesis Means for Both of Them," *Philosophical Transactions of the Royal Society, London B* (2009): 2221–28.

50. Doolittle, "Demise of Darwin's Tree."

51. Doolittle, "Attempt on the Tree of Life," 455.

52. Doolittle, "Q&A," R176.

53. W. Ford Doolittle, "Microbial Neopleomorphism," *Biology and Philosophy* 28 (2013): 351–78.

54. Doolittle, "Is Nature Motherly?," 58.

55. Gray and Doolittle, "Endosymbiont Hypothesis Proven?"

56. Doolittle and Bapteste, "Pattern Pluralism," 2048.

57. Doolittle, "Q&A," R176.

58. Doolittle, "Q&A," R177.

59. Doolittle, "Q&A," R177.

60. Erica C. Hayden, "Rosie Redfield: Critical Enquirer," *Nature* 480 (2011): 442–43.

61. Rosemary J. Redfield, "Is Quorum Sensing a Side Effect of Diffusion Sensing?," *Trends in Microbiology* 10 (1993): 365–70; Rosemary J. Redfield, "Genes for Breakfast: The Have-Your-Cake-and-Eat-It-Too of Bacterial Transformation," *Journal of Heredity* 84 (2002): 400–404.

62. W. Ford Doolittle, "Philosophy, Who Needs It?," *Current Biology* 25 (2015): R31.

BRUNO J. STRASSER

COLLECTING DREAMS IN THE MOLECULAR SCIENCES

MARGARET DAYHOFF AND *THE ATLAS OF PROTEIN SEQUENCE AND STRUCTURE*

INTRODUCTION

Possessing one of every kind has definitely not been the dream of most modern experimental scientists. Such dreams might obsess stamp collectors and other amateurs, but they disappeared from the ideals of experimentalists long ago. Naturalists in the Renaissance dreamed of collecting an example of every single living species, in particular those that were the most bizarre, to be stored in their wonder cabinets.[1] The discovery of the New World, whose new species resulted in an "information overload," made the dream hopelessly unattainable.[2] Yet still it persisted among many communities of naturalists, in a more modest form: to collect one of every kind of a smaller taxonomic group.[3] The ideal of comprehensiveness lingered on in the natural historical tradition, where it became a defining trait—despite being ridiculed by experimental scientists, starting in the mid-nineteenth century. The experimental physicist Ernest Rutherford summed up their attitude in the first decades of the twentieth century with the proclamation: "There is [experimental] physics, and the rest is stamp collecting."[4]

By 1965 many would have pronounced this collecting ideal dead, never to return: the experimental life sciences were flourishing, and the Nobel Prizes in physiology or medicine were being awarded to molecular biologists almost every year. And yet that was the year an odd collection appeared. Margaret Oakley Dayhoff, a physical chemist, published the *Atlas of Protein Sequence and Structure*: a list of every known protein amino-acid sequence.[5] A collection of one of every kind. Curiously enough, it was a direct product of modern experimentalism. A decade earlier, biochemist Frederick Sanger had devised an experimental method that made it possible to determine the sequential order of amino acids, the building blocks of proteins.[6] In 1951 he used it to determine the first protein sequence, the thirty amino acids of insulin's B chain. After this initial breakthrough, other researchers were quick to follow with sequences of other proteins from a few organisms that were easily obtainable from slaughterhouses: the ox, sheep, horse, and pig. This was the data that Dayhoff sought to collect comprehensively and turn into a published book,

Figure 8.1. Margaret Dayhoff. Photo by Ruth E. Dayhoff, MD, US National Library of Medicine.

which read like a phone book: a list of protein names followed not by phone numbers, but by amino acid sequences and a reference to the publication where they had first been described.

Dayhoff's dream stemmed from her conviction that having one of every kind of protein would yield unique insights into their evolutionary history and biochemical functions, especially if the data could be stored in a computer, another dream—computers had yet to find much use in the biomedical sciences. But Sanger's work had set something in motion: within a year of its initial publication, her *Atlas* was already outdated due to an "information explosion" in the field.[7] In 1966 Dayhoff published a new edition, twice as large as the first, in hopes of keeping pace with experimental research. In the midst of the modern experimental era, building an electronic wonder cabinet was Dayhoff's dream.

This vision seemed so strange to many experimental scientists that, for Dayhoff, it sometimes took on the character of a professional nightmare. The following decades were to cement the comprehensive collection of molecular data she had envisioned as one the most essential tools of experimental biomedical research. Today, in the form of electronic "databases" that are available online, it has become an indispensable tool that tens of thousands of researchers access daily from laboratories around the world. Around them has coalesced an entirely new scientific discipline, bioinformatics. What made Dayhoff's collection seem so novel in the 1960s was that it represented the first computerized collection of molecular data. What made it seem so archaic to many experimentalists was that it looked like just another natural history collection. Both explain why Dayhoff's vision seemed so dreamlike in the 1960s and so reasonable just few decades later, but leave open why Dayhoff thought of it as a reasonable pursuit in the postwar culture of experimental virtuosity and the slow embrace of computers by experimentalists.

COMPUTERS IN BIOLOGY

Margaret O. Dayhoff was born as Margaret Belle Oakley in Philadelphia in 1925 to a small business owner and a high school math teacher.[8] The family moved to New York City a decade later, where she attended public schools. She received her bachelor's degree in mathematics from New York University in 1945. Her career in research began at Columbia University, where she worked on her dissertation under the chemist George Kimball. She completed her PhD in quantum chemistry in 1948, at a mere twenty-three years of age.[9] As a fellow of the Watson IBM Computing Laboratory, she carried out research to calculate the physical properties of small molecules using punch-card computers, a paper technology that was a common program and data storage medium of early digital computers. After graduation she married Edward S. Dayhoff, who was still studying for a PhD in physics at Columbia University, and began working at the Rockefeller Institute (now Rockefeller University). As a research assistant in electrochemistry, her work involved measuring the density of proteins. Her husband graduated in 1951, and she followed him to Washington, DC, where he had been offered a position at the National Bureau of Standards. That same year saw the birth of their first child, Ruth E. Dayhoff, followed by Judith E. Dayhoff three years later. For the next eight years, Dayhoff gave up research to raise her children. In 1960, she returned to science with a position at a new, highly unusual research institution: the National Biomedical Research Foundation (NBRF), in Silver Spring, Maryland.

In fact, the NBRF was neither national, nor primarily biomedical; it had been founded as a small, private, nonprofit research institution by computer enthusiast Robert S. Ledley.[10] Perhaps it required a peculiar environment for Dayhoff's dream to be conceived and take root. With the creation of the NBRF, Dayhoff noted that there was "no other place where computing or engineering and biology or medicine could be combined intimately. Existing organizations," she added, were "administratively committed to just one of these areas."[11]

Robert S. Ledley was also a dreamer, driven by a desire to "computerize" biology and medicine. Armed with a technology, the digital computer, he was looking for problems to solve with it. As industrialists were hoping to replace workers with machines on the assembly line, Ledley had a vision of replacing them with computer routines. In 1963 he explained, "These routines may be thought of as analogous to the staff of a laboratory. Each routine has a function to perform just as a laboratory has people each with a job to perform: cleaning people, technicians, senior research workers, a librarian, a machinist, etc. The programmer and protein chemist have been upgraded to the chief of the computer staff."[12] At the NBRF, Ledley and his recruits developed the first whole-body computerized tomography scanner, followed by an electronic device that could calculate diagnoses from a few symptoms by applying probabilities, another that could automatically count chromosomes on microscopic images, and many more. Most of these inventions had little, if any, impact on biomedical research, at least until the 1970s. In an atmosphere where innovative technologies could be developed even if they weren't necessarily immediately marketable, Dayhoff's project didn't seem so unusual. For several years, few molecular biologists understood the full potential of her collection of molecular data.

The NBRF was also one of the rare places where experimental and computer cultures could meet and pursue biomedical dreams in the 1960s.[13] Computers were still centralized mainframes; they were available on academic campuses in the United States, but biologists rarely used them. They were too busy improving experimental systems and acquiring new data, with little interest in computerized methods to analyze the little they had. In some cases wet-lab experimentalists expressed open hostility toward computers, which they conceptually relegated to the theoretical sciences. When offered the opportunity to submit his experimental data to computational analysis, one biochemist expressed the sentiments of many in his community: "I am not a theorizer."[14] Dayhoff was well aware of the attitude; when a student proposed a project to apply computer methods to biochemical problems, Dayhoff warned him, "Make sure that the biochemists are sympathetic to the computer." Too many weren't. They hadn't profited from the environment of the NBRF, whose staff included "a medical doctor, a dentist, a chemist, and a biologist, engineers and programmers, all closely associated, understanding and appreciating each other's work."[15] It was an ideal environment to encourage Dayhoff's dream of merging biology, computers, and information science.

Dayhoff's first attempt to merge computers with biology was an algorithm to help researchers determine the entire sequence of a protein from the sequence of several overlapping fragments. Her FORTRAN programs, devised for an IBM 7090 mainframe, could take a set of partial sequences and find a solution in less than five minutes.[16] She designed other algorithms to compare protein sequences, seeking overlaps or regular patterns. To test them, Dayhoff began to collect the few known protein sequences. Without knowing it, she was assembling the embryonic material to produce the *Atlas*. Ledley and Dayhoff hoped that laboratory researchers would understand the value of their computer programs and adopt them widely, but they had no visible impact on protein sequencing. The divide that separated computing and experimental cultures was alive and well in the life sciences.

While Dayoff's efforts to build a complete collection of protein sequences was a dream, it was also a step toward one that was even more ambitious: to understand the "evolutionary history and the biochemical function" of organisms.[17] Her interest in evolution was sparked by computer simulations that she used to try to grasp the chemical transformation of the early planetary atmosphere, including the conditions under which "biologically interesting compounds, such as amino acids, were generated."[18] The work led to a collaboration with the astronomer and popularizer of science Carl Sagan, with whom she published in 1967 a study of the evolution of Venus's atmosphere.[19] Such work fit into the scientific agenda of the space age and attracted funding from NASA. While the early stages of the development of the *Atlas* were far below the radar of most biologists, an organization with its eye on the stars saw potential: it might offer insights into the earliest forms of life on Earth and thus, perhaps, on other planets and well.[20]

BRINGING BIODIVERSITY TO THE LABORATORY

One aspect of the index of Dayhoff's *Atlas* immediately stood out to its readers: it included all kinds of animals, including wild ones. Species diversity was a hallmark of natural history, while it was mostly absent in the experimental life sciences. This made sense in historical context: experimentalism had been steadily emerging since the late nineteenth century, thanks to new and better instruments, accompanied by conventions concerning the biological material most suitable for investigation. The great diversity of life that had been a blessing for naturalists was a curse for experimentalists, who hoped to derive general laws of nature from the study of single, often standardized, species.[21] When Claude Bernard studied the physiological role of the pancreas and the liver in dogs and rabbits, he had no doubt that his discoveries would hold true for all animal species, including humans. Particularly with the rise of genetics in the early twentieth century, researchers began to focus on just a handful of species they described as "model organisms."[22] A single species of mice (*Mus musculus*), flies (*Drosophila melanogaster*), worms (*Caenorhabditis elegans*), fish (*Danio rerio*), or weeds (*Arabidopsis thaliana*), for example, could represent a large class of organisms and possibly the entire living kingdom. This trend would be particularly strong among molecular biologists focusing on some of the simplest organisms on Earth: bacteria and viruses. The bacteria *Escherichia coli*, for example, became the main focus of researchers working on the molecular mechanisms of genetic regulation. The idea was the basis of a claim made in 1961 by the French molecular biologists Jacques Monod and François Jacob (who were paraphrasing Albert Kluyver): "[What is] true of *E. coli* must also be true of elephants."[23] This belief in the uniformity of nature was rather convenient, from a practical point of view; it provided scientists with an rationale for ignoring elephants, which would have been difficult to bring into the laboratory or chase down in pursuit of tissues.

Dayhoff had a different perspective, espoused by a minor, dissident branch of biochemistry called "comparative biochemistry": natural diversity could be used as a valuable tool to gain insights into fundamental principles.[24] Comparing *E. coli* and elephants could yield important knowledge about both species that could not be obtained by studying one of them. Dayhoff extended this idea to protein sequences from different organisms; comparisons, she argued, could reveal fundamental information that was "hidden in the amino acid sequence."[25] That would include information on the evolutionary history of a protein (and its host species), as Linus Pauling and Emile Zuckerkandl had suggested a few years earlier, as well as the molecular mechanisms that allowed the protein to perform its biochemical function. Two aspects of such comparisons would be informative. If a protein such as hemoglobin had versions in two species, the number of differences would give an indication of their evolutionary distance. Similarities between certain parts of the sequence would likely indicate regions in the molecule that were conserved evolutionarily and played a key role in its functional activity. Both evolutionary and biochemical research required molecular data from many species—the more, the better—rather than just a handful of model organisms. Dayhoff's all-encompassing approach was seen as original by some— and antiquated by many.

The philosophical challenges of the project were accompanied by some practical ones: it was no easy task to find information on the protein sequences of unusual species. The information was scarce, and even when someone had gone to the trouble of getting it from a species that was considered unusual by experimentalists, the results were often published in one of a wide range of journals. Thus, to find sequences from reindeers, one had to turn, unsurprisingly, to the *Acta Chemica Scandinavica*, for example.[26] Another problem was that while experimentalists exhibited great precision and care in their descriptions of experimental methods, they were surprisingly lax about the taxonomic identity of the material from which they extracted their proteins. On more than one occasion, Dayhoff had to write personally to an author, asking for specifics about the species a protein had been obtained from; she reminded a biochemist that "monkey" was not a proper scientific name (it refers to more than two hundred and fifty species).[27]

So, for the many experimentalists of the 1960s who had spent most of their careers working on a single species, or at most a few, Dayhoff's project of bringing biodiversity from the museum of natural history into the laboratory looked like a dream. Half a century later, it has been realized beyond her wildest expectations. In 2018 one successor of her *Atlas*, a database called GenBank, boasted sequences from more than four hundred thousand species. This puts the collection on the same order of magnitude as the largest museums of natural history in the world.[28]

BUILDING AN OPEN COMMUNITY

As large as Dayhoff's technological vision was, it paled in comparison to the grandeur of her social imagination. She dreamed of a community of disinterested scientists who would cooperatively share data that had cost them months or years of tedious experimental work to obtain. From the community perspective, scientists would clearly benefit from having access to a comprehensive collection of data. An individual scientist, however, saw a loophole that was just as clear: he would benefit just as much without participating—provided that everyone else did. Epistemic goals and political visions were related to each other, but did not determine each other. Collections, whether in science or some other field, have often embodied political visions, as is so obvious in the colonial collections of metropolitan museums: the objects they contain and the means by which they were obtained overtly expressed the reach and power of an empire over its colonies.[29]

Dayhoff's dream of global data collection in science stopped short of achieving empire, but she did hope to bring together researchers, irrespective of their disciplinary background or professional position, into a community that would unite efforts to make sense of nature by sharing data. To collect the initial set of data for her *Atlas*, Dayhoff and her team proceeded as did most collectors, which meant getting their hands on each item. That meant manually leafing though printed journals to find articles describing protein sequences. Unfortunately, the sequence was rarely simply printed in the article. It had to be derived from the authors' description of the experiments they performed to obtain it, which meant evaluating the methods and validating their accuracy. Even when a sequence did appear, there were often ambiguities or typographic errors. So the actual sequence usually had to be clarified through lengthy correspondence directly with the authors. When the team had finally obtained the data, they entered it manually onto punch cards, from which computers could either perform calculations or print the *Atlas*. So, from the beginning, Dayhoff was doing more than simply collecting data; she was contributing to *making* it. Collecting and curating, as every museum curator or librarian knows, is a time-consuming process consisting of many mundane tasks, from checking an incomplete reference to correcting a typographic error or resolving a contradiction between two entries. The process was unfamiliar to most experimental researchers, so they typically underestimated the amount of effort required of Dayhoff and her team.

From the beginning, the quantity of sequence data was growing faster than she could collect it, so Dayhoff hoped to establish collaborations with the researchers who were producing it. Each edition of the *Atlas* contained an invitation for researchers to submit "new data," and Dayhoff renewed the call in articles in scientific journals. But her communitarian ideal of data sharing failed to become a reality, for many reasons. Her ideal was at odds with the professional ethos of experimental scientists in the 1960s and 1970s, and embodied a contradiction between her vision of freely flowing information and her sense of ownership of the collection as a whole.

Apparently Dayhoff overlooked one of the main reasons scientists made their results "public," usually in a scientific journal: to establish priority and authorship. Authorship translated into scientific credit, the main currency by which a scientist could advance a career. Later, a molecular biologist and member of the Committee on Data for Science and Technology lamented, "Scientists are fierce individualists who consider themselves lone seekers of new knowledge. . . . The idea that they are part of an unorganized community of minds involved in a collective effort to seek knowledge may be foreign to most of them."[30] Publication in the *Atlas* conferred neither authorship nor priority, as she made clear in the introduction of each edition: "It is not our intention to become involved in questions of history or priority."[31] But asking researchers to submit data even before it was published, and thus before authorship had been confirmed in a scientific journal, neglected the fact that experimental researchers were engaged in fierce competition with each other. It surely discouraged many from giving away data they had obtained, especially when it could provide important hints that would allow their competitors to determine related sequences.

Dayhoff's dream of freely sharing information might have taken hold in the sixties, a time when the counterculture resonated with such ideals, but sharing sequences for free caused a conflict with the practical and especially financial conditions that were required to operate the enterprise. To protect the immense amount of labor and resources that the team was investing in the collection, verification, and assembly of protein sequences, Dayhoff decided to copyright her data collection. A second step she took to generate much-needed income for the project was to sell copies of the *Atlas* to cover printing costs, data collection, and its curation. In 1969, when the database was made available on magnetic tapes, she requested a payment of several hundred US dollars for a copy. And finally, adding fuel to the fire, those who acquired the tapes had to sign a contact agreeing not to redistribute the data.[32] Dayhoff reminded the buyers that "this information is proprietary."[33]

Dayhoff's "businesslike" model, as she would put it later, was directly at odds with her aim of a free sharing of data, particularly considering that an Internet-based economy with free products lay so far in the future. The scientific community, of course, noticed the contradiction. A researcher asked Dayhoff rhetorically: "you are in somewhat the position of a folksong collector who copyrights his published material; do I have to pay him if I sing *John Henry*?"[34] All of this left the collection in an ambiguous state that further hampered Dayhoff's efforts to collect sequences in a way that required the voluntary participation of researchers. They considered the *Atlas* just a collection of *their* work, while she regarded it as something of *her* own making.

Yet Dayhoff had little choice but to fund the project through its contributors and users. Science-funding agencies such as the National Institutes of Health and the National Science Foundation were reluctant to fund the database. It did not fit into the category of scientific research, from their point of view, but was merely an administrative task or at best an infrastructure. This deeply underestimated the kind of scientific expertise needed to produce the *Atlas*. To overcome this perception, Dayhoff began putting more emphasis on the results of her analyses of *Atlas* data, especially its uses in creating evolutionary trees or finding common patterns in protein sequences. Although this analytic work was widely recognized and praised, it caused further antagonism among the scientists who had produced the data in the first place—now she was taking undue advantage of her privileged access to the data collection. Dayhoff's dream was turning into a nightmare.

While Dayhoff's political dream remained elusive, her more immediate goals were starting to be realized. The second edition of the *Atlas*, published just a year after the first, contained twice as much data. The third, two years later in 1968, contained twice as much again. While shouldering the burden of most of the collecting itself and its assembly into the collection, Dayhoff still managed to produce a comprehensive collection of protein sequence data that was carefully verified and curated. It enjoyed a growing success among researchers, and sales grew rapidly as the *Atlas* became a common fixture in biomedical research laboratories. As one researcher put it, "We use your book like a bible!"[35] Reviews were also mostly positive, regarding both the results of her analysis of the data and the methods she proposed to calculate evolutionary distances, for example.

Dayhoff's *Atlas* was simply a printout of her digital database. As time went on, researchers began wanting more than a printed reference work: they wanted to access the data in a computer-readable format, which would allow them to perform the kinds of analyses that she could. Initially, she resisted sharing her punched cards, but with the transfer of the data to magnetic tapes in 1969, she agreed to sell tapes for $400 per set. By 1980, she had made subscriptions available that allowed users to connect their computers to a telephone network, using a modem, and access her database directly.

1977 saw the arrival of new methods to sequence DNA that were far more efficient than both their predecessors and even the methods used to sequence proteins. The focus of scientists was shifting from proteins to DNA. Dayhoff, who had included DNA sequences in the early editions of the *Atlas*, followed the movement by establishing a separate database for nucleic acid sequences, and she dreamed that the project represented "a mere shadow of its ultimate grandeur."[36]

It wasn't until 1982 that the NIH finally understood that sequence collections were not an archaic tool for natural historians but indispensable instruments for modern experimental scientists. This realization led to their decision to fund a national public database in the United States, partly in response to a development in Europe: two years earlier, the European Molecular Biology Laboratory had proposed the creation of a public database. Dayhoff was the most likely contender to be granted the NIH contract. Yet after a lengthy scientific, administrative, and legal battle, she lost to the Los Alamos National Laboratory, which had built a database using Dayhoff's initial collection. Dayhoff's ambiguous position with regard to the private ownership of data, and her reluctance to make it openly available for redistribution and other purposes, weighed heavily against her. Furthermore, the NBRF of the 1980s had lost the position it had held twenty years earlier at the cutting edge of computing and the life sciences. Los Alamos was able to boost its place in technological modernity and—paradoxically—openness by offering access to supercomputers such as the Cray (which were hardly needed for biological calculations), as well as to ARPANET (to which most potential users were not connected). In 1983, less than a year after losing out to her competitors, Dayhoff died of heart failure.

CONCLUSIONS

In the 1960s, molecular biologists attempted to consolidate their emerging discipline by taking over positions in biology departments in which naturalists still played a major role.[37] Part of their strategy was to define "modern" experimental molecular biology in contrast to "traditional" natural history and its highly descriptive methods. For molecular biologists, collection-based research, which especially evoked natural history museums, symbolized everything that was wrong with the naturalists' approach to biology. Dayhoff's dream seemed definitely "retro," even archaic, at a time when experimental virtuosity was guiding biology toward a bright future.

But unlike molecular biologists, who were seeking to improve their professional status, Dayhoff could sustain such dreams, partly because she had never been an experimental scientist herself and had moved across many disciplinary boundaries. While she had great intellectual ambitions, her professional ambitions remained modest. She did seek professional recognition from the scientific community sometimes—for example, through her election to professional societies. But the resistance she faced there, and her uncompromising dedication to fulfilling her dream on her own terms, led her to begin focusing exclusively on her data collection rather than seeking personal recognition. Dayhoff's professional trajectory was also influenced by her gender in many ways, including the manner in which her mainly male colleagues received her work. When asked in an interview to describe Dayhoff's personality, a leading computational biologist who had crossed paths with her in the 1970s began by saying, "She was, you know, . . ." and used his two full hands in front of his chest to mimic her breasts, "a very maternal figure,"[38] before mentioning any of her intellectual contributions, dispelling any doubts that her gender must have colored the perception of her work among the many male scientists working in the emerging field of computational biology.

That Dayhoff would attempt to build a shared resource for the scientific community rather than advance her own career did not make much sense to many (male) scientists. But for her, having a family and two children to take care of, as well as a professional home where she could pursue her intellectual dreams made perfect sense. Balancing professional and family life represented a serious challenge for any woman in the period immediately following the Second World War (and it remains so for women and men today). In 1967, in a reflection on her discontinuous career path, she said, "The 'system' does not meet the needs of women very well. I always felt that if I had not been able to get my doctorate at twenty-three, I might not have gotten it at all. Twenty-eight or thirty is much too old to permit the other activities of vital importance to women."[39] What Dayhoff attributed to reproductive physiology was just as much a reflection of marital expectations in the postwar United States.

The attitude toward data collections such as Dayhoff's changed considerably during the 1970s. No longer was the *Atlas* simply used as a printed reference for individual sequences, but increasingly, in electronic format, as a resource for the production of biochemical and especially evolutionary knowledge. Dayhoff and a growing number of molecular evolutionists had demonstrated the power of sequence comparisons in reconstructions of life's history. Her collection was often cited in the heated arguments over the neutral theory of evolution after 1969.[40] Changes in technology also played a role in the growing acceptance of electronic data collections. The increasing availability of microcomputers, such as the popular PDP-11 (still the size of a wardrobe), made the potential of Dayhoff's data collection accessible to many more researchers. A number of mathematicians and computer scientists began developing powerful methods for sequence analysis, especially for the alignment and comparison of sequences, which contributed to the growth of the new field of bioinformatics.[41]

By the twenty-first century, the genomics revolution was in full swing. Its value for the production of biomedical knowledge rested in large part on the ability to analyze vast amounts of sequence data that were stored and organized in the databases that arose in the wake of Dayhoff's initial *Atlas*. When she first dreamed of a computerized infrastructure to hold all the molecular data in the world, laboratory researchers barely used computers, and data was too scarce to warrant a collection. Half a century later, her initial collection of just seventy sequences had evolved into massive genomic databases that contain more than half a billion sequences. Today's scientists are living Dayhoff's dream.

FURTHER READING

November, Joseph A. *Biomedical Computing: Digitizing Life in the United States.* Baltimore: Johns Hopkins University Press, 2012.
Stevens, Hallam. *Life out of Sequence: A Data-Driven History of Bioinformatics.* Chicago: University of Chicago Press, 2013.

Strasser, Bruno J. "Collecting, Comparing, and Computing Sequences: The Making of Margaret O. Dayhoff's *Atlas of Protein Sequence and Structure*, 1954–1965." *Journal of the History of Biology* 43, no. 4 (2010): 623–60.

———. "The Experimenter's Museum: GenBank, Natural History, and the Moral Economies of Biomedicine." *Isis* 102, no. 1 (2011): 60–96.

———. "Collecting Nature: Practices, Styles, and Narratives." *Osiris* 27, no. 1 (2012): 303–40.

NOTES

1. Paula Findlen, *Possessing Nature: Museums, Collecting, and Scientific Culture in Early Modern Italy* (Berkeley: University of California Press, 1994); Lorraine Daston and Katharine Park, eds., *Wonders and the Order of Nature, 1150–1750* (Cambridge, MA: Zone Books, 1998).

2. Brian W. Ogilvie, *The Science of Describing: Natural History in Renaissance Europe* (Chicago: University of Chicago Press, 2006).

3. Nicholas Jardine, James A. Secord, and Emma C. Spary, eds., *Cultures of Natural History* (London: Cambridge University Press, 1996); Paul Lawrence Farber, *Finding Order in Nature: The Naturalist Tradition from Linnaeus to E. O. Wilson* (Baltimore: Johns Hopkins University Press, 2000).

4. On the history of "stamp collecting," see Kristin Johnson, "Natural History as Stamp Collecting: A Brief History," *Archives of Natural History* 34, no. 2 (2007): 244–58.

5. Margaret O. Dayhoff, Richard V. Eck, and Robert S. Ledley, et al., *Atlas of Protein Sequence and Structure* (Silver Spring, MD: National Biomedical Research Foundation, 1965). On the early history of Dayhoff's *Atlas*, see Bruno J. Strasser, "Collecting, Comparing, and Computing Sequences: The Making of Margaret O. Dayhoff's *Atlas of Protein Sequence and Structure*, 1954–1965," *Journal of the History of Biology* 43, no. 4 (2010): 623–60.

6. Soraya de Chadarevian, "Sequences, Conformation, Information: Biochemists and Molecular Biologists in the 1950s," *Journal of the History of Biology* 29, no. 3 (1996): 361–86; Miguel Garcia-Sancho, *Biology, Computing, and the History of Molecular Sequencing* (New York: Palgrave Macmillan, 2012).

7. Richard V. Eck and Margaret O. Dayhoff, *Atlas of Protein Sequence and Structure* (Silver Spring, MD: National Biomedical Research Foundation, 1966), xi.

8. Lois Hunt, "Margaret Oakley Dayhoff, 1925–1983," *Bulletin of Mathematical Biology* 46, no. 4 (1984): 467–72.

9. Margaret O. Dayhoff, "Biographical Sketch, Margaret Oakley Dayhoff," 1965, National Biomedical Research Foundation Archives, currently processed at the National Library of Medicine, Bethesda (hereafter cited as NBRF Archives).

10. On Robert S. Ledley, see Joseph A. November, *Biomedical Computing: Digitizing Life in the United States* (Baltimore: Johns Hopkins University Press, 2012).

11. Margaret O. Dayhoff to Naomi Mendelsohn, 28 June 1966, NBRF Archives.

12. "Summary Progress Report of Grant Sequences of Amino Acids in Proteins by Computer Aids," 15 January 1963, NBRF Archives.

13. On the introduction of computers in the life sciences, see November, *Biomedical Computing*.

14. Gerhardt Braunitzer to Margaret O. Dayhoff, 18 April 1968, NBRF Archives.

15. Margaret O. Dayhoff to Naomi Mendelsohn, 28 June 1966, NBRF Archives.

16. "Summary Progress Report of GM-08710," 15 January 1963, NBRF Archives.

17. Margaret O. Dayhoff to Carl Berkley, 27 February 1967, NBRF Archives.

18. Margaret O. Dayhoff, Ellis R. Lippincott, and Richard V. Eck, "Thermodynamic Equilibria in Prebiological Atmospheres," *Science* 146, no. 1461 (1964): 1461–64.

19. Margaret O. Dayhoff, Richard V. Eck, and Ellis R. Lippincott, "Venus: Atmospheric Evolution," *Science* 55, no. 3762 (1967): 556–58.

20. Margaret O. Dayhoff to George Jacobs, 12 January 1966, NBRF Archives.

21. Robert E. Kohler, *Lords of the Fly: Drosophila Genetics and the Experimental Life* (Chicago: University of Chicago Press, 1994); Karen A. Rader, *Making Mice: Standardizing Animals for American Biomedical Research, 1900–1955* (Princeton, NJ: Princeton University Press, 2004).

22. Jim Endersby, *A Guinea Pig's History of Biology* (Cambridge, MA: Harvard University Press, 2007); Rachel A. Ankeny and Sabina Leonelli, "What's So Special about Model Organisms?," *Studies in History and Philosophy of Science, Part C* 42, no. 2 (2011): 313–23.

23. Jacques Monod and François Jacob, "General Conclusions: Teleonomic Mechanisms in Cellular Metabolism, Growth, and Differentiation," *Cold Spring Harbor Symposia on Quantitative Biology* 21 (1961): 389–401.

24. Ernest Baldwin, *An Introduction to Comparative Biochemistry*, 4th ed. (Cambridge: Cambridge University Press, 1966).

25. Dayhoff, Eck, Ledley, et al., *Atlas* (1965), 2.

26. Birger Blombäck, Margareta Blombäck, and Nils Jakob Grondahl, "Studies on Fibrinopeptides from Mammals," *Acta Chemica Scandinavica* 19 (1965): 1789–91.

27. Eck and Dayhoff, *Atlas* (1966), xii.

28. On the history of GenBank, see Bruno J. Strasser, "The Experimenter's Museum: GenBank, Natural History, and the Moral Economies of Biomedicine," *Isis* 102, no. 1 (2011): 60–96.

29. Daniela Bleichmar, *Visible Empire: Visual Culture and Colonial Botany in the Hispanic Enlightenment* (Chicago: University of Chicago Press, 2012); Londa L. Schiebinger and Claudia Swan, eds., *Colonial Botany: Science, Commerce, and Politics in the Early Modern World* (Philadelphia: University of Pennsylvania Press, 2005); Lucile Brockway, *Science and Colonial Expansion: The Role of the British Royal Botanic Gardens* (New Haven, CT: Yale University Press, 2002).

30. Alain E. Bussard, "Data Proliferation: A Challenge for Science and for Codata," in *Biomolecular Data: A Resource in Transition*, ed. Rita Colwell (Oxford: Oxford University Press, 1989), 13.

31. Eck and Dayhoff, *Atlas* (1966), xiv.

32. Margaret O. Dayhoff, "LM 01206, Comprehensive Progress Report," 23 August 1973, NBRF Archives.

33. Margaret O. Dayhoff to Robert G. Denkewalter, 8 February 1971, NBRF Archives.

34. B. S. Guttman to Margaret O. Dayhoff, 10 June 1968, NBRF Archives.

35. Oliver Smithies to Winona Barker, 5 October 1970, NBRF Archives.

36. Margaret O. Dayhoff, "Technical Proposal: Establishment of a Nucleic Acid Sequence Data Bank," 1 March 1982, NBRF Archives, 12.

37. Edward Osborne Wilson, *Naturalist* (Washington, DC: Island Books, 1994). On the conflict between molecular biologists and "traditional" evolutionists, see Michael R. Dietrich, "Paradox and Persuasion: Negotiating the Place of Molecular Evolution within Evolutionary Biology," *Journal of the History of Biology* 31 (1998): 85–111.

38. Interview with X, Cambridge, MA, 16 February 2006.

39. Margaret O. Dayhoff to Russ F. Doolittle, 18 October 1968, Judith Dayhoff Personal Archives.

40. Michael R. Dietrich, "The Origins of the Neutral Theory of Molecular Evolution," *Journal of the History of Biology* 27 (1994): 21–59.

41. Hallam Stevens, *Life out of Sequence: A Data-Driven History of Bioinformatics* (Chicago: University of Chicago Press, 2013).

LUIS CAMPOS

NEANDERTHALS IN SPACE
GEORGE CHURCH'S MODEST STEPS
TOWARD POSSIBLE FUTURES

9

The author of more than 418 papers, 74 patents, and the first ever book to be encoded into DNA, George Church is a formidable example of the contemporary synthetic biologist, start-up entrepreneur, and emerging celebrity scientist. From his earliest boyhood days, fascinated with both biology and computers, Church has been keen to "start with nature's operating system, reprogram it, and collect [the] output in the form of fabulous new engineered organisms."[1] Mastering the interface between the biological and the digital promised not only to bring long-standing goals of genetics within reach—starting with novel methods for sequencing the human genome—but could even lead to a future "regenesis" beyond our wildest dreams, as we look to reinvent nature and ourselves. And Church's dreams have been pretty wild, indeed: from designing new genetic codes in "*rE. coli*" (to ensure ecosystem integrity with engineered organisms), to de-extincting mammoths (one proposed solution to our climate change problems) and bearing Neanderthal children (who will lead humanity into space when the time comes to escape a dystopian future Earth), Church's dreams of scientific futures are the stuff of science fiction.

Dreaming the future has long been a driving leitmotif of Church's work: "It's all too easy to dismiss the future," Church has claimed. "People confuse what's impossible today with what's impossible tomorrow."[2] People simply "talk themselves out of things very easily. Things that they think are a million years away or never, are actually four years away."[3] By pushing biology far beyond its ordinary boundaries and constantly promoting and redefining what it means to envision the future of life, Church has become a quintessential "visioneer."[4] And, in fact, time and again, many of Church's dreams have come true, blurring the line between speculative fantasy and laboratory reality.

While blue-sky thinking has served Church well and helped to drive technical innovation in his own laboratory, in some other cases Church's dreams have generated no end of controversy. This has most frequently occurred when he has attempted to share his dreams with larger publics, as a way of generating public discussions of possible synthetic futures. When dreams are understood no longer merely as airy visions of possible futures but as im-

Figure 9.1. George Church. Photo by Marius Bugge.

minent laboratory realities routinely brought into being through remarkable technical ingenuity, a heavy load of real dissension and concern has sometimes emerged. A careful look at some of Church's more provocative dreams thus not only reveals the tremendous power of his visionary thinking, but also offers a cautionary tale for dreamers of untrammeled innovation in biology who would fearlessly cross established boundaries. As dreams become tactics, and powerful ethical critiques are converted into questions of safety and efficacy, sensationalist visions of the future might themselves become costly dangers to be avoided. In other words, dreaming is never enough. Given that the sleep of reason produces monsters, one must always take special care that dreams not turn into nightmares.

BACK TO THE FUTURE

George Church has always been trying to get back to the future. Born at Florida's MacDill Air Force Base on August 28, 1954, George "wanted to be a fireman, construction worker or paperback-writer when [he] grew up."[5] But a visit to the 1964–65 New York World's Fair proved a formative experience: "It was bright and shiny. All the surfaces were smooth. Everything was less clunky," he recalled.[6] "It just struck me as the kind of place where we all deserved to live. And then I went back home to Florida, and I sort of waited for the future to arrive. And it didn't. I realized that if I was going to relive that moment, I was going to have to help create it."[7] Church has been trying to get back to the future ever since. "I felt like I've been to the future, part of me lives in the future, and I'm stuck here," he said, alternately describing the experience as frustrating, horrible, and even painful. "I've been trapped back in time, and I have to make the most of it."[8]

Back in Clearwater, Florida, science became the heart of George's world, at least partly due to his dyslexia ("I would focus on science books with lots of pictures"). But biology turned out to be a powerful draw. By age eight, he had become fascinated by the metamorphosis of insects, and by age nine, George idolized the famed botanist and breeder, Luther Burbank. He was thrilled to successfully repeat some of Burbank's experiments by grafting apple and pear trees in an attempt to transfer disease resistance. George also had "two greenhouses full of orchids"—in-kind payment his lawyer mother had received from a client.[9] What George learned about apples, pears, and orchids cultivated a talent to identify the low-hanging fruit in his future endeavors in biotechnology.

It was in Church's boyhood that his dreams to wed biology and computers first emerged. Not having a computer nearby, Church recalled, "I made one myself." Even as he continued experimenting with plant hormones (with the ultimate goal of creating giant Venus flytraps), Church began to work with computers in earnest after he arrived at Phillips Academy Andover in 1968 at age fourteen.[10] He had found an abandoned computer terminal in the basement of Morse Hall, which was connected to the Dartmouth College GE-635 mainframe. By age fifteen he was spending countless hours every week teaching himself BASIC, LISP, and FORTRAN, and became so involved in teaching the computer linear algebra that he stopped attending many of his classes.[11] Even though he had to repeat the ninth grade, he was already dreaming of the day when he could bring his love of biology and computers together and sequence the human genome: "I've been trying to read and write nucleic acids since I was a teenager," Church remembers.[12]

After graduating from Andover in 1972, he matriculated at Duke University where he managed to complete two BA degrees (chemistry and zoology) in only two years. Continuing with graduate studies at Duke in biochemistry (rather than at Harvard, his second choice), Church worked with X-ray crystallography to study the structure of tRNA. "It was one of the few fields in biology with any automation that had a solid physical theory behind it and used computers extensively," Church recalled.[13] Old habits died hard, though, and Church's single-minded focus in the lab—working more than a hundred hours a week—led to the neglect of his other coursework: "I was the typical obsessive scientist," Church recalled, remarking on second thought, "Maybe a little atypical."[14]

After failing one of his graduate courses, Church was automatically expelled from Duke, but was taken in as a graduate student at Harvard only a year later, finally graduating in 1984.[15] Church's PhD work, developing methods for DNA sequencing of genomes, would soon "reduce the cost of whole-genome sequencing from billions of dollars to thousands," and place Church in the center of the Human Genome Project, after which he would continue to invent the next generation of sequencing methods in his lab, bringing prices down some 10-million fold.[16] A theme of Church's career was becoming clear: he dreamed of not just "incremental improvements but transformative or sometimes positively disruptive technologies."[17]

THE CHURCH LAB: SCIENCE FICTION AS SCIENCE

After a stint working for Biogen Research Corporation and a postdoc at the University of California–San Francisco, Church followed his wife, molecular biologist Chao-Ting Wu, back to Harvard Medical School in 1986 and began climbing the ranks, rising to become Professor of Genetics in 1998.

As the director of a lab of his own, Church would be often so engaged in his work that he would forget to eat for days at a time (lab lore holds that "he lived for a year on nutrient broth from a lab vendor").[18] Disruptively dreaming the future of biology at its interface with the digital became the prime occupation of the Church lab, where researchers from a variety of disciplines from art to biology have found themselves in conversation with one another: "My particular lab depends on thinking outside the box and not dismissing things because they sound like science fiction." A low median age in the lab also helps keep things nimble, Church has noted, "because they will indulge me in my dreams. They don't yet think things are impossible."[19] The end result of Church's unorthodox approach to lab composition and management is that his lab of dreams somehow "manages to be both one of Harvard's top producers and a well-known receiving center for science's misfit toys."[20]

By the first decade of the twenty-first century, the Church lab had accomplished some remarkable achievements in next-generation sequencing. By 2004 Church had outlined a new method of DNA synthesis on microchips that would reduce costs by a thousandfold, and by 2005 he announced a new method of "multiplexing" (in analogy to signal multiplexing in electronics) that would allow for the simultaneous sequencing of millions of genomes in parallel, rather than the serial capillary sequencing that had previously been the standard. Upping the ante, by 2009 Church invented a new method of multiplex automated genome engineering (MAGE) for not merely sequencing but actually *engineering* up to fifty changes to a bacterial genome simultaneously. With his evolution machine using brute-force assaying of inserted oligos, Church reported, "in a day, you can generate a billion genomes" from "wholesale genetic changes" in an "automated fashion."[21]

Such technologies made possible still further endeavors, such as the reworking of the *E. coli* genome—labeled the "*rE. coli*" project—so that the synthesized bacteria require specially tailored nutrients to survive. "We're building genetically modified organisms that can't escape and can't influence the ecosystem because they are genetically and metabolically isolated. They're on a very short leash," he explained, addressing fears that genetically engineered organisms might escape and cause havoc in the biosphere.[22]

Church's efforts to go beyond the established boundaries of biology have been varied and legion, ranging from a privacy-free "Personal Genome Project" to new research in biofuels ("Making new petroleum should be as simple and straightforward as brewing beer").[23] He has swiftly put forward a variety of putatively "modest proposals," as he calls them: "What if it were possible to make human beings immune to all viruses, known or unknown, natural or artificial?"[24] The creation of "mirror organisms" based on an opposite stereochemistry would do the trick. By 2013, the use of the novel genome-editing technique CRISPR for the engineering of the human genome and for the simultaneous removal of up to sixty-two endogenous retroviruses in pigs (or PERVs) promised to open up new and safer possibilities for organ xenotransplantation. But perhaps the most breathtaking of Church's dreams for the future was one that surpassed the boundaries of life itself. As Church noted, "What often comes to mind is the old adage, 'Extinction is forever.' Only it isn't. . . . Genomic technology can actually allow us to raise the dead."[25]

THE RESURRECTION MAN

While science fiction visions danced in his head, Church dreamt that the reconstruction of ancient DNA sequence was not as far-fetched as it might at first seem. The successful de-extinction of the Spanish bucardo, "Celia," on July 30, 2003 marked "a turning point in the history of biology" for Church, "for on that date, all at once, extinction was no longer forever."[26] Moreover, new techniques like CRISPR were bringing the hybridization of extant and extinct species within reach: "This is real, not science fiction," he concluded, "and it's potentially applicable to . . . extinct species."[27]

At a landmark meeting on de-extinction held at Harvard Medical School in February 2012, the passenger pigeon was the focus of early discussion. But the real elephant in the room for Church was the resurrection of the mammoth: "The mammoth almost cries out for resurrection. Some specimens unearthed from permafrost are so lifelike that they appear to be merely sleeping, not dead, much less extinct." Challenging naysayers who would argue against such tales of sleeping beauties that "it is pointless to bring them back into a world in which those habitats have long since vanished," Church argued that "it is possible to bring back the habitat along with the animal itself."[28] In fact, Church's *Jurassic Park* dreams of de-extinction were inspired by Sergei Zimov's actual efforts to create a "Pleistocene Park" in northeastern Siberia.[29]

The resurrection of the woolly mammoth would be, to Church's eye, "the closest thing to time travel: a return to the flora and fauna of the Pleistocene epoch, a sort of latter-day Siberian Eden."[30] Not only could the barren northern wastes of the Pleistocene Park be transformed to "highly productive pastures" (perhaps reminiscent of Linnaeus's attempts in the eighteenth century to grow sugarcane in Lapland), but the de-extinction of mammoths could even be a "potential means of addressing anthropogenic global warming." Mammoths were "hypothetically a solution" to climate change by "getting rid of dead grass, stomping down the trees and stomping down the insulating snow in the winter so the arctic freeze can get into the permafrost. These are the sort of things that could be achieved uniquely by mammoths."[31]

However entrancing was the idea of de-extincted mammoths saving the world from climate Armageddon through carbon-sequestration, one footprint at a time, Church's claims for de-extinction rapidly descended to the more mundane: "We are not trying to make an exact copy of a mammoth, but rather a cold-resistant elephant," he noted.[32] "I'm not going to call them mammoths unless somebody insists. They're elephants with mammoth DNA."[33] By March 2015, having successfully used CRISPR to insert genes from the woolly mammoth into an Asian elephant, Church himself had become "skeptical that there will be a clean defining moment," for de-extinction.[34] At the very instant that de-extinction was becoming a real-world reality, the mammoths were dematerializing into a dream.

Other species, however, had already come to the fore. Indeed, it seemed only a small step to move from de-extincting animals to the futuristic (and yet atavistic) next frontier in re-creation science: resurrecting extinct species of humans. "The same technique would work for the Neanderthal," Church noted, "except that you'd start with a stem cell genome from a human adult and gradually reverse-engineer it into the Neanderthal genome or a reasonably close equivalent. . . . The next step would be to place it inside a human (or chimpanzee) embryo, and then implant that cell into the uterus of an extraordinarily adventurous human female—or alternatively into the uterus of a chimpanzee."[35]

Church offered several reasons why someone might dream of doing such a thing, chief among these being "to increase diversity. The one thing that is bad for society is low diversity. This is true for culture or evolution, for species and also for whole societies. If you become a monoculture, you are at great risk of perishing. Therefore the recreation of Neanderthals would be mainly a question of societal risk avoidance." Or, as he told a journalist from the German newsmagazine *Der Spiegel* in an interview that was heard around the world, "Neanderthals might think differently than we do. We know that they had a larger cranial size. They could even be more intelligent than us. When the time comes to deal with an epidemic or getting off the planet or whatever, it's conceivable that their way of thinking could be beneficial. . . . They could maybe even create a new neo-Neanderthal culture and become a political force."[36] Church's references to Neanderthals as "creatures" and the human mothers of the proposed interspecies offspring as "extremely adventurous female[s]" drew tremendous media attention. Within one week of the interview, word that Church was seeking volunteer surrogates went viral, with more than six hundred news outlets and aggregators reporting the story.[37]

While the idea of launching Neanderthals into space and coloniz-
ing the final frontier with our de-extincted hominid ancestors as a means of
saving human civilization is one of the more fantastic of Church's dreams,
it is only one of several possible space odysseys that Church has envisioned.
Escaping Earth and its implicitly dystopian future is in fact a recurring theme
in Church's long-term dreams, and he has even come to view DNA itself as a
kind of spaceship that could do amazing things—"We have received a great
gift that biology has given to us. . . . It's as if a master engineer parked a space-
craft in our back yard with not so many manuals, but lots of goodies in it that
are kind of self-explanatory." Why not imagine DNA as the ideal spaceship for
transporting life to space? In a literalization of H. J. Muller's mid-century call
for a "genetic Sputnik," Church dreamed of launching pure DNA into space,
in a grand new vision of directed panspermia. As yet another "modest step
toward these possible futures," he noted, "we need to get at least some of our
genomes and cultures off of this planet or trillions of person-years of work
will be lost. . . . We need to shoot our SCHPON (sulfur, carbon, hydrogen,
phosphorus, oxygen, nitrogen, 'spawn') into the void. . . . We will be seeding
outer space with ourselves or our descendents [sic]."[38]

Sending DNA to the moon was literally another of Church's dreams for
the future of life. In having encoded all 53,426 words and 11 images of his
2012 book, *Regenesis*, into DNA—and having made 70 billion copies on a chip
smaller than the period at the end of this sentence—Church not only realized his dream of literary immortality, but provided proof of concept of DNA
as a means of data storage with a million times the density of current disc
drives.[39] The incredible stability of DNA as a storage medium immediately
drew commercial attention from companies interested in preserving films
and recordings, with some firms even eyeing the moon as a long-term storage
location. By 2015, Church and his lab had begun to work with the French firm
Technicolor S. A. to encode an entire early silent film based on a Jules Verne
story—"A Trip to the Moon" (1902)—into DNA.[40] Scaling up his work "about a
hundredfold," Church also found ways to increase data storage with a denser
form of synthetic, "industrial-strength" DNA. His dreams of a genetic moonshot were coming together—not only could he break through undreamt-of
frontiers in the application of synthetic biology technologies, but with this
most recent work, he would literally look to "shoot it to the moon, the way
Jules Verne envisioned."[41]

While aiming for the moon has long been a feature of Church's dreams,
he's also never been afraid to shoot far beyond it. The increasing interchange-
ability of biological life with its digital representations—and the sagacious

thought of bringing together the billions and billions of possible new genomes with the vast enormity of space—even led Church to suggest the concept of remote "printing" of DNA: the idea that "fractions of the human genome could be sent to suitable planets embedded in bacteria designed for interstellar travel. Once there, the human would be reassembled."[42] Whether through Neanderthals in space, humanity practicing galactic onanism, the more mundane purposes of long-term data storage, or beaming digital life and shuffling off our mortal coils—getting life aloft has been a continuous dream. Moonshot or moneyshot, Church's dreams have always been larger than life: "I think there are some people whose job it is to cross barriers," he says. "I'm one of those people."[43]

WAKING UP TO THE CRITICS

Some may say Church is a dreamer, but they are not the only ones: Church's dreams for the future of biology are not without their critics. In the case of the de-extinction of Neanderthals, the prominent evolutionary geneticist Svante Pääbo—director of the Max Planck Institute for Evolutionary Anthropology in Leipzig and the first to successfully sequence the Neanderthal genome—publicly registered his concerns about Church's dream: "Neanderthals were sentient human beings. In a civilized society we would never create a human being in order to satisfy scientific curiosity," Pääbo wrote in an op-ed to the *New York Times* entitled "Neanderthals Are People, Too."[44] Christina Agapakis, a Harvard-trained synthetic biologist, highlighted a different important and deeply gendered dimension of Church's all-wet dream in a powerful blog post for *Scientific American*:

> In such scientific imaginings we get futuristic versions of some very retrograde cultural ideas about gender. While I know that these men don't actually think of the women in their lives and in their labs as simply vessels for DNA (some of my best friends are male synthetic biologists!), I also know that leaving these kinds of statements unexamined can lead to an environment that makes it harder for women working in these labs, harder for women to be chosen as speakers at quantitative synthetic biology conferences, and harder for women to be promoted and advance in their field. Before we discuss the potential of cloned Neandertals to boost human diversity, we must first consider our role in boosting the diversity of our labs, companies, faculty, and conferences with the humans that actually exist.[45]

For his part, Church chose to frame the controversy surrounding his inartful comments as reflective of the pathologies of contemporary science journalism. (As a follow-up article in *Der Spiegel* noted, "In the course of the past two decades, . . . he has done perhaps 500 interviews about his research and this is the first one to spiral out of control quite like this.")[46] Nevertheless, Church was forced to undertake a campaign of corrections, repeatedly saying that the de-extinction of Neanderthals was not one of his laboratory's current efforts: "Definitely not. We have no projects, no plans, we have no papers, no grants [to do that.]" And yet, he noted, still in the midst of his maelstrom, "hopefully for several years we can have a calm discussion about it. It's way better to think of these things in advance."[47] To some observers, it seemed as if the man who had successfully re-engineered life forms in countless ways had yet to discover a cure for foot-in-mouth disease.

Whether or not Church was actively looking for "extremely adventurous females" to serve as surrogates is irrelevant. The mere fact that Church has repeatedly—both in media interviews and in print in his own book—envisioned such a possibility and framed it in the way that he did highlights how his dreams of possible futures have sometimes ridden roughshod over cultural norms and expectations for proper scientific discourse. Elsewhere in his book *Regenesis*, for example, Church offers a problematic reading of a landmark American civil rights case, stating that "the trend of relaxing rules against intermarriage seems to be taking us in the opposite direction from what's needed for speciation and for the evolution of a new human species. In *Loving v. Virginia* (1967) for example . . ."[48] However much one may be in favor of continued human speciation—and however common such suggestions may have been for an earlier generation commenting on the societal implications of genetics—it is a particular variety of tone-deafness to suggest today that the landmark civil rights case *Loving v. Virginia* be viewed as an obstacle to human transformation.

Church's dreams thus function not merely as envisioned futures soon to become reality, but also as airy nothings that claim to be provocative food for thought or as mere utterances without consequence, as the situation demands. In other words, one tactic Church has deployed to attempt to avoid serious controversy is to claim either that he has been misunderstood or that his statements about life are *merely* dreams—"I am not promising anything. I am just laying out a path, so that people can see what possible futures we have." (As synthetic biology founder Drew Endy has noted, Church is "perfectly happy to spin out tons of ideas and see what might stick. It's high-throughput screening for technology and science. That's not the way most people work."[49])

In Church's hands, however, the world of the future perfect is routinely conflated with present possible. Given Church's "uncanny record" of converting his dreams into reality, it seems difficult to maintain such polite fictions to distinguish clearly between mere dreams of seemingly distant possibilities and what effectively amount to news reports from the near future.[50] In such circumstances, what counts as a reasonable interpretation of his provocative statements? For hundreds of extremely adventurous women who learned about the possibility of gestating Neanderthals, offering their bodies to science as surrogate mothers was a possible response based on an eminently reasonable interpretation. To dismiss unwanted interpretations like theirs as failures of science journalism, as Church has done, is to miss the very ways in which Church's dreams motivate larger publics to support his science.

In other circumstances, when faced with unexpected responses to his public dreaming, Church often chooses to disarm criticisms by characterizing the newest of potentially controversial techniques as mere continuations of familiar and established approaches—"Are we genetically engineering human beings? The answer is, definitely, we've been doing it for quite a while." Even such a controversial contemporary topic as human germline editing is for Church "not special with respect to permanence or consent."[51] On still other occasions, whether he is talking about how to design "mirror organisms" with reversed chirality as a safety feature (when critics have wondered whether they might dangerously outcompete natural organisms for resources) or about the controversial de-extinction of mammoths and Neanderthals, or the potential issues surrounding CRISPR and gene drives, Church chooses to acknowledge concerns that have been raised, but integrates them into his dreams for the future of biology as causes for further research: "The important thing is to actually listen to the public's concerns," he says, "and then try to visualize things that can go wrong and think of ways to guard against them."[52] Indeed, rather than serving as "deterrents to progress," Church claims that such criticisms and controversies regularly "lead to good conversations that, if anything, accelerated the field" of synthetic biology.[53] "Ultimately," Church has claimed, "our future will be what we make of it. Let us choose wisely, with carefully engineered safety and broad community engagement."[54] In such anodyne claims for the peaceful coexistence of democracy and Church's own dreams lies perhaps the most subtle use of dreaming as a form of tactics: it is not his dream alone, goes the skillful "two-step," but our shared future we are dreaming about and making together.

Although Church sometimes characterizes his interlocutors as offering "a very vague critique," more often than not he starts by agreeing with his dissenters, only to then reframe his position in terms of its superior attention to technical details.[55] In a classic technocratic utopian move (and taking a page from the 1975 Asilomar meeting on the potential biohazards of recombinant DNA research), Church frequently deploys the language of safety and effectiveness to reframe ethical questions, while presenting any failure to follow his dream as the real unconsidered risk. In the case of human germ-line editing, for instance, he characterizes some critics as having unhelpfully "entangle[d] vague unknown unknowns with very concrete safety and efficacy issues."[56] More generally, he has claimed that "the precautionary principle traditionally summarized as 'first, do no harm' should not be reduced to 'first, do nothing,' especially regarding technological fixes for our deteriorating biosphere and economy."[57]

Church has sometimes found himself at a loss when others have not shared his dreams. In a remarkable exchange at the Synthetic Biology 3.0 meeting in Zurich in 2007, Church faced off against Jim Thomas, a representative from the activist civil society organization, ETC Group. While Thomas offered a radical leftist critique of the neoliberal world order centered on the industrialization of synthetic biology (of which Church is a prime proponent), Church sat silently for a few moments before finally responding: "Let's talk about our wishlist, rather than assuming that all technology has to stop." Criticisms that could not be immediately interpreted or co-opted could be safely ignored—or used as an occasion for further dreaming.

By moving from the dream to the real world and back again, Church manages both to generate and to defuse controversy, a perpetual motion machine for his own research agenda. But this constant tacking from attention-grabbing sensationalism to seemingly responsible statements of risk assessment has not always sat well with Church's peers.[58] At a September 2014 meeting of the Forum on Synthetic Biology at the National Academies of Sciences in Washington, DC, one committee member complained, "George has the largest voice. The problem is, because George gets interested in it, here we are talking about it *as if it had happened* and were a serious threat." When talking about the very real issues likely to be faced by synthetic biology, members of the forum wondered whether Church's claims were too sensational and distracting them from other more important matters: didn't paying too much attention to his claims "risk confusing what we should be worried about because George wants to drive this, because he thinks it is out of his lab?"

Another member of the committee agreed, characterizing Church's successful media relations strategy as follows: "I am going to capture the dialogue by posing a risk that really doesn't necessarily exist, and by doing it I'm going to open my technology—which I failed to get to work—to public scrutiny just because I want to be a good citizen. The public ends up perceiving risk coming out of Harvard where the risk does not actually exist. It's fabricated as a way of dominating the gaggle, and raising money." Nor was Church alone in this method of marketing his dreams—the ways in which other federal organizations were following suit in "creating risk in order to fund their own budget" was a topic ripe for at least an entire PhD thesis, the member noted.

Such sensational claims are not without costs for some federal workers, however—following other people's dreams can incur a very real cost. The deputy director of DARPA's Biological Technologies Office was reported to have objected to the effects that such sensationalist tactics had had in drawing government attention to particular projects. The fabrication of "false threats" unnecessarily drew the attention of the Department of Homeland Security, which in some instances had slowed down some of DARPA's projects for about a year. For her part, a senior advisor for biotechnology at the FDA and an officer in charge of numerous public comment processes warned the members of the Forum of the very real costs of such risk-mongering for her and her staff. "I hope you all appreciate what happens when you all start these sorts of debates in the public. . . . I won't tell you about the death threats we get." The sheer number of bizarre requests from the public that such marketed sensationalism produced in the public comment process—such as the requests from prisoners at federal penitentiaries who were about to be released who "wanted us to clone J. Lo's ass," she noted—should be cause for significant restraint before inventing either utopian or doomsday scenarios for new biotechnologies. "When you start having these Star Wars kinds of discussions about what you could and couldn't allow, you essentially prejudice the jury pool. And what happens when you prejudice the jury pool in a 'notify and comment' process [is that] we are required by law to consider every single comment that comes in." With more than two million comments about a more mundane matter—risk assessment for AquAdvantage Salmon™—the officer noted, "We had to consider all of them. Think about what all of that means to those of you here developing technology for what your funders are going to say about projected money and timelines. I'm just urging that you understand what all of the unintended consequences of that are." Dreams not only come to life—they can also cause nightmares.

DREAMTIMES AND LIFETIMES

In a strong echo of Luther Burbank's claims to "go Nature one better," Church has announced that "today we are at the point in science and technology where we humans can reduplicate and then improve what nature has already accomplished. We too can turn the inorganic into the organic.

We too can read and interpret genomes—as well as modify them. And we too can create genetic diversity, adding to the considerable sum of it that nature has already produced."[59] Simply put, he says, "We're well beyond Darwinian limitations to evolution. Evolution right now is in the marketplace."[60] Whether it's dreaming of germ-free extraterrestrial surgery on Mars, mapping the human brain and someday "backing up my brain into another that I have in my back-pack,"[61] excising PERVs, or detecting dark matter WIMPs (weakly interacting massive particles) with a DNA detector, or synthesizing a human genome ("the Human Genome Synthesis Project," later renamed "HGP-Write"), Church has no shortage of fascinating dreams that may lead to pathbreaking science. But some of Church's dreams have never quite entirely managed to avoid causing nightmarish visions in others. ("How do you think your work will eventually destroy all mankind?" the late-night television comedian Stephen Colbert memorably asked.)

Now in his sixties, Church's dreamtime and lifetime are catching up to each other. He has matured into a genomic impresario *par excellence* whose dreams inspire young synthetic biologists, turn into powerful new technologies, boot-up companies and create entire new industries, cause journalists to froth at the pen (occasionally leading to international media scandals), tantalize the public, deeply trouble well-meaning activists, and even engender bouts of complaints by peers at the National Academies. "There are still things that I feel should have happened by now," Church has noted, "or we'd be better off if we at least cautiously explore them. There are quite a few things left on my bucket list."[62] The proleptic narcolept will doubtless keep dreaming his way to the future. But therein lies the rub—who knows what dreams may come?

FURTHER READING

Church, George M. Interview by David C. Brock in New Orleans, Louisiana, 3 March 2008. Philadelphia: Chemical Heritage Foundation, Oral History Transcript # 0408.
Church, George M., and Ed Regis. *Regenesis: How Synthetic Biology Will Reinvent Nature and Ourselves*. New York: Basic Books, 2012.
Campos, Luis. "The Biobrick Road." *Biosocieties* 7, no. 2 (2012): 115–39.
———. "Outsiders and In-Laws: Drew Endy and the Case of Synthetic Biology." In *Outsider Scientists: Routes to Innovation in Biology*, edited by Oren Harman and Michael Dietrich, 331–48. Chicago: University of Chicago Press, 2013.
———. "That Was the Synthetic Biology That Was." In *Synthetic Biology: The Technoscience and Its Societal Consequences*, edited by Markus Schmidt, A. Kelle, A. Ganguli-Mitra, et al., 5–21. Dordrecht: Springer Academic Publishing, 2009.

NOTES

1. George Church and Ed Regis, *Regenesis: How Synthetic Biology Will Reinvent Nature and Ourselves* (New York: Basic Books, 2012), 170.

2. Peter Miller, "George Church: The Future without Limit," *National Geographic*, 2 June 2014, online at http://news.nationalgeographic.com/news/innovators/2014/06/140602-george-church-innovation-biology-science-genetics-de-extinction.

3. Church's proleptic views echo those of the famed science fiction author William Gibson (who once wrote, "The future is here. It just not evenly distributed yet"); see "Welcome to My Genome," *Economist*, 6 September 2014, online at http://www.economist.com/news/technology-quarterly/21615029-george-church-genetics-pioneer-whose-research-spans-treating-diseases-altering.

4. Patrick McCray has described a "visioneer" as "a future-looking scientist who promoted bold new technological ventures that would create worlds of new possibilities." See his *The Visioneers: How a Group of Elite Scientists Pursued Space Colonies, Nanotechnologies, and a Limitless Future* (Princeton, NJ: Princeton University Press, 2012), 136. McCray has written about George Church on his *Leaping Robot* blog, "The Church of Synthetic Biology," online at http://www.patrickmccray.com/2013/01/11/the-church-of-synthetic-biology.

5. John Sundman, "Synthetic Biology Legend George Church and I Talk about Science and Civilization," *The John Sundman Blog*, online at http://johnsundman.com/2015/11/synthetic-biology-legend-george-church-i-talk-about-science-and-civilization.

6. Miller, "George Church."

7. Wyss Institute, "Disruptive: Synthetic Biology," online at http://soundcloud.com/wyssinstitute.

8. James Temple, "Meet the Time-Traveling Scientist behind Editas, the Biotech Company Going Public with Google's Help," 5 January 2016, online at http://recode.net/2016/01/05/meet-the-time-traveling-scientist-behind-editas-the-biotech-company-going-public-with-googles-help.

9. Jessica McDonald, "10 Questions for George Church, Geneticist," online at http://www.sciencefriday.com/articles/10-questions-for-george-church-geneticist/.

10. David Ewing Duncan, "On a Mission to Sequence the Genomes of 100,000 People," *New York Times*, 7 June 2010, online at http://www.nytimes.com/2010/06/08/science/08church.html; George Church, "George M. Church Personal History and Interests," online at http://arep.med.harvard.edu/gmc/pers.html.

11. Thomas Goetz, "How the Personal Genome Project Could Unlock the Mysteries of Life," *Wired*, 26 July 2008, online at http://www.wired.com/2008/07/ff-church.

12. Church, "Personal History and Interests"; Emilie Munson, "This Harvard Scientist Is Coding an Entire Movie onto DNA," *Global Post*, 10 August 2015, online at http://www.globalpost.com/article/6628554/2015/08/09/harvard-scientist-coding-entire-movie-dna.

13. Prashant Nair, "Profile of George M. Church," *Proceedings of the National Acad-*

emy of Sciences of the United States of America 109, no. 30 (2012): 11893–895, doi:10.1073/pnas.1204148109.

14. Temple, "Meet the Time-Traveling Scientist."

15. Jeneen Interlandi, "The Church of George Church," *Popular Science*, 27 May 2015, online at https://www.popsci.com/church-george-church.

16. Interlandi, "The Church of George Church."

17. Wyss Institute, "Disruptive."

18. Interlandi, "The Church."

19. Duncan, "On a Mission."

20. Interlandi, "The Church."

21. "Welcome to My Genome," *Economist*.

22. Katherine Xue, "Synthetic Biology's New Menagerie: Life, Reengineered," *Harvard Magazine*, September–October 2014, online at http://harvardmagazine.com/2014/09/synthetic-biologys-new-menagerie.

23. "Welcome to My Genome," *Economist*.

24. Church and Regis, *Regenesis*, 8.

25. Church and Regis, *Regenesis*, 133, 9.

26. Church and Regis, *Regenesis*, 136.

27. Church, "Hybridizing with Extinct Species," TEDxDeextinction, 15 April 2013, online at https://www.youtube.com/watch?v=oTH_fmQo3Ok.

28. Church and Regis, *Regenesis*, 137, 148.

29. "Pleistocene Park: Restoration of the Mammoth Steppe Ecosystem," online at http://www.pleistocenepark.ru/en/.

30. Church and Regis, *Regenesis*, 149.

31. Church, "Hybridizing with Extinct Species."

32. David Biello, "Fact or Fiction? Mammoths Can Be Brought Back from Extinction," *Scientific American*, 10 June 2014, online at http://www.scientificamerican.com/article/fact-or-fiction-mammoths-can-be-brought-back-from-extinction/?&WT.mc_id=SA_HLTH_20140610.

33. Or, as a fake George Church Twitter account would sum things up, "I make hairy elephants. #DescribeMyJobToA5YearOld." "Bored George Church" (@BoredSynBio), 2 September 2015. See also Lila Shapiro, "We May Resurrect the Mammoth Sooner Than You Think," *Huffington Post*, 18 December 2015, online at http://www.huffingtonpost.com/entry/woolly-mammoth-crispr-climate_us_567313f8e4b0648fe302a45e.

34. Biello, "Fact or Fiction?"

35. Church, and Regis, *Regenesis*, 147–48. The dream was so compelling to Church that he even described it twice in the same book: "If society becomes comfortable with cloning and sees value in true human diversity, then the whole Neanderthal creature itself could be cloned by a surrogate mother chimp—or by an extremely adventurous female human" (11).

36. "Interview with George Church: Can Neanderthals Be Brought Back from the Dead?," *Der Spiegel*, 18 January 2013, online at http://www.spiegel.de/international/zeitgeist/george-church-explains-how-dna-will-be-construction-material-of-the-future

-a-877634.html. Church expanded on this idea in *Regenesis*: "Admittedly, this will only ever happen if human cloning becomes safe and is widely used and if the possible advantages of having one or many Neanderthal children are expected to outweigh the risks" (147).

37. "Surrogate Mother (Not Yet) Sought for Neanderthal," *Der Spiegel* blog, 23 January 2013, online at http://www.spiegel.de/international/spiegel-responds-to-brouhaha-over-neanderthal-clone-interview-a-879311.html.

38. Church and Regis, *Regenesis*, 252.

39. He estimates that DNA storage could last seven hundred thousand years with minimal corruption. See Alan Boyle, "Encoding Data in DNA for Millennia? UW and Microsoft Research Are on It," *GeekWire*, 4 December 2015, online at http://www.geekwire.com/2015/encoding-data-into-dna-molecules-uw-microsoft-research-on-it.

40. John Markoff, "Data Storage on DNA Can Keep It Safe for Centuries," *New York Times*, 3 December 2015, online at http://www.nytimes.com/2015/12/04/science/data-storage-on-dna-can-keep-it-safe-for-centuries.html.

41. Boyle, "Encoding Data."

42. Erol Araf, "Space Travel Is in Our DNA, and DNA Might Be the Solution," *Montreal Gazette*, 4 August 2015, online at http://montrealgazette.com/technology/space/opinion-space-travel-is-in-our-dna-and-dna-might-be-the-solution.

43. Shapiro, "We May Resurrect."

44. Svante Pääbo, "Neanderthals Are People, Too," *New York Times*, 24 April 2014, online at http://www.nytimes.com/2014/04/25/opinion/neanderthals-are-people-too.html.

45. Christina Agapakis, "Alpha Males and 'Adventurous Human Females': Gender and Synthetic Genomics," *Scientific American*'s *Oscillator* blog, 22 January 2013, online at http://blogs.scientificamerican.com/oscillator/alpha-males-and-adventurous-human-females-gender-and-synthetic-genomics.

46. "Surrogate Mother (Not Yet) Sought," *Der Spiegel*.

47. Malcolm Ritter, "Scientist: I'm NOT Seeking a Mom for a Neanderthal," *Associated Press*, 22 January 2013, online at http://phys.org/news/2013-01-scientist-im-mom-neanderthal.html.

48. Church, *Regenesis*, 249

49. Thomas Goetz, "How the Personal Genome Project Could Unlock the Mysteries of Life," *Wired* , 26 July 2008, online at http://www.wired.com/2008/07/ff-church/.

50. Interlandi, "The Church."

51. George Church, "Encourage the Innovators," *Nature* 528 (December 3, 2015): S7, doi:10.1038/528S7a.

52. Interlandi, "The Church."

53. George Church, "The Future of Human Genomics and Synthetic Biology," presentation at Genetics and Society Symposium, North Carolina State University, 19 September 2014, online at https://www.youtube.com/watch?v=oE0a5ZaE6Gk.

54. George Church, "Safeguarding Biology," *Seed Magazine*, 2 February 2009, online at http://seedmagazine.com/content/article/safeguarding_biology/.

55. Shapiro, "We May Resurrect."

56. Wyss Institute, "Disruptive: Synthetic Biology."

57. Church, "Safeguarding Biology."

58. As McCray has noted, Church is a good example of "how today's scientist-celebrities engage, often simultaneously, in research, self-promotion, and entrepreneurship" ("The Church of Synthetic Biology").

59. Church and Regis, *Regenesis*, 12.

60. "Welcome to My Genome," *Economist*.

61. "Welcome to My Genome," *Economist*.

62. Wyss Institute, "Disruptive: Synthetic Biology."

Part IV: The Ecologists

Figure 10.1. John Todd. Still capture from *The New Alchemists*, directed by Dorothy Todd Hénaut, 1974, Canadian Film Board, 28 min.

MICHAEL R. DIETRICH & LAURA L. LOVETT

FROM NEW ALCHEMY TO
LIVING MACHINES
JOHN TODD'S DREAMS OF
ECOLOGICAL ENGINEERING

INTRODUCTION

In 1968, Stewart Brand's *Whole Earth Catalog* became a guidebook to an entire generation of "back to the land" homesteaders, communards, and intentional community founders. These modern homesteaders wanted to leave the city and suburbs for a rural life that was seen as outside of the military-industrial complex, more meaningful, more authentic, and more ecologically sustainable.[1] In the early 1970s, John Todd, Nancy Todd, and William McClarney created the New Alchemy Institute as an experimental community farm, but it was experimental in a way that only a community co-founded by ecologists could be. John Todd and Bill McClarney were young assistant professors at San Diego State University when John Todd's partner, Nancy Jack Todd, joined them in imagining a way of putting ecology into action and creating a more ecologically aware and sustainable way of living. They wanted to scientifically rethink how we live, produce food, build shelters, and generate energy.[2] It was a bold dream that put ecology into practice.

At the New Alchemy Institute, John Todd and his collaborators fashioned new ecological systems as living assemblages. These solar aquatic systems were named "living machines," or ecomachines. In Todd's words, these living machines were "engineered according to the same design principles found in nature to build and regulate the ecology of forests, lakes, prairies, or estuaries." However, their living components were "recombined in new ways" and housed within a lightweight structure where they could do their work.[3] Initially these living machines were designed to produce food as a form of traditional aquaculture. In the 1980s, however, John and Nancy Todd left the New Alchemists and radically rethought their approach to living machines: they transformed systems designed to produce food into systems designed to remove waste. As tools of bioremediation, John Todd and his partners offered living machines as ecological alternatives to chemically based industrial water treatment systems. This radical deviation from accepted practices of chemical treatment was greeted with extreme skepticism, but early successes at Cape Cod septage lagoons led to the adoption of living machines for water remediation from Providence, Rhode Island, to Fuzhou, China.[4]

THE NEW ALCHEMY INSTITUTE

John Todd and Bill McClarney met as graduate students at the University of Michigan while studying with biologist John Bardach.[5] While John Todd tried to understand how fish communicate by chemical signaling, Nancy helped lead Ann Arbor's antiwar and environmental movements.[6] As new PhDs in 1968, both John and Bill were hired at San Diego State College. San Diego was not Ann Arbor though, and the Todds realized that if they wanted to remain active in the environmental movement, they would have to create it in San Diego. The New Alchemy Institute was born in 1969 from the meetings that Nancy, John, and Bill organized at the Todd's house.[7]

Even though John's research in San Diego was going well (he had found that the insecticide DDT interfered with fish communication), he was being pushed into academic administration and moving farther away from actually doing anything about environmental problems. In 1970, McClarney and the Todd family moved to Cape Cod, where John and Bill took positions at the Woods Hole Oceanographic Institution (WHOI). John continued his work on chemical signals, while Bill joined John Ryther's group working on fish aquaculture. Ryther was chair of the Biology Department at WHOI at the time, and would be instrumental in founding the Environmental Systems Laboratory (ESL) at WHOI in 1972.[8] In many respects, the ESL and New Alchemy Institute developed in parallel in the 1970s. The ESL developed large-scale projects on aquaculture and waste treatment in marine systems, while the New Alchemy Institute created aquatic systems that relied on renewable energy sources, such as solar and wind power.

The New Alchemists also took their scientific inspiration from Howard Odum.[9] Howard and his brother, Eugene, were well-known ecologists with strong commitments to applied ecology and environmental problem solving.[10] In 1971, Howard Odum's book, *Environment, Power, and Society* reframed ecological systems for a general audience in terms of energy flows.[11] As its title suggests, Odum's book sought to situate humans, their social institutions, their history, and even their religion within systems and networks of material and energetic exchange. Of particular interest for the burgeoning New Alchemists was Odum's vision of what he called ecological engineering. In Odum's words, "The millions of species of plants, animals, and microorganisms are the functional units of the existing network in nature, but the exciting possibilities for great future progress lie in manipulating natural systems into entirely new designs for the good of man and nature."[12] Among the ways that Odum illustrates his thinking about ecological engineering is an experiment in North Carolina from 1969 and 1970 where sewage was collected in three treatment ponds, one of which was seeded with estuarine organisms that might speed decomposition. Odum declared this experiment a success and a "useful general procedure for blending the human sector with the non-human sector."[13] Odum saw waste as chemically and energetically rich, so understanding, modeling, and adapting the energy flow in such a system held special promise to him as a leading edge in ecological engineering. The solar aquaculture projects at New Alchemy were the realization of Odum's approach to ecological systems as networks of energy and materials. Their emphasis was not on waste treatment, but on sustainable food production.

The New Alchemy Institute found a home in 1971 when the Todds began renting a farm in Hatchville, Massachusetts, a few miles north of Woods Hole. Reflecting on the farm, John Todd described it as a place where they could "look the world in the eye and say sustainable food, energy, and shelter are possible."[14] Over the next ten years, the farm served as a test bed for building, food, and energy systems that integrated ecological principles. These systems were brought together in the creation of sustainable, self-contained bioshelters that the New Alchemists called Arks. The first Ark was created by the transformation of the New Alchemy farm in Hatchville. The second was created on Prince Edward Island with the support of the Canadian government. These Arks were in many ways the most visible products of the New Alchemy Institute, but proved to be expensive to create and maintain in the long run.[15]

From its beginning, the New Alchemy Institute farm had aquaculture experiments of increasing levels of complexity.[16] McClarney set up the first fish pools at the farm and enclosed two of them in translucent plastic geodesic domes (one had a recycling water filtration system; the other did not). Both ponds were stocked with tilapia (*Sarotheradon aurea*) that fed on the algae that grew in the ponds. The domes captured solar energy that allowed the fish and the ponds to survive the New England winter. From 1971 to 1974, Bill and John experimented with different configurations and different kinds of food sources, but in 1974 John began experimenting with the ecosystems in the ponds by watching them develop in five-gallon jars and then transparent fiberglass tanks that were five feet tall and eighteen inches in diameter.[17] These tanks were dubbed "solar tanks." Because they allowed solar energy to reach all of the contents of the tank, whereas solar energy only entered ponds from above, they were expected to capture more energy and so be able to produce more plant and animal growth, more food in the end. The solar tank experiments were taken up by Ron Zweig, a University of California–Berkeley graduate and a newcomer to the New Alchemists. In 1975, Zweig began experimenting with the solar tanks to see how their fish productivity compared to that of the ponds. Zweig and others compared growth in different seasons, growth with just tilapia, growth with many fish species, and differences in fish density, using tanks that were coupled together.[18] The most prolific tank turned out to be ten times as productive as the pond when controlled for size.[19] Like the ponds, some of the tanks were placed in greenhouses, where the heat they captured from solar energy would be retained and released overnight. When these tanks were linked together, they created what the New Alchemists called a Solar River. Differences among the tanks created nutrient flows that allowed for biopurification of the waste (especially ammonia) produced in the fish tanks. According to John Todd, "Our solar aquaculture tanks there raised fish productivity by a factor of 10, but we had terrible pollution problems, until by chance we started raising plants hydroponically on top of the tanks with their roots in the water." Those roots provided a place for purifying bacteria in the system, and water quality improved dramatically.[20]

In retrospect, John Todd remembers that "we started with aquaculture—the farming of fishes—and by the mid-1970s we knew that not only did these farms work, but they were elegant in their ability to self-regulate, self-purify, and produce. We were able to show that if one could just orchestrate sunlight properly in a small space, one could grow foods in *abundance* without reliance on the old energy networks."[21] The importance of energy within the aquaculture system was a crucial element of the Alchemists' approach. Their goal was to create a sustainable food system that relied as little as possible on fossil fuel energy sources. Indeed, the aquaculture program was only one dimension of the New Alchemists' experiment with solar power, wind power, and organic farming.

In the early 1970s, McClarney began dividing his time between Massachusetts and Costa Rica, where he set up a similar group.[22] He and Todd still maintained connections to WHOI and the marine aquaculture work being done at the Ecological Systems Laboratory. This connection probably helped them attract biologists to the New Alchemy aquaculture experiments. In 1977, the success of the early solar tank experiments allowed a team of researchers at the New Alchemy Institute to secure a National Science Foundation grant to monitor and document the solar aquaculture system. John Todd, Ron Zweig, and David Engstrom headed up the NSF research team and were joined by Carl Baum, who led the hydroponic plant experiments, and Joel Wolfe, who created a computer model that simulated the entire solar aquatic network.[23] The group knew that the solar tank systems were productive. The grant allowed them to model and monitor the system and find out why the system worked. Despite the success of the solar aquatic research and the productivity of the system as a food source, when funding ended in the early 1980s, the New Alchemy Institute made the controversial decision to ends its aquaculture program. Some members saw this as a breaking point in the community, as a retreat from their mission. Ron Zweig left New Alchemy as did the Todds.[24]

SEWAGE, SEPTAGE, AND SOLAR AQUATICS

In the 1980s, the Todds were ready to move into a new phase of their lives.[25] John Todd describes the New Alchemy research and experience as giving rise to new endeavors ranging from building design to sustainable food production. The Todds, however, chose to focus on water as they moved forward.[26] The next phase of their lives was marked by the creation of two new organizations, Ocean Arks International and Ecological Engineering Associates (EEA). What Ocean Arks International developed as experiments, EEA turned into commercial realities.[27] Part of the motivation for this move toward nonprofit and for-profit corporations was the slow pace of change in academia. In John Todd's words, "The old idea of just letting this information be assimilated slowly by the next generation of students, from whom it would spread into the academic world, engineering firms, and eventually into society, just wasn't fast enough—that takes at least twenty years. The only way I could think of compressing the process was through the corporate arena. So, Ecological Engineering Associates was set up, with the kind of backing and management that was needed."[28] That said, famed anthropologist Margaret Mead was an equally important source of inspiration for the creation of Ocean Arks International in 1981.

Mead challenged the Todds to engage their system of living with those in the developing world. In 1976, John and Nancy Todd were invited to a celebration of Mead's seventy-fifth birthday.[29] A few months later, Mead invited them to join her in Bali for the Pacific Science Congress on Appropriate Technology. While traveling with Mead, the Todds came to appreciate how carefully local farmers interconnected and integrated their systems of housing, water use, plant cultivation, and animal care. Mead saw a parallel between the sustainable systems and technologies being designed at New Alchemy and the systems that had been in use for years in Balinese villages. Before she passed away, Mead urged the Todds to find a way to make their technologies available in the developing world.

The idea of taking the New Alchemy Ark on the road spurred the design of Ocean Arks, sailing ships that would "provide life support for environmentally damaged coastal regions around the world."[30] They built two ships that were used to help fisherman in Guyana and Costa Rica, but, as they recalled, the "seagoing Ark proved to be an idea well ahead of its time," and initial funding from the Canadian International Development Agency did not lead to further support.[31] So, in the mid-1980s, the Todds shifted their attention to water quality. This decision, they recalled was motivated by the realization that "the water quality in our town was rapidly eroding. Cancer rates were going through the roof, and here we were raising kids. So we went out and bought spring water. That was a symbol to me that we had fundamentally violated our environment."[32] Solar aquatics offered a possible solution.

The Todds' new venture in water restoration cut its teeth on two waste treatment projects, beginning in 1986: a sewage treatment project in Vermont and a septage treatment project on Cape Cod. Both projects were designed for year-round use in a cold climate, which presented an energetic challenge for a sustainable waste treatment system. The solution was the adaptation of New Alchemy's solar aquaculture design for waste water treatment.

At the time, less than thirty miles from the Stowe, Vermont, Mountain Lodge created by the Von Trapp Family after they fled Austria, "the hills were alive" with microbes at a competing ski resort. Sugarbush, Vermont, was under orders to clean up its waste treatment facilities by officials from the Green Mountain State after effluent from its leachfield, downstream from the ski condos, was determined to contain high levels of ammonia. Open to alternatives, the management at Sugarbush proposed to compare two sewage treatment plants and see which worked better. As journalist and environmental activist Donella Meadows noted in her 1988 essay, the disparate facilities represented two approaches to sewage treatment and to the world. One, turned to chlorine gas, a tool for disinfecting that had been discovered in the 18th century and first put to widespread use to purify water in the US in the late 19th century to combat cholera. The other, turned back to the system for refining first put into play by mother nature, creating a solar aquatic system that John Todd adapted from the system for food production at the New Alchemy Institute. As Meadows noted, the chlorine gas, first named by Sir Humphrey Davies as a gas, was dangerous: "Hanging on a wall [of the square, windowless structure] is a gas mask and instructions IN CASE OF CHLORINE EMERGENCY." The other plant, a plastic greenhouse, held tanks of water, whose contents meandered through floating bamboo, cattails, and swamp irises. The sign at that facility jokingly read: "NO DIVING."[33]

The Sugarbush system was a single series of translucent fiberglass tanks arranged in a series inside of a greenhouse. Bacteria, algae, Daphnia, some snails, and fish live below a layer of floating plants. Waste that entered on one end of the solar river exits five days later and is filtered in an artificial marsh where cattails, bulrushes, bamboo, and irises finished the process. In the end, the water produced was virtually free of ammonia and bacteria.

The first challenge for the solar aquatic system was energetic. During the spring and summer, the solar-powered system worked well with the low levels of sewage being processed. Winter, however, saw an increase in the amount of sewage to be processed and, of course, a drop in temperature and sunlight (there was no direct sunlight on the treatment facility after around 2:00 p.m. in mid-winter). As the biological oxygen demand increased, the need of the solar aquaculture system to remove ammonia increased, yet the system still removed 99% of the ammonia.[34]

The second challenge of the solar aquatic approach was its administration. The practices of monitoring that were used first under New Alchemy's solar aquatic NSF grant were crucial components of the ecomachine, because they provided the evidence that skeptical regulators needed to understand the novel system in the context of water treatment. Vermont had the strictest water standards in the United States, and the move from a chemically based treatment system to an ecologically based treatment system was greeted with extreme skepticism.[35] Elizabeth Walker, the compliance officer for Sugarbush, noted that "state engineers view the greenhouses in the same manner that doctors view chiropractors."[36] Although the system worked, there was a crucial gap in data collection during the winter, and the system did not respond quickly to intervention. For an industry trained on quick chemical reactions that could be turned on and off, the complex ecosystem of the solar aquatic treatment required a different expertise and a different expectation regarding its management. In the end, Sugarbush had to go with the conventional chemical treatment regime. Nevertheless, the Sugarbush experiment was an important incubator for John Todd's ecomachines that led him to understand the regulatory climate and the need for redundancy in the series of tanks as reassurance should one fail.

The second experiment in adapting solar aquatics for waste remediation came in Harwich, Massachusetts. Like 25% of the homes in the United States, the homes of Harwich residents used septic tanks to process home sewage.[37] The septage that has to be pumped out of these tanks every few years was dumped in lagoons. Typically fifty to one hundred times more concentrated than sewage, septage lagoons are susceptible to leaks. Harwich had been collecting septage in outdoor lagoons for years.[38] With its water table only twenty-five feet below the bottom of these lagoons, the town faced a significant threat of groundwater contamination, and the Massachusetts Department of Environmental Protection ordered it to find a better solution by 1992.[39]

A Harwich resident, Hunter Craig, had heard of the Sugarbush experiment and arranged for Harwich town officials to pay a visit. They agreed to let Todd develop a septage system for Harwich in the summer of 1988.[40] The newly founded Ecological Engineering Associates, directed by Susan Peterson, an anthropologist by training, who had worked with John Todd at WHOI, created the twenty-tank solar aquatic system. Together with John M. Teal, a senior scientist at WHOI, Peterson wrote technical evaluations of the Harwich experiment and its development into a commercial treatment facility.[41]

The Harwich system consisted of a series of twenty connected, fiberglass solar aquatic tanks that together comprised what the Todds called an eco-machine. Septage was pumped into the first series of ten tanks, and then most of the effluent was drained into an artificial marsh. Effluent from the marsh was then pumped into the second series of eleven tanks. All of the tanks were individually aerated. Water from a local pond was used to start the tanks, and marsh, plants, and a mix of commercially available microbes were added to create the initial ecosystems. On June 15, EEA began pumping septage into tank one from ten hours a day, five days a week, at a rate of 0.13 liters per second, or up to 4.6 cubic meters of septage per day.[42] It took around fifteen days for effluent to run through the entire system. Measured levels of volatile organic compounds and heavy metals were very significantly reduced and gave Teal and Peterson evidence that the system could produce water that met drinking-water standards.[43]

The Harwich experiment gave Todd and Ecological Engineering Associates confidence that the ecomachine approach worked for both sewage and septage.[44] In 1989, the town of Harwich agreed to build a pilot plant for septage treatment based on the solar aquatic experiment. The pilot solar aquatic treatment system opened in March 1990, and after a training period of several months, moved into full processing.[45] In 1992, after the pilot plant at Harwich had been processing septage for one year, Massachusetts certified Ecological Engineering Associates' ecological approach to waste treatment as a legitimate treatment method.[46] That same year, Todd patented the solar aquatic waste treatment process and assigned it to Ecological Engineering Associates.[47] According to Teal and Peterson, "The innovation in SAS [Solar Aquatic System] technology involves concentrating the system, optimizing the mix of biological components, controlling the process, and treating a concentrated waste through all seasons."[48] Similar reasoning justifies the patent application, which also emphasizes the difference between this system and pond-based systems, such as the one that Odum had experimented with as early as 1969. The solar tanks located in a greenhouse allowed greater energy transmission and retention than a pond ever could. Combined with

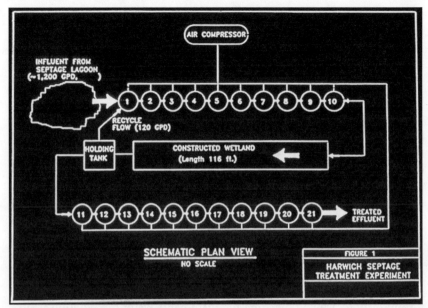

Figure 10.2. The ecomachine for processing septage in Harwich, MA, c. 1987. Courtesy of John Todd.

agitation to keep solids in suspension and controlled cycling of tank water to maintain biodiversity, the solar aquatic system provided a structure for a sustainable artificial ecosystem capable of treating both sewage and septage. Without chemicals, without producing sludge, Ecological Engineering Associates had developed a machine that was economical, effective, and ecological.

CONCLUSION

Commenting on John Todd's path from New Alchemist to ecological engineer, journalist and activist Donella Meadows described him as "an ecologist, a visionary, a whole systems thinker."[49] Indeed, Todd and his colleagues embody W. Patrick McCray's idea of a "visioneer"—someone who is a combination of researcher, futurist, and promoter; who can imagine how a technology might radically alter the future, conduct the research to create that technology, and effectively take that tool and vision to the public.[50]

The nature of Todd and his partners' innovation, as in the systems he helped design, is one of adaptive assembly. He and his collaborators sought the best assembly of ideas, technologies, people, and organizations to further their vision of a truly sustainable way of life. Faced with the challenge of sewage treatment, Todd and his colleagues borrowed solar aquatic technology developed for sustainable food production and reimagined it as a means of sewage treatment. The success of that adaptation relied on their ability to create new ecological systems or assemblies that would function as self-sustaining systems in different local contexts where demands could be radically different.

Todd's movement from academia to WHOI, the New Alchemy Institute, Ocean Arks International, and Ecological Engineering Associates traces a path through a series of institutions that allowed him greater freedom to experiment with complex ecological systems with an increasing emphasis on their application to pressing social and environmental problems. These structures supported the innovation of Todd and his collaborators, but Todd and his fellow travelers pushed the limits of these institutions. Indeed, when the academy was seen as overly restrictive, they moved on to different organizations that allowed them to engage in different forms of scientific work and different applications of that science.

Taking a solar aquatic system designed for food production and transforming it into a living machine for sustainable water remediation is profoundly creative. What makes the work of John Todd and colleagues much more radically innovative is that they realized that their scientific and technological innovations would flourish best in different social and organizational structures. When academic life proved too constraining, they moved to the research labs at WHOI and created the New Alchemist farm. When the farm ran its course, they invented a combination of profit and nonprofit companies to advance their vision of engineering with living machines. As masters of creating new, assembled systems, John Todd and his colleagues created ecological systems in their living machines and the human systems that would best see their vision realized.

FURTHER READING

Ausubel, Kenny. *The Bioneers: A Declaration of Independence*. White River Junction, VT: Chelsea Green, 2001.

Todd, Nancy Jack. *A Safe and Sustainable World: The Promise of Ecological Design*. Washington, DC: Island Press, 2005.

Todd, Nancy Jack, and John Todd. *From Eco-Cities to Living Machines: Principles of Ecological Design*. Berkeley, CA: North Atlantic Books, 1993.

Trim, Henry. "A Quest for Permanence: The Ecological Visioneering of John Todd and the New Alchemy Institute." In *Groovy Science: Science, Technology, and American Counterculture*, edited by David Kaiser and W. Patrick McCray. Chicago: University of Chicago Press, 2016.

1. Jeffrey Carl Jacob, *New Pioneers: The Back-to-the-Land Movement and the Search for a Sustainable Future* (College Station: Pennsylvania State University Press, 1997). For earlier "back to the land" movements, especially irrigation settlements, see Laura L. Lovett, "Rooted in the Soil: Family Ideals, Land Reclamation and Irrigation Resettlement as Welfare in the United States, 1897–1933," in *Families of a New World: Familialism and the Process of State-Making*, ed. Lynne Haney and Lisa Pollard (New York: Routledge, 2003), 85–98.

2. Henry Trim, "A Quest for Permanence: The Ecological Visioneering of John Todd and the New Alchemy Institute," in *Groovy Science: Science, Technology, and American Counterculture*, ed. David Kaiser and W. Patrick McCray (Chicago: University of Chicago Press, 2016); Nancy Jack Todd, *A Safe and Sustainable World: The Promise of Ecological Design* (Washington, DC: Island Press, 2005). Todd's book and Trim's essay richly document the history of the New Alchemy Institute. Here our focus is not on the institute per se, but on the development of "living machines" in the 1970s and their application by John Todd and his colleagues in the decades after the New Alchemy Institute to the problem of treating waste water.

3. John Todd and Beth Josephson, "The Design of Living Technologies for Waste Treatment," *Ecological Engineering* 6 (1996): 109–36.

4. Molly Farrell, "Purifying Wastewater in Greenhouses," *BioCycle* 37, no. 1 (1996): 30–33.

5. Todd, *Safe and Sustainable World*.

6. John Todd, J. Atema, and J. E. Bardach, "Chemical Communication in the Social Behavior of a Fish, the Yellow Bullhead (*Ictalurus natalis*)," *Science* 158 (1967): 272–73; J. Bardach, J. H. Todd, and R. Crickmer, "Orientation by Taste in Fish of the Genus *Ictalurus*," *Science* 155 (1967): 1276–78; John Todd, "The Chemical Languages of Fishes," *Scientific American*, May 1971, 98–108.

7. Nancy Jack Todd and John Todd, *From Eco-Cities to Living Machines: Principles of Ecological Design* (Berkeley, CA: North Atlantic Books, 1993).

8. "In Memoriam: John H. Ryther," Woods Hole Oceanographic Institute, July 10, 2006, online at http://www.whoi.edu/mr/obit/viewArticle.do?id=14526&pid=14526; John E. Huguenin, "Development of a Marine Aquaculture Research Complex," *Aquaculture* 5 (1975): 135–50.

9. Todd, *Safe and Sustainable World*, 61.

10. Joel B. Hagen, "Teaching Ecology during the Environmental Age, 1965–1980," *Environmental History* 13 (2008): 704–23; and *An Entangled Bank: The Origins of Ecosystem Ecology* (New Brunswick, NJ: Rutgers University Press, 1992).

11. Howard T. Odum, *Environment, Power and Society* (New York: John Wiley, 1971).

12. Odum, *Environment, Power and Society*, 279.

13. Odum, *Environment, Power and Society*, 289.

14. Donella Meadows, "The New Alchemist Turns Tycoon—John Todd's Wastewater Treatment Plant," box 6, folder 41, Donella Meadows Collection, Rauner Special Collections Library, Dartmouth College, Hanover, NH.

15. Trim, "Quest for Permanence"; Todd, *Safe and Sustainable World*.

16. Earl Barnhart, "A Primer on New Alchemy's Solar Aquaculture," New Alchemists Publications (21 February 2006), online at https://newalchemists.files.wordpress.com/2015/01/solar-aquaculture-primer-by-eab1.pdf.

17. Todd, *Safe and Sustainable World*, 43–45.

18. Ron Zweig, "Solar Aquaculture," *Journal of the New Alchemists* 6 (1979): 93–95.

19. Todd, *Safe and Sustainable World*, 45; Ron Zweig, "The Saga of the Solar-Algae Ponds," *Journal of the New Alchemists* 4 (1977): 63–68.

20. Meadows, "New Alchemist Turns Tycoon."

21. Robert Gilman, "Restoring the Waters: An Interview with John and Nancy Todd," *In Context: A Quarterly of Humane Sustainable Culture* 25 (1990): 42.

22. William McClarney, "New Alchemy—Costa Rica," *Journal of the New Alchemists* 4 (1977): 17–23.

23. The New Alchemy Institute, *Solar Aquaculture: Perspectives in Renewable, Resource-Based Fish Production* (results from a Workshop at Falmouth, Massachusetts, 28 September 1981); David Engstrom, John Wolfe, and Ron Zweig, "Defining and Defying the Limits of Solar-Algae Pond Fish Culture," *Journal of the New Alchemists* 7 (1981): 83–87; John Wolfe, Ronald Zweig, and David Engstrom, "A Computer Simulation Model of the Solar-Algae Pond Ecosystem," *Ecological Modeling* 34 (1986): 1–59.

24. Todd, *Safe and Sustainable World*, 50.

25. John Todd and Nancy Jack Todd, *Tomorrow Is Our Permanent Address: The Search for an Ecological Science of Design as Embodied in the Bioshelter* (New York: Harper & Row, 1980).

26. Kenny Ausubel, *The Bioneers: A Declaration of Independence* (White River Junction, VT: Chelsea Green, 2001).

27. Farrell, "Purifying Wastewater in Greenhouses."

28. Gilman, "Restoring the Waters," 42.

29. Todd, *Safe and Sustainable World*, 145–47.

30. John H. Todd, "History," online at http://oceanarksint.org/index.php?id=a-history.

31. Todd, *Safe and Sustainable World*, 147.

32. Gilman, "Restoring the Waters," 42.

33. Meadows, "New Alchemist Turns Tycoon"; Donella Meadows, "Ecology vs. Engineering: A Clash of Values on a Mountain in Vermont," *Los Angeles Times*, 10 July 1988, online at http://articles.latimes.com/1988-07-10/opinion/op-9355_1_sewage-treatment-systems.

34. Todd, *Safe and Sustainable World*, 149.

35. Björn Guterstam and John Todd, "Ecological Engineering for Wastewater Treatment in New England and Sweden," *Ambio* 19 (1990): 173–75; Gilman, "Restoring the Waters," 42.

36. William Burke, "Restoring Water Naturally," *Technology Review* 94 (1991): 16–17.

37. John M. Teal and Susan Peterson, "The Next Generation of Septage Treatment,"

Research Journal of the Water Pollution Control Federation 63 (1991): 84–89; Ausubel, *Bioneers*.

38. Todd and Josephson, "Design of Living Technologies."

39. Laura Van Tuyl, "A 'Living Machine' Purifies Waste," *Christian Science Monitor*, 13 February 1991.

40. Todd, *Safe and Sustainable World*, 152.

41. John M. Teal and Susan Peterson, "A Solar Aquatic System Septage Treatment Plant," *Environmental Science and Technology* 27 (1993): 34–37.

42. Teal and Peterson, "Next Generation"; Guterstam and Todd, "Ecological Engineering."

43. Teal and Peterson, "Next Generation."

44. Todd, *Safe and Sustainable World*, 155.

45. Ocean Arks International, Past Projects, online at http://www.oceanarksint.org /?id=past-projects; Robert Spencer, "Lower Cost Way to Septage Treatment," *BioCycle* 33, no. 3 (1992): 64–68.

46. Todd, *Safe and Sustainable World*, 156.

47. John Todd and Barry Silverstein, "Solar Aquatic Apparatus for Waste Treatment," US Patent Number 5,087,353 (11 February 1992), online at https://www.google .com/patents/US5087353.

48. Teal and Peterson, "Solar Aquatic System," 37.

49. Meadows, "New Alchemist Turns Tycoon."

50. McCray, *Visioneers*. Henry Trim identifies Todd as an ecological visioneer in "A Quest for Permanence."

PHILIPPE HUNEMAN

STEPHEN HUBBELL AND THE PARAMOUNT POWER OF RANDOMNESS

INTRODUCTION

Stephen Hubbell became the father of the "neutral theory" in ecology when he published his masterwork, *The Unified Theory of Biodiversity and Biogeography* in 2001.[1] In this book, Hubbell suggests a systematic explanation of many patterns of biodiversity with which ecologists are familiar, such as species richness in a community, species-abundance distribution, and species-area curves; it is as simple as it is counterintuitive, and those two features constitute together its groundbreaking character. Imagine going into a forest—a tropical forest, which is a classical object of study for ecologists like Hubbell. You find thousands of species of different trees, and then you wonder: Why are there these species here? What accounts for the fact that three or four species are very abundant, while many of them are sparse and many of them are very rare? A natural answer comes from our familiarity with Darwinian biology: some trees are good at exploiting wet soils, some are excellent in dry soils, some can flourish with little light while others need more light, and so on. Hence, each species is found in the places where it thrives, its *niche*, and the distribution of these places ultimately accounts for the distribution of the tree species, each species living in the area to which it is adapted. The accounting process for all that is the essential *explanans* of adaptation, as Darwin taught us—namely, natural selection.

Hubbell suggests that this intuitive answer is, if not wrong, at least not to be taken for granted. He built a sophisticated and wide-ranging theory showing that if you suppose that natural selection is not acting—that is, if the effects of competition, the major dimension through which natural selection acts, cancel out at the scale of the community and beyond—then the distribution of tree species can be accurately predicted. In other words, assuming that fitnesses of all individuals are the same (i.e., that species have the same birth and death rate per capita), then you can build a model that predicts patterns of diversity very close to the ones recorded, especially concerning the distribution of abundances. In a tropical forest such as the one on Barro Colorado Island in Panama, where Hubbell gathered data for decades, beginning in the 1970s, the neutral theory of ecology does better than the models using the concepts of niche, competition, and natural selection.

Figure 11.1. Stephen Hubbell. Photo by Reed Hutchinson, UCLA Newsroom.

Hence the paradox: Hubbell articulates a theory that is simpler—in terms of its parameters and assumptions—than the existing ones and has at least an equal or better predictive accuracy. However, far from being unanimously positively received, it raised an intense controversy regarding deep philosophical issues about model testing: How could such a simple and apparently unrealistic theory be so accurate?

Neutral ecology is simple (aside from its technical apparatus, which is very sophisticated)—so simple that few ecologists genuinely considered it as a serious alternative. The natural inclination among most ecologists was to complicate extant niche theories in order to fit the data and include many processes that increasingly seemed to be relevant. It is fair to say that Hubbell's dream of a neutral theory for ecology revolutionized theoretical community ecology. So, how did Hubbell come to this theory; how did it meet, in a radically novel manner, the issues faced by ecologists; what were its ambitions; how was it received and then incorporated into ecological theory?

THE THEORETICAL BACKGROUND: NICHES, COMPETITIVE EXCLUSION, AND LIMITING SIMILARITY

Ecology has been called the "science of the struggle for existence" (Haeckel) or the "science of the relations between organisms and their environments."[2] This duality indicates that competition and selection may receive more or less emphasis in the conception of ecology. Community ecology is interested in accounting for the diversity of species composing the community and its variation (especially, the conditions for its stability). In the 1960s, community ecology and biogeography were deeply affected by the theoretical ambitions of MacArthur and Wilson's *Theory of Island Biogeography* (1967). Together with Richard Levins and Richard Lewontin, those biologists and ecologists wanted to introduce into ecology mathematical rigor and systematic modeling. Many of them were students of George Evelyn Hutchinson, who achieved in the 1950s and 1960s a major synthesis of ecological thought.[3] Ecology was also divided according to the scales one considers: biogeography considers regions, which comprise various communities that may exchange species through dispersal.

A major issue raised by both biogeography and community ecology concerned *patterns of biodiversity*. A crucial pattern, studied by biogeography, was the "species-area curve," which relates the area of a territory to the number of species it includes. MacArthur and Wilson proposed the so-called island-mainland model, which models the biodiversity dynamics yielding species-area curves. Simply said, in an island-mainland model, species occupy a mainland and can colonize islands. The amount of species to be found on each island depends upon its size, its distance from the mainland, and the number of species on the mainland and other islands; the amount is given at the equilibrium between extinction and immigration.[4] This simple model allows predictions using species-area curves, their differences in the continuous mainland and in archipelagoes, as well as the evolutionary fate of some characters of those species (convergence or divergence between characters).

In turn, traditional *community ecology* asks: how are species distributed *according to their abundances*?[5] What processes account for these patterns? Many of the species-abundance distributions are log-series, as indicated by Fisher; yet Preston contradicts him, asserting that some are also log-normal curves, but mainly at a smaller scale.[6] Regardless of scale, there is a regularity here that calls for explanation, as well as an explanation of the differences.[7]

Keywords here are *competition* and *niches*. Gause in 1935 established the "competitive exclusion principle": two species with exactly the same requirements (such as amount of water and light required to survive and thrive, distance from predators, optimal temperature, amount of light for trees, etc.) can't coexist, since one will always drive the other extinct because of competition. The set of requirements that is proper to a species, represented, according to Hutchinson, as a subspace of the hyperspace of ecological parameters (such as temperature, drought, pH, light, etc.), is what is called the "niche."[8] From this "fundamental niche," one should distinguish the "realized niche," that is, the result of competition between two species whose fundamental niches overlap. The better competitor will occupy its proper fundamental niche, whereas the lesser competitor will occupy the part of its fundamental niche that is not part of the fundamental niche of the better competitor. Competition therefore partitions the niche space into non-overlapping realized niches that correspond to each species after competition, and this explains the coexistence of species at equilibrium. Such an explanation is called "limiting similarity" (since each species may limit the fundamental niches of others when they share niche parameters).

Yet Hutchinson was aware that this explanation did not entirely capture the facts of biodiversity. As he famously argued in several papers,[9] if you consider plankton species, there are very few parameters along which the environment differs in ocean: light, pH, temperature, and so on. Therefore, there should be few realized niches, and we expect few species; but in fact we find thousands of phytoplankton species. Hutchinson coined the phrase the "paradox of the plankton," asserting that modeling realized niches is not enough to account for biodiversity. Hutchinson suggested other explanations: for in-

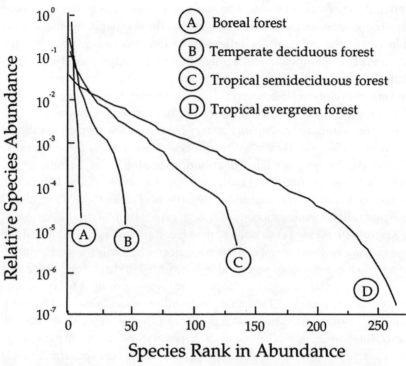

Figure 11.2. Some species-abundance distributions in various ecological regions. Dominance-diversity curves for tree species in four closed-canopy forests, spanning a large latitudinal gradient. The four curves seem to represent a single family of mathematical functions, suggesting that a simple theory with few parameters might capture the essential metacommunity patterns of relative species abundance in closed-canopy forests. Republished with permission of Princeton University Press, from Stephen P. Hubbell, *The Unified Neutral Theory of Biodiversity and Biogeography*, MPB-32 (Princeton, NJ: Princeton University Press, 2001), 116, fig. 5.1; permission conveyed through Copyright Clearance Center, Inc.

stance, that the environmental parameters are quickly changing, so that the directions of competition constantly shift, and the best competitors don't have time to competitively exclude others.

Competition theory developed and grew in sophistication in order to account for facts and patterns of biodiversity discovered through censuses in tropical forests, mangroves, coral reefs, and natural parks. Theoretically, community ecology revolved around this axiom: species coexistence should always be based on niche differences, be they obvious differences or tiny, periodical, chaotic, nondetectable differences. But Hubbell proposed a radically new conception in opposition to this axiom: in his view, explaining species coexistence and various diversity patterns does not in principle or a priori need to consider any difference between species biological requirements.

STEPHEN HUBBELL AND THE
ELABORATION OF THE NEUTRAL THEORY

The neutral theory in ecology appeared first as a book in 2001. Instead of summarizing the teachings of already extant and discussed papers (Hubbell had been an active researcher in this field since the 1970s), it elaborated into a single theory the insights he gained from three decades of reflections upon community ecology theory and data gathering in Barro Colorado Island, Panama.

Hubbell was born in 1942. His mother was a statistician, and his father was an entomologist and evolutionary biologist (director of the Museum of Zoology at the University of Michigan) who gave his son the taste for fieldwork by taking him along on collecting trips. He wanted to be an architect, but lacked a gift for drawing, and he turned to something closer to his parental background (i.e., biochemistry), after having graduated at Minnesota. As a grad student in Berkeley, he turned to ecology, working on isopod crustaceans. (He now suggests that this could have been part of a general drive toward empirical science and application, initiated by the fear of the Sputnik as a major soviet technical achievement.) Then, as a researcher at the University of Michigan, he was sent in Costa Rica to teach tropical ecology. This experience turned him toward his major topic, tropical plant ecology. He returned to a position at the University of Iowa, but remained attached to the tropical forest, especially by getting involved in the Smithsonian program on tropical forests. Beginning in the late 1970s, Hubbell gathered data on Barro Colorado Island, Panama, and in 1980, together with Robin Foster, he took charge of a program of monitoring exhaustively a plot of fifty hectares in Barro Colorado, and then followed it year after year (noting the growth of propagules, their dispersion, the fate of trees of all species, etc.).[10]

The interest Hubbell took in establishing the Barro Colorado Island data also shows that he elaborated the neutral theory through a very tight intertwining of data gathering (i.e., fieldwork) and mathematical modeling. Concern for tropical forests is a recurrent feature of his scientific activity, and he never stopped publishing about their specific features—mostly with co-authors Robin Foster and Richard Condit—from the 1970s to his most recent works, which include a paper trying to evaluate the number of species in the tropical forest.[11] Some papers are about conservation;[12] others compare Barro Colorado Island to Pasoh, a fifty-hectare plot in Southeast Asia that matches it.[13] Hundreds of scientists have visited the Barro Colorado Island Center and used it as a basis for theoretical work in ecology and conservation: for example, Theodore Schneirla has resided there for thirty-five years, studying ecology of ants, and Egbert Leigh wrote two books about the island on community ecology. The emergence of the neutral theory in ecology relied heavily on this institutional structure.

Before he began researching tropical forests, Hubbell was comfortable with the theory of limiting similarity; however, at the Barro Colorado Island Center, he realized that such models don't easily apply (personal communication). Because of the number of species and their proximity, the distance between competitors is very small; species compete in general with their neighbors, but in a tropical forest, given the vast number of individuals from common species, these to some extent don't "see" their neighbors and so don't actually compete. Such a realization underlay his view of a neutral theory.

Hubbell's first paper introducing ideas of what would develop into the neutral theory was published in 1979, entitled "Tree Dispersion, Abundance, and Diversity in a Tropical Dry Forest."[14] It discusses the Janzen-Connell hypothesis,[15] according to which dispersal and seed predation prevent any given species from dominating a forest as a monoculture by lowering its probability of self-replacement. Hubbell's paper tests such ideas about density dependence by confronting them with data about intergenerational turnover in Barro Colorado Island. In the discussion of the paper, Hubbell suggests a view of communities that is the core of the neutral model to come, but this view was a side hypothesis and is not developed extensively.[16]

The neutral theory itself had to wait almost twenty years more before Hubbell wrote the book articulating the new framework. In the meantime, Hubbell published three texts that added some of the core ideas, but no publication in a major ecology or biology journal spelled out his views on the neutral theory before the book came out. As a result, many community ecologists— especially tropical ecologists or coral reef ecologists—were acquainted with parts of Hubbell's message, as ecologist Mark McPeek recalls (personal communication), but they did not put the pieces together as a major theoretical proposition before the book.

While at Princeton University, Hubbell started to write his monograph on the neutral theory, entitled *The Unified Theory of Biodiversity and Biogeography*. He intended to provide a novel view of community ecology and biogeography. Even though many ideas had been laid out before, his book was both a systematic elaboration of a worldview and a set of models that articulated it in a way that was testable—as well as a series of empirical corroborations of the theory, including some using data from Barro Colorado Island.[17]

The fate of the theory—its latency for almost twenty years—is not common, even leaving aside the personal history of the author himself, which accounts in part for this delay. Hubbell clearly had in this 1979 paper the vision of what would be a neutral theory: namely, an explanation of biodiversity patterns that didn't rely on anything except, quickly said, random replacement of equivalent dying individuals (notwithstanding the species they belong to). This vision has been completed in the 2001 book.

Such a gap seems intriguing, and explaining it is significant for the meaning of Hubbell's neutral theory. The notion of an account of biodiversity that does not appeal to natural selection and focuses mostly on dispersion has been around for some time, especially with MacArthur and Wilson's *Theory of Island Biogeography*. (Hubbell in his book calls theories of this kind "dispersal assembly theories.") But those authors were focusing on biogeography and species-area curves, not community ecology and species-abundance distributions.

At the time of Hubbell's first paper, community ecology was marked by a heated debate over the role of null models—namely, models with no selection, to which any hypothesis about the role of selection should then compare its prediction. Jared Diamond argued that when ecosystems display a checkerboard distribution of species, for instance in an archipelago, species being present on an island only when another is absent locally and reciprocally, then this pattern is the result of competition.[18] Daniel Simberloff and John Connor argued in response that this pattern means nothing, since, first, one should randomly shuffle the data, producing all patterns that could happen randomly, and then compute among them the probability of the checkerboard distribution.[19] This probability being high, they concluded that nothing could be said regarding the effects of competition from a checkerboard pattern. It was a rude controversy, which for some commentators reflected a difference between field-oriented and mathematically oriented ecologists, and culminated in an issue of *American Naturalist* entitled "A Round Table on Research in Ecology and Evolutionary Biology" (vol. 122, no. 5 [1983]). Hubbell's hesitance to publish on neutrality after his first paper was motivated by a desire to stay out of this controversy (personal communication), and, because of the possible hostility that neutral models in ecology might trigger among some field ecologists, Hubbell grouped himself with those ecologists who tended toward a synthesis of fieldwork and mathematics. Hubbell delayed articulating the neutral theory until this "null models war" calmed down a bit.

The extent of Hubbell's theory—reaching to both community ecology and biogeography—distinguishes it from earlier theoretical elaborations. Even though neutrality, in the sense of fitness equality, was not new, Hubbell's idea was that it could yield a very general theory likely to encompass biogeography and community ecology. His dream was that a rigorous formulation of the idea of fitness neutrality in metacommunities could lead to a very simple theory accounting for all recurrent biodiversity patterns—instead of following the natural practice of ecologists—namely, complicating the niche theory of limiting similarity.[20] Neutrality, indeed, is the simplest hypothesis, since it assumes a minimal number of processes. And the simplest theory is more testable than a theory in which a wealth of parameters could always be fine-tuned to fit the data. Thus, Hubbell was dreaming that one would address biodiversity in a radically different manner: namely, by no longer refining the possible ways natural selection could yield all those patterns, and starting anew with what would happen randomly—that is, without selection—and modeling it in detail.

Simplicity and testability were the key epistemic values he advanced, as well as the unifying character and parsimony of the theory. Testability has a Popperian ring; Hubbell says that, as a student, he was impressed by a paper published by biologist Joseph Platt in *Science*, "Strong Inference," which advocated a strong stance on testability. This hypothetico-deductive model of science favoring predictions and mathematical simplicity was a first inspiration for his methodology, and a continuous one, whereas few philosophers of science have been explicitly included in his references. More generally, Hubbell advocated from the start an attitude inspired by physicists, especially with regard to their use of mathematics, which impressed him when he took those classes as a student (personal communication). He was impressed by the way they began with simple theories, checked with experiments, changed parameters, and so forth, all within a mathematical framework.

The simple idea that underlies this dream of a grand theory, unifying disciplines and joining simplicity with testability, is the notion of "ecological equivalence": this is how randomness—in the sense of the absence of any determinate or directed process—can be turned into a resource for modeling ecological communities across scales and used to build a universal theory.

ECOLOGICAL EQUIVALENCE AS THE
CORNERSTONE OF NEUTRAL THEORY

The ecological equivalence assumption is the foundation of the models explored in the neutral theory: it states that the per capita birth and death rates and migration or speciation probabilities for all species are equal. That may be not true, or realistic, but Hubbell's first question is: How does a model of biodiversity dynamics built upon this assumption behave—which obviously eliminates niche effects, or the effects of selection, since it states that the specific nature of an individual does not in principle affect its relative chances of birth and death. Then, species in the community—understood as "groups of trophically similar, sympatric species that actually or potentially compete in a local area for the same or similar resources"[21]—undergo several processes: *dispersal* occurs as the appearance of an offspring at some distance from its parent individual; there is *speciation* (i.e., appearance of new species), a process that was absent from the core idea in 1979 and that will prove crucial for Hubbell's unifying project. In most of the book, speciation occurs instantaneously relative to the other species in the community. Finally, there is *"ecological drift,"* meaning stochastic birth and death: individuals die randomly and are randomly replaced. This concept corresponds exactly to the process of frequency change of equal-fitness alleles in population genetics, named "random genetic drift" by Sewall Wright.[22] Even if the parallel was not clear from the beginning to Hubbell and therefore does not really belong to its context of discovery, Motoo Kimura's neutral theory, focusing on genetic drift, constitutes an exact parallel to the neutral theory in ecology: alleles in the former are species in the latter; mutation becomes speciation; and so forth (some formulae and key parameters are common to both theories, as Hubbell will notice while writing his book).[23] Papers by Hubbell after 2001 make this similarity explicit.[24]

Hubbell's model is a "zero-sum game." Suppose that an individual of a given species in the community dies and is then replaced by an individual of a randomly chosen species. The assumption of the zero-sum game is one way to model ecological equivalence; empirically, a zero-sum game is plausible to the extent that biotic communities are generally saturated.[25] Ecological drift combines with dispersal, which is represented by an individual that dies in a community and is then replaced by an individual from another community that has a different species pool. So, at each time step, randomly dying individuals are replaced by randomly chosen individuals of the same or other species (fig. 11.3) Hubbell's first model then determines the time of fixation of one species by pure drift (exactly as population genetics models start by computing the time for fixation of one allele by drift alone).

Embedding local communities in a regional set of communities called a *metacommunity* is crucial. This term was derived from the term *metapopulation*, a word introduced by Richard Levins[26] and developed by, among others, Ilkka Hanski, who elaborated a "metapopulation ecology" that has been developed in the 1990s especially and is concerned with the fragmentation of populations in distinct habitats.[27] Metacommunities are pools of communities exchanging individuals through colonization but otherwise separated

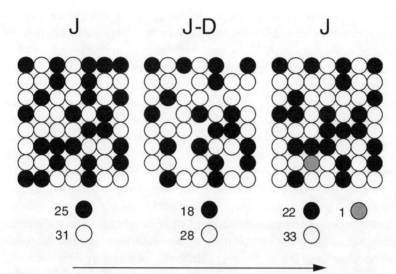

J   J-D   J

25 ●   18 ●   22 ●   1 ◐
31 ○   28 ○   33 ○

Figure 11.3. Simulation of the Markov process of a zero-sum game in a community. Cartoon of one disturbance cycle in a model community undergoing zero-sum ecological drift. At the beginning of the cycle are two species whose individuals occupy all sites or resources (*left*). Immediately after the disturbance, which killed several individuals of both species, vacant sites or unutilized resources are opened up (*middle*). These are occupied by recruits from the two species in the local community, and by an immigrant individual of a third species from the metacommunity source area (*right*). Republished with permission of Princeton University Press, from Stephen P. Hubbell, *The Unified Neutral Theory of Biodiversity and Biogeography*, MPB-32 (Princeton, NJ: Princeton University Press, 2001), 77, fig. 4.1; permission conveyed through Copyright Clearance Center, Inc.

(hence noncompeting). The relation between community and metacommunity instantiates the relation between a *local* scale (explored by community ecologists interested in relative abundance patterns) and a *regional* scale (explored by MacArthur and Wilson's biogeography). Hubbell's theory first explores the neutral dynamics in communities and then in metacommunity, while their relation defines the unity across scales that's aimed at by the theory.[28] Biodiversity patterns result from this coupled dynamic.

In such a model, in a community, because of drift, all species would go extinct except one, picked at random. In a metacommunity, "because all species ultimately go extinct, diversity is maintained, in the last analysis, solely by the origination of new species in the metacommunity,"[29] and this speciation process in metacommunity proves essential to understanding the maintenance of diversity.

A major feature of this theory is that species richness is derived from the model, as well as species-abundance distributions (SAD)—there is no need to assume the number of species to derive the abundance distribution, as was usually done in niche theories. A crucial result is that across those equations emerges a dimensionless number, which Hubbell called the "fundamental biodiversity number," occurring both in equations determining the number of species and in those determining SAD (and isomorphic to the parameter discovered by Kimura, which is similarly compounded by effective population size and mutation rate, and governs neutral evolution).

Moreover, from Hubbell's models, one can also derive species-area curves in the metacommunity. So, biogeography as a study of species-area curves and community ecology as a study of relative abundance can be both merged in a single mathematical theory centered on ecological equivalence, modeling metacommunities undergoing speciation, dispersal, and drift—and governed by a single parameter, which in turn, can be estimated by abundance data from each single metacommunity. And at each step, Hubbell compares some data to the resulting patterns of the model—mainly Barro Colorado data—and show that they fit as well or better than the predictions of niche theories.

Now, like any post-Darwinian biologist/ecologist, Hubbell knows that natural selection is powerful in shaping the traits of species and in governing the dynamics of allele frequencies among populations. Why then the ecological equivalence? First, dispersal—especially dispersal of seeds—may delay competitive exclusion by minimizing the effects of competition upon close neighbors. In such a way, competitive exclusion is trumped, and individuals appear as having invariant fitnesses.[30] Ecological equivalence, or near equivalence, can therefore be assumed even if individuals, when considered for themselves, have varying fitnesses. For reasons pertaining to dispersal limitation, ecological equivalence is therefore a very likely outcome of any evolutionary dynamic in ecosystems. Moreover, as Hubbell explains in chapter 10, trade-offs between life-history traits of individuals of various species will be increasingly numerous as species and traits are numerous, and this may result in equalizing fitnesses across species: "Niche differentiation along life-history trade-offs is the very mechanism by which per capita relative fitnesses are equalized among the coexisting species in a community."[31]

THE HERITAGE FROM ISLAND BIOGEOGRAPHY: NICHE VS. DISPERSAL ASSEMBLY

Hubbell's neutral theory was intended as a major theoretical alternative. Where ecologists were looking for niche differences as ultimate accountants, the neutral theory started instead with ecological equivalence.

Thus, Hubbell expected to meet some resistance, and, in fact, his neutral theory triggered a debate that is not yet settled. Yet, as Hubbell indicates in the first chapter of his book, the neutral theory is an extension of MacArthur and Wilson's *Theory of Island Biogeography*. They begin their treatise by considering only immigration and extinction, and therefore they model island-mainland dynamics without including considerations about species-fitness differences. To this extent, the *Theory of Island Biogeography* is a "dispersal-assembly" model of communities, whereas prior community ecology mostly favored a niche-assembly model of biodiversity dynamics. Yet their ecological equivalence is precisely thought at the *level of species*, whereas Hubbell's neutral equivalence is defined as *individual* equivalence. The latter therefore extends MacArthur and Wilson's framework in a manner that allows Hubbell to model the variations of individuals' frequencies within species—namely, the *abundances* of species.

In this sense, Hubbell can predict not only species-area curves, as the *Theory of Island Biogeography* does, but also species-relative abundances—a program proper to community ecology, which, however, used to assume the number of species in order to compute SADs. As Hubbell says, by shifting the level of equivalence, his theory unifies the "theory of island biogeography and relative species abundance."[32] By doing so, it drastically reduces the number of parameters of the theory, which makes it more testable and less prone to ad hoc calibration than current sophisticated niche theories involving a huge parameter space.[33]

But "unification" in the neutral theory also means a unification of *niche*- and *dispersal*-assembly views of diversity. For Hubbell, *dispersal*-assembly had to first be elaborated—in the neutral theory—in order to consider any possible unification. What appears ultimately is that ecological equivalence may be created by niche differences themselves in some complex way: "Life-history trade-offs and fitness invariance rules potentially decouple niche differentiation from control of the species richness and relative species abundance of communities." Niche assembly therefore allows ecological equivalence to be set, which yields dispersal-assembly processes, and hence a unifying perspective is permitted. Given that dispersal-assembly studies have been pursued at large regional scales, while niche-assembly views have been elaborated at a local, community scale, the two kinds of unification overlap with each other.

This long-standing concern with unification makes sense in the context of the history of community ecology. When Hubbell started his career, subdisciplines proliferated within ecology: after Eugene Odum's ecosystem ecology, a focus on trophic networks gave rise to functional ecology; community ecology was supplemented by landscape ecology, considering objects at larger scales of time and space; population biology gave rise to metapopulation biology; conservation biology and biogeography arose, which shared some of the tenets of theories in those fields but not all of them. This questioned the status of community ecology, which did not have any absolutely proper method or object. In this sense, Hubbell's neutral theory in ecology is a response to the suspicion raised against the isolation of community ecology: in his own view, the dynamics of biodiversity at all scales (hence, biogeography, landscape ecology, and community ecology) should be understood through the same theoretical framework, since the assumption of ecological equivalence is likely to be construed across all these scales.

One of the first reviews of Hubbell's book, by community ecologist Peter Abrams in *Science*, was entitled "A World without Competition";[34] it expresses the shared feeling that Hubbell was shaking up the key term of *competition* cherished by ecologists for decades—from Gause to McArthur's Marlboro group, through Hutchinson. That said, other aspects of Hubbell's dream of an extensively equivalent nature were already in the air: neutral evolutionary genetics, of course, outside ecology, and island biogeography within ecology. The radical novelty of Hubbell's neutral theory comes by extending such theoretical elements: a new kind of theory emerged through radical redeployment of extant concepts within a new vision, focused on unification of scales and formal simplicity.

But what newly emerges here with Hubbell is an ecology of nonequilibrium—which was already indicated as an empirical result in the 1979 paper: "The available circumstantial evidence suggests that the forest is in a nonequilibrium state."[35] Major theories in ecology were equilibrium theories, considering the state of affairs from the viewpoint of equilibrium reached: competitive exclusion; equilibrium between immigration and extinction in the theory of island biogeography; optimal strategies reached by selection in behavioral ecology.[36] In many cases, natural selection brings about those equilibria, which entitles biologists to consider traits and phenotypes from the viewpoint of achieved selection. This is not available in Hubbell's neutral theory; on the contrary, the zero-sum game models a system that is not oriented toward equilibrium (as fitness maximization would be). "Our most familiar theories in ecology all concern the population dynamics and community ecology of specific, named, or labeled species, each of which has an assigned dynamical equation or set of equations. However, because the dynamics of any given set of species in the metacommunity obey an absorbing process (all species eventually go extinct), no fixed, nontrivial equilibrium dominance-diversity distribution can exist for any set of named species in the metacommunity. Thus, the analysis of metacommunity dynamics is qualitatively different from most classical ecological theory."[37] Since the "balance of nature" has been such a powerful metaphor in ecology,[38] the disruption Hubbell introduces was rightly seen as deeply provocative.

FATE AND DEVELOPMENTS

The neutral theory in ecology had a mixed reception, even though most reviews acknowledged the novelty of the book. Sophisticated discussions pervaded ecology and general journals, mostly triggered by the obvious, counterintuitive fact that an assumption as unrealistic as ecological equivalence sounded as good as the most refined niche models to account for known biodiversity patterns. Several aspects of the theory have been especially addressed in those discussions: first, the empirical attestations, the fit with the extant data; second, the legitimacy of the notion of ecological equivalence; third, the logical status of the theory itself.

Since 2001, Hubbell himself has made decisive advances on the two first points. As always, writing with many collaborators—papers about Barro Colorado Island, as expected, often involved more than ten people—he applied the neutral theory to Barro Colorado data and other tropical forest plots; he also developed various other neutral models and, with Volkov and Maritan,[39] formulated an analytic solution for species distributions under ecological equivalence, whereas he had often done simulations.[40] He suggested an evolution of the ecological equivalence,[41] and he recently elaborated the hypothesis that parasitic relation between fungi and trees was affecting competitive relations between trees in a way that made ecological equivalence plausible.[42]

The current work on the neutral theory is much concerned with what Hubbell highlighted in his 2006 paper on the evolution of ecological equivalence: namely, the fact that many processes, often niche processes, may lead to neutral patterns. Therefore, instead of trying to falsify the neutral theory, another path for research could be concerned with integrating niches into the neutral theory,[43] in a framework where the effect of niche difference should be progressively added to a baseline neutral model.[44] Whatever the next development, it seems fair to assert that, as in the case of neutral evolution, an idea that originally sounded so disruptive regarding the almost century-long teaching of ecology has proved to be part of the theoretical apparatus of the science and clearly fuels empirical research programs—even though its core was a shift in mathematical modeling.

Other authors, after the publication of *The Unified Neutral Theory* and subsequent neutralist papers, criticized claims for the widespread validity of the theory (e.g., for coral reefs). Yet a major issue raised about the status of the theory was the *place for null hypotheses*. Some ecologists indeed consider that the neutral theory is a null hypothesis,[45] against which any hypothesis about species coexistence or SAD should be measured, but not a hypothesis by itself. Yet the neutral theory is not exactly a null hypothesis, in the sense of the models proposed by Simberloff and Connor against Diamond in the early 1980s,[46] because it is a parsimonious description of processes rather than the randomization of a pattern.[47]

CONCLUSION

The neutral theory today belongs to the corpus of theoretical ecology; Morin's community ecology textbook, published eight years after the publication of *The Unified Neutral Theory*, includes a section on the neutral theory.[48] To some extent, the fate of this theory parallels its predecessor and analog within evolutionary biology. While Kimura triggered new questions, not only about the extent of neutrality but also about new objects, such as the molecular clock, Hubbell brought up a new problem for ecologists: namely, the extent of ecological equivalence, the detection of equivalence-ecology/neutrality patterns in extant ecosystems, and the estimation of fundamental biodiversity number.

Neutral views in ecology have preceded Hubbell: papers by Caswell and Levinton in the 1970s were also obvious stand-alone forerunners of the idea of individual ecological equivalence as a base for model-building.[49] The historical question was therefore, "Why did the neutral theory, as such, emerge in 2001 with Hubbell?" The answer pinpointed Hubbell's proper trajectory and intellectual interests, as well as his epistemic stance toward systematicity and generality; but it's also about the embedding of the theory within thorough fieldwork of measuring and monitoring trees in a tropical forest, which provided a source for data that continuously fueled mathematical modeling and provided patterns that fit predictions.

Granted, Hubbell's neutral theory was entrenched within earlier theories, especially the theory of island biogeography—exactly like Kimura's theory, which developed well-established conceptions about random drift and extended them. Yet Hubbell's neutral theory extended early theories with a new concern for parsimony and a thorough mathematical treatment, to the point where they could make manifest the paramount power of randomness regarding ecological diversity—in a way that does not exactly contradict the "paramount power of selection" as demonstrated by Darwin, but provides us with new tools to understand the levels, scales, and situations where it can be overwhelming.

FURTHER READING

Chave, J. "Neutral Theory and Community Ecology." *Ecology Letters* 7 (2004): 241–53.
Hubbell, S. P. "The Neutral Theory of Biodiversity and Biogeography and Stephen Jay Gould." *Paleobiology* 31 (supp.) (2005): 122–32.
Marquet, P., A. Allen, J. Brown, et al. "On Theory in Ecology." *Bioscience* 64 (2014): 701–10.

NOTES

1. S. Hubbell, *The Unified Neutral Theory of Biodiversity and Biogeography* (Princeton, NJ: Princeton University Press, 2001).

2. G. E. Hutchinson and P. Deevey, "Ecological Studies on Population," *Survey of Biological Progress*, vol. 1, ed. G. Avery (New York: Academic Press, 1949), 325–55.

3. See Jay Odenbaugh, "Searching for Patterns, Hunting for Causes: Robert MacArthur, the Mathematical Naturalist," in *Outsider Scientists: Routes to Innovation in Biology*, ed. O. Harman and M. Dietrich (Chicago: University of Chicago Press, 2013), 181–200.

4. R. H. MacArthur and E. O. Wilson, *The Theory of Island Biogeography*, Monographs in Population Biology (Princeton, NJ: Princeton University Press, 1967).

5. The abundance of a species in a community means the number of individuals of this species.

6. R. A. Fisher, A. S. Corbet, and C. B. Williams, "The Relation between the Number of Species and the Number of Individuals in a Random Sample of an Animal Population," *Journal of Animal Ecology* 12 (1943): 42–58; F. W. Preston, "The Commonness and Rarity of Species," *Ecology* 29 (1948): 254–83.

7. "Some are steeper, and some are shallower, but all of the distributions basically exhibit an S-shaped form, bending up at the left end and down at the right end. Is there a general theoretical explanation for all of these curves?" Hubbell, *Unified Neutral Theory of Biodiversity*, 4.

8. A. Pocheville, "The Ecological Niche: History and Recent Controversies," in *Handbook of Evolutionary Thinking in the Sciences*, ed. T. Heams, P. Huneman, G. Lecointre, et al. (Dordrecht: Springer, 2015).

9. See G. E. Hutchinson, "Homage to Santa Rosalia; or, Why Are There So Many Kinds of Animals?," *American Naturalist* 93 (1959): 145–59; and "The Paradox of the Plankton," *American Naturalist* 95 (1961): 137–45.

10. For more on ecological research in Panama, see Megan Raby, "'The Jungle at Our Door': Panama and American Ecological Imagination in the Twentieth Century," *Environmental History* 21 (2016): 260–69; and "Ark and Archive: Making a Place for Long-Term Research on Barro Colorado Island, Panama," *Isis* 106 (2015): 798–824.

11. S. Hubbell, F. He, R. Condit, et al., "How Many Tree Species Are There in the Amazon, and How Many of Them Will Go Extinct?," *PNAS* 105, no. 1 (2008): 11498–504.

12. S. P. Hubbell, "Species-Area Relationships Always Overestimate Extinction Rates from Habitat Loss," *Nature* 473 (2011): 368–72.

13. R. Condit, P. S. Ashton, N. Manokaran, et al., "Dynamics of the Forest Communities at Pasoh and Barro Colorado: Comparing Two 50-ha Plots," *Philosophical Transactions of the Royal Society B: Biological Sciences* 354 (1999): 1739–48.

14. S. P. Hubbell, "Tree Dispersion, Abundance, and Diversity in a Tropical Dry Forest," *Science* 203 (1979): 1299–1309.

15. D. H. Janzen, "Herbivores and the Number of Tree Species in Tropical Forests," *American Naturalist* 104 (1970): 940; J. H. Connell, "On the Role of Natural Enemies in Preventing Competitive Exclusion in Some Marine Animals and in Rain Forest Trees," in *Dynamics of Population*, ed. P. J. Den Boer and G. R. Gradwell (Wageningen: Pudoc, 1970).

16. "Suppose that forests are saturated with trees, each of which individually controls a unit of canopy space in the forest and resists invasion by other trees until it is damaged or killed. Let the forest be saturated when it has K individual trees, regardless of species. Now suppose that the forest is disturbed by a wind, storm, landslide, or the like, and some trees are killed. Let D trees be killed, and assume that this mortality is randomly distributed across species, with the expectation that the losses of each species are strictly proportional to its current relative abundance. Next let D new trees grow up, exactly replacing the D "vacancies" in the canopy created by the disturbance, so that the community is restored to its predisturbance saturation until the next disturbance comes along. Let the expected proportion of the replacement trees contributed by each species be given by the proportional abundance of the species in the community after the disturbance. Finally, repeat this cycle of disturbance and re-saturation over and over again. In the absence of immigration of new species into the community, or of the recolonization of species formerly present but lost through local extinction, this simple stochastic model leads in the long run to complete dominance by one species. In the short run, however, the model leads to lognormal relative abundance patterns, and to geometric patterns in the intermediate run." Hubbell, "Tree Dispersion," 1306.

17. The same year, Graham Bell, ecologist at Montreal, published a long paper in *Science* entitled "Macroecology," where he presented roughly similar ideas about ecological equivalence—also referring to Hubbell's forthcoming book; see his "Ecology—Neutral Macroecology," *Science* 293 (2001): 2413–18.

18. J. M. Diamond, "Assembly of Species Communities," in *Ecology and Evolution of Communities*, ed. M. L. Cody and J. M. Diamond, 332–445 (Cambridge, MA: Harvard University Press, 1975).

19. E. F. Connor and D. S. Simberloff, "The Assembly of Species Communities: Chance or Competition?," *Ecology* 60, no. 6 (1979): 1132–40.

20. Consider, for example, the powerful R* ("resource-ratio") theory, developed in D. Tilman, *Resource Competition and Community Structure* (Princeton, NJ: Princeton University Press, 1982).

21. Hubbell, *Unified Neutral Theory of Biodiversity*, 5.

22. Sewall Wright, "Evolution in Mendelian Populations," *Genetics* 16 (1931): 97–159.

23. M. Kimura, *The Neutral Theory of Molecular Evolution* (Cambridge: Cambridge University Press, 1985).

24. For instance, see X.-S. Hu, F. He, and S. P. Hubbell, "Neutral Theory in Macroecology and Population Genetics," *Oikos* 113 (2006): 548–56.

25. Hubbell, *Unified Neutral Theory of Biodiversity*, 53.

26. R. Levins, "Some Demographic and Genetic Consequences of Environmental Heterogeneity for Biological Control," *Bulletin of the Entomology Society of America* 71 (1969): 237–40.

27. I. Hanski, "Metapopulation Dynamics," *Nature* 396 (1998): 41–49. A major driver of fragmentation is human activities; hence, metapopulation ecology directly connects to conservation biology.

28. Hubbell, *Unified Neutral Theory of Biodiversity*, 111.

29. Hubbell, *Unified Neutral Theory of Biodiversity*, 113.

30. G. C. Hurtt, and S. W. Pacala, "The Consequences of Recruitment Limitation: Reconciling Chance, History, and Competitive Differences between Plants," *Journal of Theoretical Biology* 176 (1995): 1–12.

31. Hubbell, *Unified Neutral Theory of Biodiversity*, 325.

32. S. P. Hubbell, "A Unified Theory of Biogeography and Relative Species Abundance and Its Application to Tropical Rain Forests and Coral Reefs," *Coral Reefs* 16 (1997): S9–S21.

33. See, for instance, the R* theory.

34. P. Abrams, "A World without Competition," *Nature* 412 (2001): 858–59.

35. Hubbell, "Tree Dispersion," 1305.

36. See, for example, R. H. MacArthur and E. R. Pianka, "On Optimal Use of a Patchy Environment," *American Naturalist* 100 (1966): 603–9.

37. Hubbell, *Unified Neutral Theory of Biodiversity*, 115.

38. S. Pimm, *The Balance of Nature? Ecological Issues in the Conservation of Species and Communities* (Chicago: University of Chicago Press, 1991); D. Simberloff, "The 'Balance of Nature': Evolution of a Panchreston," *PLoS Biology* 12, no. 10 (2014): e1001963.

39. I. Volkov, J. R. Banavar, S. P. Hubbell, et al., "Neutral Theory and Relative Species Abundance in Ecology," *Nature* 424 (2003): 1035–37.

40. Hubbell, *Unified Neutral Theory of Biodiversity*. He did simulations because he had no time to get involved in the analytic treatment before handing the book manuscript to the publisher, even though he wanted ultimately an analytic solution, because it would provide an indubitable result about the model (personal communication).

41. S. S. Hubbell, "Neutral Theory and the Evolution of Ecological Equivalence," *Ecology* 87 (2006): 1387–98.

42. A. Barberan, K. L. McGuire, J. A. Wolf, et al., "Relating Belowground Microbial Composition to the Taxonomic, Phylogenetic, and Functional Trait Distributions of Trees in a Tropical Forest," *Ecology Letters* 18 (2015): 1397–1405.

43. P. B. Adler, J. Hillerislambers, and J. M. Levine, "A Niche for Neutrality," *Ecology Letters* 10 (2007): 95–104; R. D. Holt, "Emergent Neutrality," *Trends in Ecology and Evolu-*

tion 21 (2006): 531–33; B. J. McGill, "Towards a Unification of Unified Theories of Biodiversity," *Ecology Letters* 13 (2010): 627–42.

44. M. Vellend, "Conceptual Synthesis in Community Ecology," *Quarterly Review of Biology* 85 (2010): 183–206.

45. B. McGill, B. A. Maurer, and M. D. Weiser, "Empirical Evaluation of Neutral Theory," *Ecology* 87 (2006): 1411–23; C. P. Doncaster, "Ecological Equivalence: A Realistic Assumption for Niche Theory as a Testable Alternative to Neutral Theory." *PloS ONE* 4 (2009): e7460.

46. E. F. Connor and D. S. Simberloff, "The Assembly of Species Communities: Chance or Competition?," *Ecology* 60, no. 6 (1979): 1132–1140; E. F. Connor and D. S. Simberloff, "Interspecific Competition and Species Co-occurrence Patterns on Islands: Null Models and the Evaluation of Evidence," *Oikos* 41 (1983): 455–65. The controversy started with J. M. Diamond, "Assembly of Species Communities," in *Ecology and Evolution of Communities*, ed. M. L. Cody and J. M. Diamond (Cambridge, MA: Harvard University Press, 1975), 332–445. The second paper by Connor and Simberloff responded: see J. M. Diamond and M. E. Gilpin, "Examination of the 'Null' Model of Connor and Simberloff for Species Co-occurrences on Islands," *Oecologia* 52 (1982): 64–74.

47. For more about the epistemic status of the neutral theory, see F. Munoz and P. Huneman, "From the Neutral Theory to a Comprehensive and Multiscale Theory of Ecological Equivalence," *Quarterly Review of Biology* 91, no. 3 (2016): 321–42.

48. P. J. Morin, *Community Ecology* (London: John Wiley, 2008).

49. Caswell built an ecological model in an explicit correspondence with Kimura's neutral theory; though Hubbell discusses in detail its model, it did not much influence community ecologists, and Hubbell overlooked it at the time; see H. Caswell, "Community Structure—Neutral Model Analysis," *Ecological Monographs* 46 (1976): 327–54. Levinton proposed a model based on ecological invariance to account for paleontological patterns of marine diversity; see J. S. Levinton, "A Theory of Diversity Equilibrium and Morphological Evolution," *Science* 204 (1979): 335–36.

ACKNOWLEDGMENTS
The author warmly thanks Mark McPeek for personal communications about the neutral theory, and above all Steve Hubbell for deep and rich conversations. He is grateful to Mike Dietrich and Oren Harman, to Arnaud Pocheville and Sébastien Dutreuil, and to anonymous reviewers for fruitful comments and criticisms.

JANET BROWNE

RACHEL CARSON

PROPHET FOR THE ENVIRONMENT

12

Rachel Carson speaks most directly to us today through her environmental book *Silent Spring*, published in 1962. She is revered as an exceptional writer and literary naturalist in the tradition of John Muir and Aldo Leopold—a romantic populist, knowledgeable, scientifically accurate, visionary, holistic, and passionate. Her personal "dreaming" was of the pristine seashores and untouched rocky landscapes of Northeast America that she came to love so profoundly. At another level, simultaneously more organic and more transcendent, she dreamed of an interconnected world in which the natural and human were closely interwoven in harmony, an early vision of a balanced living system that helped generate the modern ecological way of thought. Her dismay at the prospect of the destruction of this harmony by thoughtless human interference turned her into an exceptionally focused writer who crystallized existing scientific knowledge and gave powerful new direction to the growing environmental movement. In *Silent Spring* she warned about the ecological devastation that would inevitably appear through the indiscriminate use of pesticides. Carson was suffering from cancer while she wrote the book, and died at age fifty-seven in April 1964, but not before making forceful political interventions relating to the chemical industry in the United States. *Silent Spring* generated a storm of interest and was the foundation for important legislative action regulating the use of chemicals in the environment. Her dream became the dream of many others, and her role in shaping the environmental movement was crucial.

Many scholars have asked how Carson's transformative dream emerged.[1] She was born in Springdale, Pennsylvania, in 1907, and much of her life was dominated by the conflicting demands of a struggle for financial security and strong literary aspirations. She was the third and youngest child of Maria Frazier McLean, a former schoolteacher, and Robert Warden Carson, an insurance salesman and then an employee at West Penn Power. Her father was mostly unable to support the family, and his death in 1935 left them with very limited means. From her young adulthood, Rachel Carson took financial responsibility for her mother, sister, and her sister's children.

Carson was greatly influenced by a succession of exceptional women,

Figure 12.1. Rachel Carson. Photo courtesy of the National Digital Library of the United States Fish and Wildlife Service.

of which no doubt her mother was the first. Throughout her life she held a deeply affectionate and nearly exclusive relationship with her mother. Maria Carson kept house, edited and typed Rachel's manuscripts, and encouraged her daughter until her death in 1958. Carson retrospectively credited her mother with instilling in her an early love for nature. Her mother raised her, it could be claimed, in the American "nature studies" movement that fostered active learning outdoors.[2] Her mother also introduced her to the world of literature, and at the age of eleven, she began sending short fiction stories from her home in Pennsylvania to the children's section of *St. Nicholas Magazine*, several of which were published. Carson believed that she gained from her mother an uncompromising Calvinistic morality, in which waste and intellectual laziness were abhorred. She was in early life a devout Christian, a point of view that she abandoned in middle age. Yet even without an obvious religious faith, these personal commitments were refocused into an intense belief in the natural world as the holistic setting for human life, and she adopted a personal credo in which human beings had a moral duty to protect, understand, and appreciate nature in all its forms. This form of secular religion provided Carson with a framework of meaning that was as compelling as any more traditional faith.

Carson was educated at the Pennsylvania College for Women (now Chatham University), where she majored in biology and met Mary Scott Skinner, a biology teacher who became her mentor. She studied marine biology during one summer at Woods Hole Marine Biological Laboratory, graduated in 1929, and joined Johns Hopkins University, aiming for a PhD in zoology. Financial difficulties obliged her to cut short her studies at Johns Hopkins with an MA in 1932. During this time she became an assistant in Raymond Pearl's laboratory, where she worked with rats and *Drosophila* to earn money for tuition. Her own research was on the embryonic development of fish. It is perhaps not sufficiently appreciated that Carson was therefore exceptionally well educated in genetics at a time when it was at the cutting edge of biological science. At the urging of Skinner, with whom she subsequently kept in touch, she took a position with the US Bureau of Fisheries to write radio scripts on biological matters during the Depression and supplemented her income by providing articles on natural history for the *Baltimore Sun*. Her family relocated to Maryland to be with her, and she supported the household. When the opportunity emerged, she sat the civil service exam, outscored all other applicants, and, in 1936, became only the second woman employed professionally at the Bureau of Fisheries. She rose to become editor-in-chief of all publications for the bureau (subsequently the US Fish and Wildlife Service). Her career has, in fact, been a beacon for subsequent women in the sciences and is the topic of a great deal of insightful writing by feminist scholars. Her life marked a turning point in the way women participated in science, and she was instrumental in shifting the prevailing view of women as contributing to science through a variety of supporting roles to one in which women were recognized as engaging in fully active scientific careers.[3]

In these early years, above all else, Carson remembered her first youthful encounter with the sea. This was during the summer of 1928 at Woods Hole Marine Biological Station, in Cape Cod, Massachusetts. For her, the sea held a mystical quality. In an undated letter (c. 1941), she mentioned that she imagined herself under the water "until I could see the whole life of those creatures as they lived them in that sea world."[4] A sense of holism and identification with other, nonhuman, organisms was developing that came to underpin her understanding of ecosystems. These ideas took permanent root during her continuing work in the US Fish and Wildlife Service, where she encountered—and edited—much of the latest oceanographic research. This was how she came across a series of reports on the new pesticide DDT (dichloro-diphenyl-trichloroethane), which would feature so much in her later work. Carson proposed writing a popular account on pesticide research for the *Reader's Digest* but was turned down.

At this point in her life, her family's financial situation was dire. She decided that she could help resolve the crisis by writing a natural-history best seller. In the 1930s, she therefore began to write with great determination and dedication in the hours after her work in the bureau, aiming for a Pulitzer Prize.[5] Her first book was *Under the Sea Wind* (1941). It began as an article on animal migration for the *Atlantic*. Encouraged by an editor from the publishing house of Simon and Schuster, and the author Hendrik Willem Van Loon, she turned it into a book, presenting her material from the perspective of individualized sea and land creatures as they moved through their respective life cycles. It was based on numerous firsthand observations of American fauna, as in the central narrative of a sanderling migrating from Patagonia to Alaska, finding a mate, hatching out chicks, encountering predators, and so forth. Despite the anthropomorphism, the book is full of real nature: tough, ruthless, and subject to chance. She included humans as part of this global ecosystem—for example, describing fishermen netting a shoal of mullet. She already showed deep understanding of the organic cycles of life: "For in the sea, nothing is lost. One dies, another lives, as the precious elements of life are passed on and on in endless chains." This first book was initially well received, and two chapters were reprinted by William Beebe in his influential anthology of America's greatest naturalist writers, published in 1944, but any hope of commercial success was cut off by the beginning of war. It did not sell in any great numbers and was remaindered by the publisher.

Success came ten years later with her widely praised book *The Sea Around Us* (1951). This too was first published in a periodical, in three parts in the *New Yorker*. The book won Carson recognition as a gifted writer and brought her much-needed financial security. She quit her job at the US Fish and Wildlife Service, purchased a two-tone Oldsmobile, and built a small house in Maine, on the coast, near Southport, where she and her family spent their summers for many years. The book received the US National Book Award, although no Pulitzer ever came. In it, her purpose was to give a sense of the sea as an environmental whole, and she divided the book into sections, describing sea life according to depth, the movement of winds and currents, geography of the sea floor, fish stocks and other commercial resources, and the ocean's effect on climate. The intent was larger than this however. She wrote of the eternal rhythms and cycles of life, of the fact that mankind cannot control the sea, about continuities and interrelatedness. There was a biblical tone to the opening pages, although no creative force was cited. All life, she wrote, depends on the oceans. The book was deeply informed by a secular evolutionary understanding of nature, and she spoke in many places about the fierce competitive forces at the heart of biology. She had adopted contemporary Darwinism long before, back at Johns Hopkins, and by the time she published this book, it seems, she was no longer a practicing Christian.

But the moral authority of *Silent Spring* was still a long way off. Her next book was called *The Edge of the Sea* (1955). During this period she became close friends with a Maine neighbor, Dorothy Freeman. Released from the need to continue in a job, she took long nature walks with Dorothy, collected organisms from pools on the tide line, and examined specimens under an expensive new microscope—all of which contributed to the text of the new book. While Carson had other dear friends who gave professional and literary support, she increasingly relied on Dorothy Freeman for emotional nourishment. She also became friends with Bob Hines, who illustrated *The Edge of the Sea*. The structure of the book was more or less geographical, embracing various types of shoreline: sandy, rocky, and subtropical (coralline). This third book was published to great acclaim. Her publishers correspondingly set out to reissue all three books together as a major sea trilogy that explored the whole of ocean life from the shoreline to the oceanic depths.

After this hard-won success, Carson became temporarily adrift in literary terms. Several years beforehand, she had been planning a popular book on evolution—a timely thought with the centenary of Charles Darwin's *Origin of Species* on the horizon in 1959. Yet Julian Huxley forestalled her with his *Evolu-*

tion in Action (1953). She tried writing for national television, and delivered an attractive script on clouds that was aired in 1956. It was therefore only in early 1957 that she turned to a project on ecology that she tentatively called *Remembrance of the Earth*. For this project, unlike her others, she did not yet have a sense of her storyline. Here, however, the dreamer in her soul emerged.

She was writing a short article for *Women's Home Companion* called "Help Your Child to Wonder" that featured her grandnephew Roger, who lived with her, and the advice offered to Roger revealed much of her innermost passion for the natural world. Slowly, she began to articulate her subject. It would be life itself, and the relationship of living beings to the physical environment in the modern, atomic age. A shift in her thinking had begun with the wartime bombing of Hiroshima. Now, some twelve years later, in a letter to Dorothy Freeman, she said that she found it difficult to be confident in the capacity of life to withstand such human assaults. She was starting to question the scientific progress that she had formerly taken for granted. Mankind "seems likely to take into his hands . . . many of the functions of God. . . . He must do so with humility rather than arrogance."[6]

What pushed her to focus so courageously on pesticides and the agricultural chemical industry? Carson was not the only one expressing concern about chemicals in the environment. Fears about pesticides had been raised as early as 1933 by Arthur Kallett and Frederick Schlink in their book *100,000,000 Guinea Pigs: Dangers in Everyday Foods, Drugs, and Cosmetics*. The one hundred million guinea pigs in the title referred to the size of the human population of the United States at the time. Widely read periodicals such as the *Atlantic Monthly, Reader's Digest,* and *Harper's Magazine* occasionally voiced concern. Products used in agriculture containing heavy metals, such as lead, copper, and arsenic, were denounced, especially by Ruth Deforest Lamb, the chief information officer of the US Food and Drug Administration, who wrote about arsenic in spray residue on crops.[7] Even man-made fertilizers were criticized by Rudolf Steiner, the educationalist and organic farmer. The dangers of DDT were described in the *Atlantic Monthly* in 1945 by the British entomologist Vincent Wigglesworth in an article called "DDT and the Balance of Nature." More expressly, Clarence Cottam, a personal friend of Carson's in the US Fish and Wildlife Service, published a report on DDT in 1946. He became Carson's mentor on the subject of pesticides and conservation. And John Kenneth Terres, a soil biologist and birder, wrote of the possibility of pesticides moving up the food chain. Carson called on the studies of these and other concerned individuals when she began her research.

Carson was not alone, either in evocative or politically pointed environmental thinking. William Vogt, a prominent member of the Conservation

Foundation and an advocate for studying the relationship between climate, population, and resources in Latin American countries, published his challenging text, *Road to Survival*, in 1948. The Sierra Club was founded in 1892, the National Audubon Society in 1905, the Conservation Foundation in 1947, and the Nature Conservancy in 1951. She fitted into a well-established American nature-writing trajectory from John Muir and Aldo Leopold to Mary Stoneman Douglas, author of *The Everglades: Rivers of Grass*. These writers saw in each living thing its own particular beauty and place in the natural world.

So it seems likely that it was moral urgency that pushed Carson onward. Judging from the words of *Silent Spring*, her profound secular morality generated the conviction that modern technology was creating in human beings an arrogant hubris that ought to be replaced by a new humility toward nature. The eternal truths of environmental harmony and ecological interrelatedness that she wrote about in her first three books were threatened. The tipping point appears to have been information from an acquaintance, Olga Owens Huckins, who alerted her to the effects of pesticides on wildlife, as seen in a mosquito eradication program near Huckins's home in Duxbury, Massachusetts. The aerial spraying of some unidentified pesticide led to the death of dozens of birds and the disappearance of bees in Huckins's garden. Once Carson knew the extent of disruption caused by poisons, she felt she had to speak out. Inspired by her dreams of an unspoiled nature, reinforced by her childhood memories and all-absorbing fascination for the intricacies of the web of life, Carson thus came to recognize that her existing fame as a popular nature writer and her strong scientific background could be used politically to make a real difference. She committed herself to a crusade, relinquishing her sense of scientific impartiality, and took on the role of a spokeswoman for nature.

Carson immediately began collecting information about pesticides. At first she focused on the United States Department of Agriculture's war against fire ants that was, in Carson's words, "an outstanding example of an ill-conceived, badly executed, and thoroughly detrimental experiment in the mass control of insects." Fire ants had entered the United States from Brazil in the 1920s and spread across the Southwest, building nests as much as five feet high, as strong as concrete. Multiple ant bites could kill a small animal. Nevertheless, fire ants and farmers managed to coexist. But in the mid-1950s, the USDA initiated an eradication program that called for the spraying by airplane of some twenty to thirty million acres of agricultural land. The department chose a pesticide that included dieldrin and heptachlor, both highly potent chlorinated hydrocarbons (it was later shown that heptachlor causes liver damage, and dieldrin is a neurotoxin). In 1959, the USDA's Agricultural Research Service responded to criticism by Carson and others with a public service film, *Fire Ant on Trial*. Carson concluded that the department had probably never investigated the possible toxicity of these substances or, if it did, had ignored the results. Her conviction deepened in 1959 when US cranberries were found to contain high levels of the herbicide aminotriazole, and the sale of all cranberry products was halted. Carson attended the ensuing Food and Drug Administration hearings on pesticide regulation. According to her letters of the time, she was deeply dejected by the tactics of chemical industry representatives. She reported to Dorothy Freeman after the hearings that the testimony of the agricultural industry's representative, Edward B. Astwood, "can be shot so full of holes as to be absolutely worthless, and the disheartening thing is that he must know this full well."[8]

A second case concerned attempts to control the gypsy moth, another non-native species that was ravaging large areas of forest in the Northeast corridor. Inadequacies of the program became a hot public issue when the properties of wealthy inhabitants of Long Island were sprayed up to fifteen times with DDT in 1957. Residents objected vigorously on a variety of grounds and brought legal action to halt the spraying. Carson engaged with this protest, became politically active by supporting John F. Kennedy in his run for the presidency, and urged Democrats to take up issues of pollution control. She joined the National Resources Committee of the Democratic Advisory Council and then the Women's Committee for New Frontiers. The work of National Cancer Institute scientist, Wilhelm Hueper, who classified many pesticides as carcinogens, was helpful to her at this time, as was the Washington, DC, chapter of the Audubon Society, which actively opposed chemical spraying programs and asked Carson to help publicize the US government's practices. By now, she was sufficiently famous as an environmental writer to be able to recruit nearly everybody she needed. In so doing, she built up a formidable network of allies and colleagues in high places that stood her in good stead when controversy flared after publication of *Silent Spring*.

Pesticides were far grimmer than the celebrations of natural harmony that Carson had previously produced. Her legacy perhaps rests in part on creating a new kind of nature book: one that deals with the gloomy consequences of human actions upon the earth. Her friend Dorothy Freeman called *Silent Spring* the "poison book." Even so, the title remained undecided until Carson lifted it from her chapter on birds. She prefaced the work with lines from "La Belle Dame sans Merci" (1819) by the English poet John Keats:

The sedge is wither'd from the lake
And no birds sing.

Silent Spring was mostly about DDT. It was first published as three substantial essays in the *New Yorker* in June 1962. Carson struck a fair balance between technical detail and the narrative about chemicals in the environment that she wished to make, but the nature of the subject matter meant that the book was not nearly so lyrical as her former publications. Unlike many modern pesticides, whose effectiveness is limited to destroying one or two types of insects, DDT is capable of killing many different organisms at once, especially water-based animals such as aquatic invertebrates and fish. The insecticidal action of DDT was discovered in 1939 by the Swiss chemist Paul Hermann Müller, who was awarded a Nobel Prize in recognition of the possibility of using DDT to eradicate mosquitoes in the global fight against malaria. The substance was immediately relevant to the US war effort and was used as a spray distributed by airplane during World War II for clearing South Pacific islands of mosquito larvae (these occur in swampy places). It was also used by Allied troops and by some civilian populations to control the ticks and lice that carry typhus. After 1945 it was used extensively in the United States as an agricultural insecticide and in Europe as a domestic fly spray. The proprietary formulation Flit, contained in a popular household spray gun, contained 5% DDT in the late 1940s and early 1950s, before the negative environmental impact of DDT was widely understood.

Carson described how DDT enters the food chain and accumulates in the fatty tissues of animals, including human beings, and can cause some forms of cancer. She thought DDT caused genetic damage (i.e., was mutagenic rather than merely toxic), and had good reason to believe so at that time, but the current evidence is ambiguous on this. Her text nevertheless made clear that DDT accumulates in the food chain. Thus, while DDT might appear only mildly toxic to one organism, if the carrier is eaten by another, the chemical will accumulate and remain active in body tissues. Her scientific case studies often read as miniature narratives about the great interrelatedness of life. One vivid example that she included concerned the spraying policy in Michigan to control beetles that carried Dutch elm disease. Researchers discovered that earthworms living in the forest litter under the sprayed trees were accumulating DDT from their vegetable diet and that the robins eating them were being poisoned. Another study showed that DDT alters a bird's calcium metabolism in a way that results in thin eggshells, sometimes so thin that they break when the female incubates them. These, and other scientific studies, proved to be powerful arguments. A single DDT application on a crop, she wrote, killed insects for months and remained toxic in the environment even after dilution by rainwater. Most of the book is devoted to the effects of pesticides on natural ecosystems, but she allocated space to describing cases of human illnesses attributed to pesticide poisoning. In the final chapter she offered alternatives to the practices she condemned; and ended with the hope that mankind would learn to live in harmony with the environment, "rather than in combat."

Not all her case studies came from the experts. Carson respected the observations of nonprofessionals, a democratic vision of expertise that ran counter to most contemporary ideas about the scientific enterprise at large. The fact that she relied on non-expert witnesses made her book significant in the history of public debate about science: after *Silent Spring*, public confidence in the superiority of scientific expertise could no longer be taken for granted, and environmentalism would develop as a political movement that did not necessarily trust scientists or their scientific results. To write such a book in the politically active 1960s was seen as a direct attack on the edifice of professional science. Yet it is important to remember that Carson was not anti-science. It would be more accurate to say that she believed in expertise without elitism.

Most famously, the book's opening chapter, "A Fable for Tomorrow," is noted as a literary classic. It depicts an imaginary American town where all life—from fish to birds, apple blossoms, and human children—has been "silenced" by the effects of DDT. "It was a spring without voices," she wrote: "It is our alarming misfortune that so primitive a science has armed itself with the most modern and terrible weapons, and that in turning them against the insects it has also turned them against the earth." On its publication in 1962, *Silent Spring* shocked readers across America and, not surprisingly, brought a howl of indignation from the chemical industry and other interested parties.[9] Agribusinesses and a number of chemical industry bosses went on the attack. Their response can now be seen as consistent with the tactics used by business concerns that deny the findings of science in order to promote economic interests.[10] One prominent adversary was Robert White-Stevens, director of American Cyanamid, a major agricultural chemical company whose activities had been restricted by pesticide regulation after the 1959 cranberry scare. White-Stevens invited the public to imagine the calamity of a world without chemicals or medicines, a position that Carson never advocated. Others, such as William Darby, head of biochemistry at Vanderbilt School of Medicine, dismissed Carson as overexcited. He suggested that if Americans accepted her ideas, they would face hunger and the end of all human progress in the health sciences: "It means disease, epidemics, starvation, misery, and suffering."[11] Ezra Taft Benson, secretary of agriculture to President Eisenhower, belittled her engagement with genetics, asking why "a spinster with no children" should be so concerned about heredity. Economic entomologists understandably also felt themselves under direct attack. It should be said that economic entomologists had grown rapidly in professional status with the development of the pesticide program, and that the discovery and application of DDT was the equivalent for them of the atomic bomb project at Los Alamos National Laboratory, a project that brought public celebrity and national relevance to physicists. Large-scale pesticide projects were doing the same for entomologists. Entomologists, however, believed that DDT posed no threat to humans and minimal danger to wildlife.

In 1962, President John F. Kennedy was quick to assign Jerome Weisner, his science advisor, to set up a panel of the President's Science Advisory Council to investigate Carson's claims. The report concluded that there should be an orderly reduction of pesticide use in the environment. At the same time, public response to the book was high, and letters protesting indiscriminate pesticide usage poured into the political offices of senators and congressmen. Individual states began rapidly to introduce bills to limit pesticide use. The furor over *Silent Spring* ultimately led to the creation of the Environmental Protection Agency, the passage of the Clean Air Act, the Clean Water Act, the Endangered Species Acts, and the banning of a long list of pesticides, including DDT and dieldrin.

Carson's political goal was clear. Yet her years in government also made her pragmatic. She wanted her arguments to bring about improvements but also to be politically feasible. Appearing on a CBS documentary about *Silent Spring* in April 1963, she remarked, "Man's attitude toward nature is today critically important simply because we have now acquired a fateful power to alter and destroy nature. But man is a part of nature, and his war against nature is inevitably a war against himself. . . . I think we are challenged, as mankind has never been challenged before, to prove our maturity and our mastery, not of nature but of ourselves."[12] Robert White-Stevens also appeared in the documentary, predicting starvation and disease across the globe without agricultural chemicals. Carson's calm and quiet manner stole the show. At the end of May she appeared at a Senate hearing chaired by Abraham Ribicoff, at which she presented the recommendations for policy change that had been in her mind ever since her research into pesticides began. A year later she died in the Cleveland Clinic from the cancer she had been fighting for years.

But her dream did not die. Carson initiated a fundamental shift in the way we regard the environment. The impact of this shift can hardly be overstated. Its influence can be seen in the way that controversies still linger about whether Carson's writing should be regarded as legitimate science, and indeed some commentators from the scientific and business communities define science in a way that relegates her book to the category of "popular science"—a label that deliberately diminishes the power of her evidence and makes it easier to ignore her message. This anti-Carson rhetoric has evolved over the decades, to the point that it has become the object of scholarly analysis in its own right.[13] In early criticisms, Carson's gender was frequently mentioned, often in order to disparage her arguments. The *New York Post* called on Abraham's Lincoln's apocryphal words to Harriet Beecher Stowe: "So you're the little woman who wrote the book that made this great war!" The parallel implied that Carson's abilities were not equal to the outcomes that she initiated. Recent work indicates that a distinguishing feature of Carson's achievement was her reliance on a network of female support that lay outside the normal boundaries of the predominantly masculine science of the day. Something of this female powerbase persists in the recent rise of ecofeminism.[14] In sum, her work generated heated antipathy and equally heated support that continues today.

Carson was an outstanding writer and a powerful inspirational force. Perhaps some of the changes she envisaged might have happened anyway. The environmental movement would probably have emerged without her—she has been wrongly credited with single-handedly launching American environmental politics. Nevertheless, she was the primary catalyst for events waiting to happen and imparted to them a special character. Carson concluded that DDT and other pesticides irrevocably harmed living beings and had contaminated the world's food supply. Subsequent research has established the point beyond any doubt. Due to the persistence of DDT and its metabolites in the environment, very low levels still continue to be detected in foodstuffs, even though it was banned from use in the United States in 1972. Her contribution to the making of the modern world was also much more than an alarming warning about pesticides. Her landmark book had an immediate and lasting impression on the public, and alerted readers of all backgrounds—political, scientific, and literary—that our choices and decisions about the environment, and how we conduct our relationship with nature, genuinely matter.

FURTHER READING

Dunlap, Thomas R. *DDT: Scientists, Citizens, and Public Policy*. Princeton, NJ: Princeton University Press, 1981.

Freeman, Dorothy. *Always, Rachel: The Letters of Rachel Carson and Dorothy Freeman, 1952–1964*. Edited by Martha Freeman. Boston: Beacon Press, 1995.

Gottleib, Robert. *Forcing the Spring: The Transformation of the American Environmental Movement*. Washington, DC: Island Press, 1993.

Lear, Linda J. *Rachel Carson: Witness for Nature*. New York: Henry Holt, 1997.

Lytle, Mark Hamilton. *The Gentle Subversive: Rachel Carson, "Silent Spring," and the Rise of the Environmental Movement*. Oxford: Oxford University Press, 2007.

Murphy, Priscilla Coit. *What a Book Can Do: The Publication and Reception of "Silent Spring."* Amherst: University of Massachusetts Press, 2005.

Pimente, David, and Hugh Lehman, eds. *The Pest Question: Environment, Economics, and Ethics*. London: Chapman & Hall, 1993.

Waddell, Craig, ed. *And No Birds Sing: Rhetorical Analyses of Rachel Carson's "Silent Spring."* Carbondale: Southern Illinois University Press, 2000.

NOTES

1. One of the most authoritative biographies is by Linda J. Lear, *Rachel Carson: Witness for Nature* (New York: Henry Holt, 1997). See also Robert Gottleib, *Forcing the Spring: The Transformation of the American Environmental Movement* (Washington, DC: Island Press, 1993); and Mark Hamilton Lytle, *The Gentle Subversive: Rachel Carson, Silent Spring, and the Rise of the Environmental Movement* (Oxford: Oxford University Press, 2007).

2. Sally Gregory Kohlstedt, *Teaching Children Science: Hands-On Nature Study in North America, 1890–1930* (Chicago: University of Chicago Press, 2010).

3. Rebecca Raglon, "Rachel Carson and Her Legacy," in *Natural Eloquence: Women Reinscribe Science*, ed. Barbara T. Gates and Ann B. Shteir (Madison: University of Wisconsin Press, 1997), 196–211.

4. Lytle, *Gentle Subversive*, 35.

5. Lytle, *Gentle Subversive*, 56.

6. Dorothy Freeman, *Always, Rachel: The Letters of Rachel Carson and Dorothy Freeman, 1952–1964*, ed. Martha Freeman (Boston: Beacon Press, 1995), 204.

7. Roger Meiners, Pierre Desrochers, and Andrew Morris, eds., *The False Crises of Rachel Carson: Silent Spring at 50* (Washington, DC: Cato Institute, 2012), 42; and Ralph H. Lutts, "Chemical Fallout: Rachel Carson's *Silent Spring*, Radioactive Fallout, and the Environmental Movement," *Environmental Review* 9 (1985): 210–25.

8. Lear, *Rachel Carson*, 342–44, 358–60.

9. Priscilla Coit Murphy, *What a Book Can Do: The Publication and Reception of "Silent Spring"* (Amherst: University of Massachusetts Press, 2005).

10. Naomi Oreskes and Erik M. Conway, *Merchants of Doubt: How a Handful of Scientists Obscured the Truth on Issues from Tobacco Smoke to Global Warming* (New York: Bloomsbury Press, 2010), 216–23, 226–27.

11. Lytle, *Gentle Subversive*, 174.

12. Quoted from Lear, *Rachel Carson*, 450.

13. David K. Hecht, "How to Make a Villain: Rachel Carson and the Politics of Anti-Environmentalism," *Endeavour* 36, no. 4 (2012): 149–55.

14. See Carolyn Merchant, *Radical Ecology: The Search for a Liveable World* (New York: Routledge, 2005), 193–222.

Part V: The Ethologists

Figure 13.1. Dr. Jane Goodall at the US Department of State in Washington, DC, on October 27, 2015. State Department photo/Public domain.

DALE PETERSON

JANE GOODALL
SHE DREAMED OF TARZAN

13

> *High up among the branches of a mighty tree she hugged the*
> *shrieking infant to her bosom, and soon the instinct that was*
> *as dominant in this fierce female as it had been in his tender*
> *and beautiful mother—the instinct of mother love—reached*
> *out to the tiny man-child's half formed understanding, and he*
> *became quiet. Then hunger closed the gap between them, and*
> *the son of an English lord and an English lady nursed at the*
> *breast of Kala, the great ape.*
> —*Edgar Rice Burroughs,* Tarzan of the Apes *(1914)*

Those words, written by American author Edgar Rice Bur-
roughs in a fantasy-adventure tale for young readers, *Tarzan of the Apes* (1914),
may have been read by millions of children and young adults during the first
half of the twentieth century.[1] The fictional hero Tarzan was the only child of
an English couple who were marooned by mutineers and left to die on the
forested coast of Africa. Had he grown up in England, the baby would have
gone on to enjoy the privileged life of an English aristocrat. Raised by the
apes, he became their aristocrat, blessed with human intelligence and a forest ape's superhuman athleticism. Instead of becoming civilized, he became
feral: freed to sleep in the trees and to live as close to nature and to animals as
any imaginative young girl might desire in her wildest dreams.

One dreamy reader of that mythic tale was a young English girl named Valerie Jane Morris-Goodall. She loved many of the standard children's stories
of the time, especially *The Story of Dr. Dolittle*, which described an eccentric
physician who had decoded the secret language of animals and traveled to
Africa to speak to them. But after outgrowing Dr. Dolittle, the girl moved on
to Tarzan. In the household garden were many trees, and she had a special
favorite, a beech tree called Beech. In warm weather she would take a blanket
and a book up into the top branches of Beech, where she could sway in the
wind and listen to the birds while reading about Tarzan. Her fantasy was not
to have a romantic relationship with the fictional Tarzan. It was to *be* Tarzan.
That was the dream.[2]

■

Valerie Jane Morris-Goodall was born in London on April 3, 1934. Her mother, Margaret Myfanwe Joseph, or Vanne, was the daughter of an intellectual minister who preached at a Congregational church near Bournemouth, a seaside resort town in southern England. The Reverend Joseph died in 1921, leaving Vanne's mother with one boy and three girls to raise on a tiny widow's pension and some help from others in her family.[3]

Valerie Jane's father, Mortimer Herbert Morris-Goodall, was part heir to a fortune generated by the playing-card company founded by his grandfather. Mortimer worked as a technician for a telephone-cable testing company, while the playing-card inheritance provided a supplementary, if dwindling, income. He loved to drive expensive cars and became a well-known amateur racecar driver for the Aston Martin team. By the time his first daughter was one year old, the family had moved out of London and into a house in suburban Weybridge, where they lived in convenient proximity to the famous Brooklands race track.[4] Valerie Jane and her sister Judith, four years younger, were safe and comfortable during the Weybridge years, loved by a nanny, attended to, and amused by a dog and a tortoise. The paternal family's money may have contributed a sense of freedom and creative possibility. The maternal family's tradition of scholarship in the service of a Christian ministry may have added a bracing sense of discipline and moral clarity. Those two interest clusters joined in the psyche of this promising child to produce a competent and confident grownup who also happened to possess an unusual capacity for focused concentration.

That capacity first became apparent when, in the fall of 1939, she disappeared. Family and friends were enlisted in the search, followed by members of the army (who happened to be billeted nearby in preparation for a developing war in Europe). The disappearance took place in the countryside, incidentally, where Mortimer's mother and stepfather maintained a home that was surrounded by fields with cows, sheep, and horses, and also had henhouses and hens, as well as a flock of geese and a pack of dogs.

She had been missing for hours. It was getting dark. People were about to telephone the police when at last the missing child wandered serenely into view with bits of straw adhering to her clothes, explaining that she had just discovered where eggs came from. She often gathered eggs from the hen houses, and she knew that hens laid them. But how? It was a mystery she solved by hiding quietly behind a bale of straw inside the hen house for hours until a hen came in. As she recalled years later, "[The hen] was about five feet away from me, . . . and she had no idea I was there. If I moved I would spoil everything. So I stayed quite still. So did the chicken." After some time passed, the hen "raised herself from the straw. She was facing away from me and bending forward. I saw a round white object gradually protruding from the feathers between her legs. It got bigger. Suddenly she gave a little wiggle and—plop!—it landed on the straw. I had actually watched the laying of an egg." That was, the older Goodall concludes, her "first serious observation of animal behavior."[5]

Of course, no one thought of it as a serious observation at the time, and the child never imagined herself being a scientific observer. Being a scientist was not part of the childhood dream, which was purer and simpler. It was simply the dream to be intimately close to animals in the style of Tarzan with his apes in Africa. The dream, then, was an emotional-narrative complex, a private, personal story that showed a young child how she might most fully embrace her emotional self, which was centered on a deep interest in animals and nature.

The crisis happened around the same time as the henhouse episode: the autumn of 1939. That was when, after England declared war on Germany, her father enlisted in the army. Mortimer's inheritance was gone or nearly so by then, and his army pay enabled him to send home only around twenty pounds a month. Within a short time, Vanne had taken the two girls back to her mother's home in Bournemouth, a red-brick Victorian known as The Birches, where they would stay, along with Vanne's two unmarried sisters, for the duration of the war and the remainder of the girls' growing-up years.

During the war, life at The Birches was marked by severely diminished finances compounded by wartime rationing. They lived frugally: reusing envelopes, saving string, mending their clothes, patching the holes in their shoes with pieces of cardboard. The children were treated like responsible members of the family, expected to contribute, reasonably indulged without being coddled.[6] With no men around, the family organized itself into a benevolent matriarchy led by a grandmother of Victorian sensibilities who was backed up happily, if on occasion rebelliously, by her three adult daughters. In that environment, Valerie Jane was, as she later recalled, "never, ever told I couldn't do something because I was a girl."[7]

Likewise, her unusual fascination with animals was never discouraged. She watched birds, learned to identify them, trained them to perch on the sill of her bedroom window and, in good weather, to fly inside. At one point, she maintained a stable of five racing snails, which were raced on Sundays across an open stretch of church pew and at other times encouraged to exercise in the garden accompanied by three guinea pigs on leashes. There were Hamlette the hamster, Jacob the turtle, and Peter the canary.[8] There were horses—and riding lessons, which she paid for by working in the stables.

And dogs, most especially a black spaniel with a white blaze on his chest who belonged to the couple managing the hotel across the street. Rusty would dash over to The Birches the instant he was let out in the morning, barking sharply at the front door to be let in. Perhaps their greatest mutual pleasure involved going out for walks onto the cliffs and beaches of Bournemouth and enjoying the fresh air and the smells, sounds, and sights of the natural world.

■

The girl finished school, shortened her name to Jane Goodall, and entered the adult world. The childhood vision of what she wanted her adulthood to look like now had one serious obstacle: reality. Jobs were hard to find, and women faced the additional constraint of having very few obvious career choices.

Nor was university an option. The family could not afford it. And, in any case, what would she study? There were few field biologists in those days and no female ones; and the rare men who studied animal behavior were mostly parsing the acts of insects, birds, or fish in Europe. No one went to Africa. No one studied the behavior of large and possibly dangerous wild animals. Her mother remained positive, however, telling her that she could accomplish great things with steadfast determination but that she should first acquire a practical skill. Following Vanne's advice, Jane took a flat in London and enrolled at Queen's Secretarial College in South Kensington. Classes began on May 4, 1953.

A year later, secretarial degree in hand, she took a job at the Oxford University registrar's office for a year and worked for a second year at a London film studio that specialized in shorts for advertising and education. She enjoyed her time in London and was attracted to some of the men she met and dated, but she found no one who could match her own fierce emotional life, while her job at the film studio was stultifying. Then came a letter from an old school friend, Marie-Claude Mange (or Clo). Clo's father had recently bought a farm in Kenya Colony, intending to try his hand at farming. Knowing Jane's often-stated dream of going to Africa, Clo invited her to come out to Kenya and stay with the family for several months. Jane quit the film studio job in the spring of 1956, returned to live with the family in Bournemouth, and earned money as a waitress at a local hotel until at last she had saved enough to purchase a ticket to East Africa on a passenger steamship.[9] She arrived in Nairobi on the morning of her twenty-third birthday, April 3, 1957.

Being in Africa was the first part of the dream. Perhaps the one person in all of East Africa who could properly connect her to the second was Dr. Louis Seymour Bazett Leakey, the son of English missionaries, a man who had grown up with black Africans as his closest friends and playmates before enduring the English weather and food long enough to acquire a doctorate from Cambridge University. Dr. Leakey made his career as a paleoanthropologist in Africa. By 1957 he had written eight books, been elected president of the third Pan-African Congress on Prehistory, and appointed curator of the Coryndon Museum in Nairobi (now the National Museums of Kenya).[10]

Jane stayed with the Mange family for a few weeks before moving down to Nairobi, where she found a job as typist at an engineering firm, and then—on the advice of someone who knew of him by reputation—she telephoned Dr. Leakey at the Coryndon and asked for an appointment. Their meeting on the morning of Friday, May 24, was an exciting moment for both of them, during which they discovered shared interests in animals and natural history as well as congenial personal styles. By the end of the morning, "Dr. Leakey" was "Louis" and had offered her a secretarial job at the museum.[11]

Their contact that summer included work at the museum; a visit to meet his wife, Mary, at their home outside Nairobi; and several weeks camping in Olduvai Gorge, where Louis and Mary were excavating ancient fossils and artifacts from early humans and hominids. By the end of August, as their stay at Olduvai was coming to an end, Louis described his plan to Jane. He wanted her to travel to a remote forest, live for some months in a tent in a wilderness where there were wild chimpanzees, and learn as much as she could about them.[12] The fact that she had no scientific training and no credentials, he assured her, was irrelevant.

Primatology, in fact, was only beginning in 1957 to develop as a distinct discipline. It is true that many scientists were interested in studying primates by then, but they were still likely to identify themselves as psychologists, animal sociologists, physical anthropologists, zoologists, or ethologists.[13]

American psychology was just then emerging from its behaviorist phase, an extended time when many in the discipline experimented with rats and other small animals with the hope of understanding how the mechanics of animal behavior might reflect on human behavior.[14] Primates had begun to enter psychologists' laboratories during the early 1930s, however, after a young psychologist named Harry Harlow, recently hired by the University of Wisconsin, was unable to persuade that institution to give him a decent rat laboratory. Finally, accompanied by a few of his students, Harlow began walking from the campus to a nearby zoo that kept some primates—two orang-utans and a baboon—he could experiment on. Soon, however, the baboon developed a passionate attachment to one of Harlow's students, a young woman remembered as "Betty." That and other surprising experiences led the psychologist to conclude that primates possessed a psychological complexity he had never seen in laboratory rats.[15] Harlow went on to develop a series of experiments that manipulated the conditions of monkeys in cages in order to study attachment, affection, depression, and other psychological states or experiences, and in 1958 he was honored for his research by being elected president of the American Psychological Association.[16] Primates in cages had been accepted as appropriate—even ideal—models for studying human psychology.

A decade earlier, at the end of World War II, a Japanese naturalist named Kinji Imanishi created the discipline of animal sociology, promoting the close study of animal social behavior starting with that of Japanese deer, rabbits, and feral horses. In December of 1948, two of his students, while watching feral horses, happened upon a troop of Japanese monkeys, and the monkeys proved interesting enough that Imanishi and several of his students began to concentrate on them. The construction of the Japan Monkey Center in 1956 established a physical and institutional home for Imanishi's animal sociology, which was by then a distinctively Japanese primatology, recognizing primates as the ideal models for the study of human social behavior.[17]

Before the war, Harvard University's Earnest Hooton had argued, in his books *Apes, Men and Morons* and *Man's Poor Relations*, for the importance that primate field studies could have in advancing physical anthropology. Such ideas "did not fit the extant paradigm,"[18] however, until physical anthropologist Sherwood Washburn declared in a seminal essay of 1951 that his discipline should re-conceptualize itself as "the new physical anthropology" that would turn to the vision of humans as a single species and focus on the puzzle of human evolution.[19] Washburn traveled to Africa in July of 1955 to attend the third Pan-African Congress on Prehistory in Northern Rhodesia (now Zambia). After the conference was over, he stayed on for a few days. Always interested in comparative anatomy, he occupied himself by dissecting dead baboons for part of the day, then relaxing on the veranda of his hotel for another part. One of his students from that time, Irven De Vore, has recalled that the distinguished American anthropologist, while sitting on the hotel veranda with a drink in hand, became fascinated by the drama created by live baboons raiding the hotel gardens—and soon announced his interest in comparative behavior over comparative anatomy. By 1958, another Washburn student,

Phyllis Jay, had begun the first study of common langurs in India. By 1959, Washburn and De Vore were back in Africa to study baboons in Nairobi National Park. Free-ranging primates, Washburn thought, would provide ideal models for thinking about the behavioral evolution of humans.[20]

In sum, the study of primate behavior was becoming an important fashion during the 1950s, while a small community of scientists had started to recognize the emerging subdiscipline of primatology. A few of those people also understood that if primates were to be considered models for the study of human psychology, social behavior, or behavioral evolution, then, logically, a scientist would want to concentrate on that small cluster of primate species known to be evolutionarily closest to *Homo sapiens*: the great apes. But the great apes—gorillas, chimps, pygmy chimps (bonobos), and orangutans—presented an accessibility problem. Baboons you could find romping about openly in a savanna environment and watch them, often, while sitting inside a car. Great apes, living deep within their tropical forest homes, were hard to locate in the first place and notoriously elusive in the second. And then, of course, there was the danger problem. Everyone knew that apes are much larger than monkeys and many times stronger than any human researcher could ever be. Everyone also knew—from reading the reports of early hunters and explorers, and from the knowledge acquired by zookeepers and others dealing with apes in cages—that those animals were not only amazingly strong but also crafty, often hostile, and temperamentally volatile. What, on a bad day, would keep a wild ape from ripping off someone's face?

Additionally, two young researchers had already tried studying wild apes with decidedly unimpressive results, both of them sent out to the field by an advanced-thinking psychologist from Yale University named Robert Yerkes. The first was Harold C. Bingham, who, accompanied by as many as forty African assistants, trekked into the forests of eastern Belgian Congo in the summer of 1929. Because he was afraid of being attacked by gorillas, and therefore carried a gun, his occasional sightings of panicked apes were seriously interrupted by the killing of one.[21] The second expedition took place in 1930, when Henry W. Nissen went to French Guinea, West Africa, to learn about wild chimpanzees. He returned to the United States after a little more than two months and wrote up his report, which amounted to a stylistically elaborate recitation of the obvious, the mundane, and the mistaken. The mistakes included a couple of basic ones: "The chimpanzee is nomadic, having no permanent home" and "The chimpanzee group is composed of from four to fourteen animals." Nissen also failed to challenge the reigning false orthodoxy about wild chimpanzees: that they were vegetarians.[22]

In short, as much as an ambitious young scientist might have hoped during the 1950s to study great apes in the wild, it was not clear that a person actually could—although Louis Leakey certainly thought so.

■

Leakey was aware that great apes would be prize subjects for thinking about
human behavioral evolution. It is possible that he and Sherwood Washburn
met and talked about this very subject during the 1955 Pan-African Congress
on Prehistory, in which they both were important participants, and it is quite
likely that Leakey had entertained plans for a great ape project much earlier
than that. He once claimed to have sent someone out to study chimpanzees
in the mid-1940s, although the man "failed utterly."[23] We also know that in
1956 Leakey sent his previous secretary at the Coryndon, Rosalie Osborn, out
to Uganda to try habituating mountain gorillas. Osborn began sighting the
apes regularly, and she was interested in developing her early observations
into a scientific study. Unfortunately, her mother back home in Scotland,
after reading in the newspaper that her twenty-two-year-old daughter was
watching gorillas in a Ugandan forest instead of typing letters in a Nairobi
museum, put an end to all that.[24]

But Leakey was still convinced that studying great apes would result in
significant evolutionary and anthropological insights—and perhaps chimps
would be even better in that regard than gorillas. Since modern chimpanzees and modern humans share a recent common ancestor, there was the
potential for a theoretical kind of triangulation: that is to say, any important
behavioral commonalities between modern humans and modern chimps
would suggest similar behaviors among their shared ancestors who were alive
roughly six million years earlier. That was the structural logic of the puzzle he
hoped to solve, or at least to address. As someone with practical experience
in organizing wilderness expeditions in Africa, Leakey also had good reason
to believe that most wild animals will not attack unless they feel threatened.
The right person, someone who could project a calm and non-threatening
presence, and who had patience and determination, might overcome the
natural fear most wild animals have of humans and get close enough to the
supposedly dangerous apes in order to study them. Finally, Leakey believed
that the best person to do such a thing would be a woman. He liked women
and tended to fall in love with them, but he also had a more reasoned bias. He
believed that women were generally less aggressive and therefore less threatening to animals than men were, and that they were also superior observers.[25]

By virtue of a life spent in Africa and his position as curator of the Coryndon, Louis Leakey had some good ideas about where an enterprising scientist

might find great apes. In fact, he believed he knew of the perfect spot to study wild chimpanzees: a piece of forest situated on the edge of Lake Tanganyika in the British-run Tanganyika Territory. The apes had long been protected both by an extremely rugged terrain and by the local Ha people, who may well have traditionally regarded the forest as sacred territory.[26] During the last several decades, moreover, the forest had been officially protected by the German colonialists and then by the British, who designated it the Gombe Stream Chimpanzee Reserve.

Gombe Stream was where Leakey planned to send Jane Goodall in order to start a research project involving wild chimpanzees. He knew her well enough by August of 1957 to recognize that she was perfectly qualified in all the important ways. She was competent around animals of all sorts. She understood dogs intuitively and was a superb observer who had trained herself to watch many kinds of creatures. She had none of the predictable fears—of, for example, spiders, scorpions, and snakes. She liked living in a tent and camping. She had all the skills and qualities needed to survive gracefully on an extended safari under rough conditions. She had, in addition, demonstrated the focus and determination necessary to succeed at a very difficult task. And finally, she had great confidence in herself and a guiding passion and vision of what she wanted to do: a dream. That was qualification enough—in one way. In another, it was not.

While Leakey felt that having a university degree would be meaningless for such an enterprise, he also recognized that without it, Goodall would remain forever, in the view of the scientific community, an amateur. This problem became increasingly clear as, during the remainder of 1957 and into 1958, he searched in vain for funding to send his secretary out to watch chimps. Leakey at last turned to an eccentric American millionaire, Leighton Wilkie, who had already helped to fund the excavations at Olduvai. When, in early 1959, he asked Wilkie for an additional grant to support a "Chimpanzee Project" that would send two "research workers" out to Tanganyika for four months, a check for $3,000 was quickly written.[27] According to the application for that grant, Leakey's research workers would start in September of 1959.

It took nearly a year longer, partly because in July of 1959, the Leakey team uncovered at Olduvai a fossilized skull that seemed to belong to a new hominid genus and was, as Louis soon announced, "the oldest yet discovered maker of stone tools."[28] That discovery gave both Louis and Mary Leakey sudden fame, which was accompanied by the profitable opportunity for him to give a series of triumphal lectures across the United States. A second reason for the year-long delay may have been, once again, Miss Goodall's lack of academic credentials. During his lecture tour in the States, Leakey happened to meet an intelligent and eager young woman, Cathryn Hosea, who had just earned a bachelor's degree in anthropology and wanted to work in Africa. With a characteristic impulsivity, he soon offered the chimpanzee job to her. As she would recall the words years later, Louis spoke of "a young girl in my office who wants the job in the worst way. But she doesn't have the credentials."[29] To replace Goodall at that point would have been an astonishing betrayal, of course, but in any case Hosea, who had never been to Africa, eventually balked at the thought of going so far into the unknown. She turned down Leakey's offer late in the spring of 1960, at which point he informed Jane that it was time for the expedition to start.

Considered from a careerist perspective, Leakey's dithering for almost a year was serious, since by then a real race was on to see who would first complete a successful study of great apes. It might be said, indeed, that a zoology graduate student from the University of Wisconsin had just won it. Funded by the National Academy of Sciences, George Schaller and his wife, Kay, had by the summer of 1959, set up housekeeping in a cabin within the Belgian Congo's Albert National Park, where Schaller went out daily looking for gorillas. He soon found them and eventually was able to get close enough to conduct a year-long preliminary study that marked the auspicious start of a brilliant career. Schaller wrapped up his work at around the same time Goodall was unfolding her tent.[30] A few months before that, meanwhile, the Dutch ethologist Adriaan Kortlandt had discovered a large banana and papaya plantation in the eastern Belgian Congo where chimpanzees were tolerated as crop raiders, emerging frequently from an adjacent forest in order to pilfer fruit. Kortlandt built five blinds of various heights at strategic locations on the plantation, and then he spent nine weeks hiding inside the blinds each day in order to watch, film, and photograph the thieving apes. He finished at the end of June, 1960.[31]

In sum, George Schaller was the first scientist to conduct a successful year-long study of great apes, and Adriaan Kortlandt was the first to study wild chimpanzees up close for almost two months.

■

Jane Goodall arrived at Gombe on July 14, 1960, or about two weeks after Kortlandt had finished his project in the Belgian Congo. She was accompanied by her mother (an official chaperone as required by the colonial authorities) and a recently hired African cook. Aside from those two essential camp companions, however, Goodall found herself blissfully alone in what seemed like a dreamer's paradise. As she wrote home to her family in England that first week, "It is the Africa of my childhood's dreams, and I have the chance of finding out things which no one has ever known before."[32]

Within a few days she was seeing chimpanzees occasionally. Within several weeks, she was observing them regularly. By the end of October, she had discovered that wild chimpanzees eat meat, thus helping to end the established error that all primates are vegetarians.[33] By November, she had documented that chimpanzees fashion and use simple tools, thus overturning the accepted truism that humans could be defined as the tool-using species.[34] By the time she had finished her first five months at Gombe (accompanied by her mother to satisfy the requirements of the chief warden, who was concerned for her safety), she was already the world's foremost expert on wild chimpanzees—and on her way to being accepted as a top scientist in an exciting new field. At this stage, Leakey arranged for funding so that Goodall could pursue formal studies at Cambridge University, where, as a member of Newnham College, she became only the eighth person in university history to be admitted for doctoral work without first having obtained a BA or BSc degree. Under the tutorship of Robert Hinde, Goodall wrote a thesis entitled "Behaviour of the Free-Ranging Chimpanzee," largely based on her first few years of field work. After receiving a PhD in ethology in 1966, Dr. Jane Goodall stayed actively involved in the primate research at Gombe for another two decades, while the research center that she founded continues to support field studies at Gombe to this day.

Why was Jane Goodall so successful in unlocking the secrets of ape behavior, when others before her had failed or barely begun? What allowed her to see and describe chimpanzee behavior in ways that had not been done before? These questions, it seems to me, can be answered most simply through a few comparisons with the approaches of her immediate predecessors in the field, Adriaan Kortlandt and George Schaller.

Kortlandt was a bold and imaginative scientist who in later years developed the habit of unfavorably comparing Jane Goodall's way of studying chimpanzees with his own, complaining of her "parochialism" and "isolationism" and the inefficiency of her "St. Francis of Assisi approach."[35] It is certainly true that Goodall's approach was different from Kortlandt's, and perhaps the clearest difference had to do with efficiency. Out of the reasonable urge to get quick results (and also, perhaps, from the fear of being attacked), the Dutch ethologist approached his subjects from a safe place, placing them on one side of a visual barrier and himself on the other, establishing a physical and psychological remove while maintaining the stance of objective observer. Yes, hiding in blinds made some sense, and the fact that the apes were already being provisioned by fruits from the plantation meant they were relatively easy to find. At the same time, Kortlandt was forced to watch the chimps in a single place and situation (raiding fruit trees in the comforting circumstance of open land rather than closed forest) and from a limited physical perspective; as a consequence, he was unable to witness the full repertoire of their behavior both in and out of the forest or the details of their most intimate moments. That physical limitation alone meant that Kortlandt, even as late as September of 1962 (when his article "Chimpanzees in the Wild" was published in *Scientific American*) still insisted that chimpanzees were vegetarians and wondered why they did not use tools.[36]

Schaller's approach to studying the gorillas in eastern Congo was, by contrast, much less efficient than Kortlandt's. It required a substantial commitment of time and patience, as well as the willingness to take real physical risks. Schaller refused to carry a gun, believing that such a weapon and the self-assurance gained by carrying one was likely to be threatening to the apes. Although he had been instructed by the Belgian colonial government to travel with an assistant, Schaller usually asked the assistant to stay back and went by himself as he tried to approach the gorillas, reasonably presuming that two people were more threatening than one. He also made a point of moving openly, showing himself in the forest rather than trying to hide since hiding was the predictable act of a predator. And it worked. Over time, the gorillas seemed to conclude that he was not a threat. They became used to him fully enough that he could move in close. Once that happened, Schaller learned to recognize individuals, based on distinctive physical features and styles of behavior or personality; and, perhaps largely as a mnemonic device, he gave them names. Altogether, Schaller's approach can be described as *intimate immersion*, a non-threatening relationship between scientist and subject that places both in close proximity within the same physical and psychological field.

Jane Goodall's approach was also one of intimate immersion. Like Schaller, she never carried a gun or any other kind of weapon. Although she was supposed to be accompanied by an assistant, as Schaller was, she did her best to explore the forests alone; and upon finding the apes, again as Schaller had with his gorillas, she generally made a point of showing herself: moving openly rather than secretively. She also began to recognize individuals, and she gave them names partly as a mnemonic device and partly because names seemed intuitively appropriate for animals with such obvious differences in personality or character. Like Schaller's, her method of data gathering was comparatively simple and direct. She made handwritten notes, typically marking the times when some new event or behavior began and ended, and she typed up those notes every evening in a field journal that thus presented a time-marked narrative of the day's observations. And since virtually everything she saw in chimp behavior had not been seen or recorded before, virtually everything was potentially relevant, and therefore details were especially important. On September 9, 1960, for example, she had her first comparatively good observation of a chimpanzee making a night nest. She was looking through binoculars and still not close enough to identify whether the nest maker was male or female, but, she saw,

> It squatted in a leafy tree, near the top. It then rapidly pulled small leafy branches towards it, from each direction, treading on them to hold them in place. It then sat down for a moment: stood up & pulled off a branch from higher up which it incorporated into the nest. This it did 4 times, with about 1/2 minute between each picking. It then lay down, hardly visible. Another couple of minutes & it reached out & picked a very small bunch of leaves which it appeared to place under its head. Then it stretched right out so that its feet projected beyond the structure of the nest.[37]

George Schaller might be conceptualized as a dreamer like Jane Goodall, someone who followed his own Tarzan sort of dream to study apes. However, I would regard him more as an inspired pragmatist. He was a trained zoologist who, working in a discipline that provided no obvious models for studying large and dangerous wild animals of high intelligence, pragmatically followed common sense and a bit of intuition. And the fact that Jane independently developed a similar technique for studying a similar species was testimony to the good sense she used in following her own intuition as well as her childhood experiences of watching and sometimes trying to tame wild animals. But George Schaller was a scientist whose primary ambition was to do zoological science, not to live among the animals. He might easily have spent his time studying any other species—lions, for example, or birds or giant pandas or sea otters—which indeed he went on to do. Jane Goodall was a dreamer whose primary ambition was to be close to wild animals, to live among them, and since science gave her a way to do that, she went through the necessary training to become accepted as a scientist.

The difference between the two ultimately might be considered a matter of intensity and commitment, which can in turn be measured in time. Schaller stayed with the gorillas of eastern Congo for a year. That was the length of his study and thus commitment. Goodall worked with and regularly lived among the chimpanzees of Gombe for twenty-five years before turning her research station over to other scientists who continued the work. And the longevity of the study itself should be considered an important aspect of her method and mark of her contribution. She learned many important things about the lives and society of chimpanzees during that time that no one dreamed of knowing after the first year or two, five, or even ten years.

Jane Goodall was a dreamer who demonstrated that a person can break through many of the ordinary barriers that separate humans from nonhumans; she also showed that the best research requires an open mind, a flexible approach, and a very long-term commitment. Because she opened a window onto the lives of humanity's closest relatives, her discoveries have contributed to a growing sense of ourselves as part of the natural world and thus potentially sharing social, behavioral, emotional, and cognitive commonalities with a broad swath of the animal kingdom.[38] Over time, she became recognized as one of the great pioneers in primatology, and she stands today as one of the two or three most celebrated women scientists in history.

FURTHER READING

Goodall Jane. *The Chimpanzees of Gombe: Patterns of Behavior*. Cambridge, MA: Harvard University Press, 1986.

———. *My Life with the Chimpanzees*. New York: Pocket Books/Simon and Schuster, 1988.

———. *In the Shadow of Man*. Boston: Houghton Mifflin, 1971.

Kortlandt, Adriaan. "Chimpanzees in the Wild." *Scientific American* 206, no. 5 (1962).

Morell, Virginia. *Ancestral Passions: The Leakey Family and the Quest for Humankind's Beginnings*. New York: Simon and Schuster, 1995.

Peterson, Dale. *Jane Goodall: The Woman Who Redefined Man*. Boston: Houghton Mifflin, 2006.

Schaller, George B. *The Year of the Gorilla*. Chicago: University of Chicago Press, 1964.

Strum, Shirley C., and Linda Marie Fedigan, eds. *Primate Encounters: Models of Science, Gender, and Society*. Chicago: University of Chicago Press, 2000.

NOTES

1. "Tarzan of the Apes," *Wikipedia*, online at https://en.wikipedia.org/wiki/Tarzan _of_the_Apes.

2. Dale Peterson, *Jane Goodall: The Woman Who Redefined Man* (Boston: Houghton Mifflin, 2006), 38, 46.

3. Peterson, *Jane Goodall*, 29–31.

4. Peterson, *Jane Goodall*, 3–11.

5. Jane Goodall, *My Life with the Chimpanzees* (New York: Pocket Books/Simon and Schuster, 1988), 1, 2.

6. Jane Goodall and Philip Berman, *Reason for Hope: A Spiritual Journey* (New York: Warner, 1999).

7. Peterson, *Jane Goodall*, 29, 30.

8. Peterson, *Jane Goodall*, 39.

9. Peterson, *Jane Goodall*, 67–91.

10. Sonia Cole, *Leakey's Luck: The Life of Louis Seymour Bazett Leakey, 1903–1972* (New York: Harcourt Brace Jovanovich, 1975); Virginia Morell, *Ancestral Passions: The Leakey Family and the Quest for Humankind's Beginnings* (New York: Simon and Schuster, 1995).

11. Peterson, *Jane Goodall*, 100–102.

12. Peterson, *Jane Goodall*, 117, 18; Jane Goodall, *Africa in My Blood: An Autobiography in Letters: The Early Years* (New York: Houghton Mifflin Harcourt, 2000), 114.

13. Shirley C. Strum and Linda M. Fedigan, "Changing Views of Primate Society: A Situated North American View," in *Primate Encounters: Models of Science, Gender, and Society*, ed. Shirley C. Strum and Linda Marie Fedigan (Chicago: University of Chicago Press, 2000), 3–49; Thelma Rowell, "A Few Peculiar Primates," in ibid., 57–70; Alison Jolly, "The Bad Old Days of Primatology?," in ibid., 71–84.

14. For an engaging discussion of this complex topic, see Deborah Blum, *Love at Goon Park: Harry Harlow and the Science of Affection* (New York: Basic Books, 2002), 61–73.

15. Blum, *Love at Goon Park*, 78.

16. Blum, *Love at Goon Park*, 170.

17. Personal communications from Michael Huffman, Takayoshi Kano, and Toshisada Nishida. See Hiroyuki Takasaki, "Traditions in the Kyoto School of Field Primatology in Japan," in Strum and Fedigan, *Primate Encounters*, 85–103.

18. Robert L. Sussman, "Piltdown Man: The Father of American Field Primatology," in Strum and Fedigan, *Primate Encounters*, 89.

19. Sherwood Washburn, "The New Physical Anthropology," in *The New Physical Anthropology: Science, Humanism, and Critical Reflection*, ed. Shirley C. Strum, Donald G. Linburg, and David Hamburg (Upper Saddle River, NJ: Prentice Hall, 1951), 1–5.

20. Personal communication from Irven De Vore.

21. Harold C. Bingham, *Gorillas in a Native Habitat* (Washington, DC: Carnegie Institution, 1932).

22. Henry W. Nissen, "A Field Study of the Chimpanzee," in *Comparative Psychology Monographs* 8 (1931–32): 13, 25, 73.

23. Morell, *Ancestral Passions*, 239.

24. Peterson, *Jane Goodall*, 118, 119.

25. Donna Haraway famously examined how female and male researchers have differed in their observations of apes in *Primate Visions: Gender, Race, and Nature in the World of Modern Science* (New York: Routledge, 1990), and gender and primatology has been a blossoming field.

26. Michele Wagner, "Nature in the Mind in Nineteenth- and Early Twentieth-Century Buha, Tanzania," in *Custodians of the Land: Ecology and Culture in the History of Tanzania*, ed. Gregory Maddox, James L. Giblin, and Isaria N. Kimambo (Athens: Ohio University Press, 1996): 175–99.

27. Peterson, *Jane Goodall*, 151–55.

28. L. S. B. Leakey, "A New Fossil Skull from Olduvai," *Nature* (August 1959): 493.

29. Peterson, *Jane Goodall*, 160.

30. George B. Schaller, *The Year of the Gorilla* (Chicago: University of Chicago Press, 1964).

31. Adriaan Kortlandt, "Chimpanzees in the Wild," *Scientific American* 206, no. 5 (1962): 128–38.

32. Peterson, *Jane Goodall*, 179.

33. Peterson, *Jane Goodall*, 206–7; Solly Zuckerman, *The Social Life of Monkeys and Apes* (1932; repr., London: Routledge and Kegan Paul, 1981).

34. Peterson, *Jane Goodall*, 207–11.

35. Adriaan Kortlandt, "Some Comments on American Teaching Programs in Primatology and Evolutionary Anthropology," a circulated preliminary draft (10 March 1998).

36. Kortlandt, "Chimpanzees in the Wild."

37. Peterson, *Jane Goodall*, 199.

38. See, for example, Donald R. Griffin, *Animal Thinking* (Cambridge, MA: Harvard University Press, 1984); Jaak Panksepp, *Affective Neuroscience: The Foundations of Human and Animal Emotions* (Oxford: Oxford University Press, 1998); Frans B. M. de Waal and Peter L. Tyack, eds., *Animal Social Complexity: Intelligence, Culture, and Individualized Societies* (Cambridge, MA: Harvard University Press, 2003); Marc Bekoff, Colin Allen, and Gordon M. Burghardt, eds., *The Cognitive Animal: Empirical and Theoretical Perspectives on Animal Cognition* (Cambridge, MA: MIT Press, 2002).

14

FRANCIS CRICK AND THE
PROBLEM OF CONSCIOUSNESS

Born to a boot maker near Northampton, Francis Crick's prewar scientific work was in physics—in particular, fluid viscosity. During the war, Crick worked on naval mines, and developed clever ways for mines to overcome anti-mine measures.[1] But "at the end of the war" Crick tells us, "I found my thoughts turning increasingly towards biology. Two major problems fascinated me—the distinction between living things and non-living matter, and the distinction between self-conscious animals and machines."[2] That he was drawn to these two problems in particular is not an accident. Crick once advised that "in approaching a new discipline it is a useful exercise to attempt to separate those topics that, although far from being understood, appear at least capable of explanation by familiar approaches of one kind or another from those for which no ready explanation, even in outline, seems available at the present time."[3] Crick, like many others, took the "familiar approaches" to be the approaches defined by mathematical physics and chemistry. And as late as the end of the nineteenth century, two major phenomena were not in the category of phenomena that seemed capable of explanation in terms of mathematical physics and chemistry: life and consciousness—precisely the two to which Crick found himself drawn.

As to rendering life in terms of chemistry and physics, Watson and Crick's work on the double-helix structure of DNA was the headline grabber, but the muscle behind the headline was the development of the entire field of molecular biology—the theory, techniques, equipment, and social/institutional structures, all of which Crick was centrally involved with constructing as well.

It speaks to Crick's level of intellectual boldness that, having contributed to research that fostered our understanding of life in terms of nonliving matter, he turned his attention to the second of the two big problems, the nature of consciousness. And, as it happens, his overall approach in this case was quite similar to his approach to molecular biology.

Crick's significant change of research focus went along with a significant geographical change. While his work in molecular biology took place largely while he was in the UK at the Cavendish Laboratory at Cambridge, in 1976 Crick moved to the Salk Institute of Biological Studies (where he had been a nonresident fellow since 1960) in La Jolla California. There he fully reori-

Figure 14.1. Francis Crick in his office. Behind him is a model of the human brain that he inherited from Jacob Bronowski. Photo: Marc Lieberman, https://commons.wikimedia.org/wiki/File:Francis_Crick.png.

ented his research to neuroscience—in particular, the mechanisms of consciousness.

I do not provide here anything like an overview of Crick's approach to the study of consciousness. There are many excellent sources for such overviews, including Crick's own.[4] Rather, I want to explain what I take to be visionary about Crick's approach. I first describe the large-scale structure of the approach Crick took to the study of both life and consciousness. That approach, while bold in a certain way, is also fairly conservative in key respects. So, what set Crick apart was not the overall structure of his approach. I then address a common view as to what made Crick's approach bold and visionary—the fact that he took consciousness to be his target. Though this was indeed his target, he was not as much of a maverick in this regard as many (including Crick himself) thought. Consciousness was a topic that was widely addressed in scientific and philosophical circles. Rather, as I show in the final part of this essay, what genuinely made Crick a visionary was his ability to see that the conceptual hardware that was ubiquitous at the time was not suitable for the task; instead, his exploration and development of alternate frameworks proved more enlightening.

But first we turn to the overall structure of Crick's approach to the big questions of life and consciousness. Though Crick's choice of questions in both cases was quite bold, his approach to both was markedly conservative. What I mean is this. Questions that resist being answered in terms of a familiar framework (such as that provided by physics and chemistry) can be approached in a number of ways. First, one might just christen a special branch of proprietary ontology for the recalcitrant phenomenon. We can call such approaches *dualist*. Second, one might explain the phenomenon as the result of a functional or mechanistic structure (or just collective properties) of entities and processes provided by the familiar, typically physical, ontology. We can call this approach, very broadly, *mechanist/functionalist*. Third, one can deny that there is such a phenomenon, initial convictions notwithstanding. This can be called *eliminativist*. Finally, one might take the phenomenon to be a relational feature. We can call this approach, broadly, *relationalism*. Examples of all four approaches are common. Cartesian dualism and various forms of vitalism would be examples of dualistic approaches. Functionalism in philosophy of mind that identifies mental states only through their functional roles, and the kinetic theory of heat would be examples of mechanist/functionalist approaches. Eliminativism in the philosophy of mind, and perhaps moral nihilism, would be examples of eliminativist approaches that deny entire categories of entities. And finally Dennett's "intentional stance,"[5] and the relational theory of color would be examples of relationalism.

The approach I am here calling mechanist/functionalist is the most conservative in two respects. First, unlike dualism, it does not stray beyond highly defensible ontological boundaries. It is conservative in a straightforward, ontological sense. Second, unlike eliminativism and relationalism, it need not involve any significant measure of conceptual realignment. The eliminativist's claims that there are no beliefs (or no consciousness, or no moral properties) are conceptually revisionist to say the least. Ditto for a relationalist approach, according to which my having beliefs is no more than someone attributing beliefs to me. In both of these cases our pre-theoretic understanding of the phenomenon is sharply challenged in one way or another. But the mechanist/functionalist approach is, typically anyway, conservative in both these respects—it attempts to explain the puzzling phenomenon as it is more or less already understood, in terms of entities and processes whose ontological bona fides are not in dispute.

Now while this approach is conceptually conservative, it is of course pragmatically bold. If it were easy to understand the recalcitrant phenomenon in terms of the familiar and unquestioned ontology, there wouldn't be a problem to begin with. One is signing up for a long haul of difficult work in uncertain conditions by taking this approach. Setting oneself the task of explaining phenomena such as life and consciousness in terms of a physical ontology is bold in the same way that setting oneself the task of swimming the English Channel is bold. It is conservative in the sense that it's conceptually straightforward, and plenty of people have thought of it, but it is unquestionably a pragmatically bold undertaking nevertheless.

We can see that Crick's approach to understanding the natures of both life and consciousness are examples of the mechanist/functionalist approach. In the case of life, he was positing no conceptual reorientation with respect to what life was. He understood it, like most everyone else, to be something intrinsic to the things we pre-theoretically understood to be living things. And the extension of the term *living thing* was more or less as anyone took it to be. He was neither eliminativist nor relationalist. He was not, like the vitalists, positing any special entity (e.g., *elan vitale*) that was not among the items usually recognized as part of the physical world.[6] Analogously, with respect to the problem of consciousness, Crick's approach did not promote any major conceptual reorientation concerning consciousness of the sort that, for example, Dennett does,[7] nor was he an eliminativist with respect to consciousness. Moreover, Crick did not promote any augmentation of reality as currently understood by the physical sciences, as did John Eccles for instance.[8]

In any case, if we are looking for what is visionary in Crick's approach to consciousness, we have to dig deeper. For, though bold, there is nothing groundbreaking about adopting the most conservative strategy possible to understanding a puzzling phenomenon.

The outline of Crick's mechanist/functionalist approach starts with an analysis of the puzzling phenomenon. Concepts used for phenomena such as life or consciousness tend to be less precise than those involved in physics and chemistry, and Crick saw the imprecision as a barrier to progress. The point of the analysis is to provide at least some initial conceptual traction. Ideally, it takes the form of some sort of functional analysis. In the case of life, a preexisting focus on *trait inheritance* was enough. For a nice, historically important, example of the reasoning behind understanding the mechanisms of *inheritance* as the key to understanding the physical mechanisms of *life*, see Erwin Schrödinger's *What Is Life?*[9]

The basic folk understanding of consciousness is perhaps even more vague than that of life. And while he warned against premature definitions, Crick quite rightly worried that characterizing the phenomenon of interest by the term *consciousness* would render any hope of progress dim. The meaning of the expression is vague at best, and quite variable from field to field and researcher to researcher. The other side of this, of course, is that any attempted "clarification" of the concept will almost certainly leave some people dissatisfied, since they will feel that it omits one or another feature or nuance they deem critical—even if in some inchoate way it is the vagueness itself whose removal triggers the dissatisfaction. Unlike the case of life, there was no de facto analysis of the key components of consciousness such that if those components were understood in terms of a physical mechanism, there would be broad agreement that the phenomenon would be approachable in terms of a physical ontology.

Note that this is *not* because nobody was interested in consciousness. A common misunderstanding about Crick's investigations into consciousness is that, when he began his investigations, the topic of consciousness was taboo. Indeed, as late as 1990, Crick and his collaborator Christof Koch state that "it is remarkable that most of the work in both cognitive science and the neurosciences makes no reference to consciousness (or 'awareness')."[10] The belief that the topic was generally avoided made possible the idea that a large part of what made Crick's approach bold and groundbreaking was his willingness to dive headfirst into a topic that was shunned in polite scientific company.

The problem is that this isn't quite correct. Or, at minimum, it is misleading. True, there was a sense in which behaviorists distanced themselves from "consciousness" talk. And while, at the time of Crick's turn to consciousness, behaviorism was being challenged, it was still a significant force in the relevant scientific circles. We can see evidence of this in Eric Kandel's 727-page *Cellular Basis of Behavior*, which has nearly half a page of index entries for *behavior*, and fifteen subcategories of index entries for *conditioning*, but not a single entry for *consciousness*.[11]—not to mention the presence of the word *behavior* in the title. So there is no denying that, for at least some researchers, the topic of consciousness was shunned to some extent or other.

Nevertheless, consciousness was far from being a *quaestio non grata* at the time. It was commonplace for behaviorists to provide an analysis of "consciousness" to yield a plausible reduction in terms of processes that were co-pacetic from the behavioristic point of view. To take a few examples, Calvin Hall (in a 1960 psychology textbook very much in the behaviorist tradition) says, "A person is rarely, if ever, fully conscious of everything taking place. We might liken being conscious to a spotlight on a darkened stage. Although the spotlight illuminates only a small section of the stage and leaves the rest in darkness, it may be moved around to illuminate other parts of the stage."[12] And Robert Silverman (in a 1978 undergraduate psychology textbook) has an entire chapter on consciousness, and also links it to, *inter alia*, attention: "Consciousness in all its various forms and functions also allows selective attention to the most important parts of our environment. We filter out infor-mation that is distracting or irrelevant. We do not respond to every stimulus around us; we respond to the stimuli we need to respond to."[13] And not just psychologists—including those flying the behaviorists' flag—but many neu-roscientists as well were very interested in consciousness. For example, the *Handbook of Clinical Neurology* (published in 1969) includes an entire chap-ter on consciousness, and many of the other chapters discuss consciousness openly and with no overt signs of shame.[14]

So it is simply not true, Crick's own protestations to the contrary, that by taking consciousness as the topic of study, he was somehow bucking a unani-mous trend.[15] This isn't to say that Crick wasn't groundbreaking. It is just to point out that it was not *this particular ground* that was being broken.

In 1979, after claiming (not entirely accurately) that contemporary psy-chologists ignored the problem, Crick looks back to late nineteenth-century psychology, including William James, for inspiration and finds three "basic ideas":[16]

1. Not all operations of the brain correspond to consciousness.
2. Consciousness involves some form of memory, probably a very short-term one.
3. Consciousness is closely associated with attention.

These points are soon after reemphasized by comparison to the theories of more contemporary cognitive psychologists. And then the phenomenon of interest gets narrowed down even more by focusing on vision, because "it seems more accessible to direct experiment."[17] The result is an analysis of the vague concept of consciousness in terms of a focus on vision—in particular, visual attention and its connection to short-term working memory.

Note that Crick's starting point here is actually quite similar to the start-ing point adopted in the behaviorist psychology textbooks that I quoted from above. Hall was so behaviorist that he identified consciousness as a type of behavior before going on to refine his position to be that the sort of behavior in question is *attention*—precisely Crick's third point above. Hall also pointed out that much of what happens in the person, and the person's nervous system, was not available to consciousness—Crick's first point above.[18]

This is notable for the following reason. If any group would have been responsible for making consciousness a taboo topic, it would have been the behaviorists. But it turns out that many of them not only dealt with the topic, but dealt with it in exactly the same way Crick did—by analyzing it as something amenable to a certain kind of investigation and, indeed, with a generally similar analysis. But behaviorists were the ones being criticized for *ignoring* consciousness! Whether or not that charge is accurate, it remains true that if these sorts of behaviorist analyses of consciousness counted as an attempt to dodge the issue, then Crick's analysis would have to count as just such a dodge.[19]

Now, obviously, Crick's *investigations* of consciousness were quite unlike those of behaviorists, even if his *analysis* of consciousness would have been accepted by many of them. The difference is in what happens with the conceptual analysis. Having isolated attention and working memory (for example) as the hallmarks of consciousness, behaviorists then turn to understanding these components in behavioristic terms. Crick wanted to understand them in terms of operations of the brain.

But while the focus on the brain set Crick apart from behaviorists, it did not set him apart from many other psychologists and neuroscientists. There were many who were quite interested in consciousness and relating it to brain operations and structures. In fact, there were those who were identifying some of the same brain areas as central to consciousness as those Crick eventually identified. For instance, in his 1972 textbook *Introduction to Neuroscience*, Jeff Minckler claims that the thalamus is crucial for consciousness and provides some reason for thinking so.[20] And Frederiks mentions the reticular formation in particular.[21] These are among the very regions that Crick would converge upon.

So again, we are driven to dig yet deeper to find what is groundbreaking about Crick's approach. Perhaps the most insightful part of Crick's approach—something worthy of the appellation *visionary*—was his particular choice of modeling frameworks for understanding what neural hardware was doing. What I mean here is the following. Whatever one's functional breakdown might be for some mental phenomenon, the neural basis of these functions is going to be large and complex. The idea that one could just describe the low-level physical properties and causal interactions of the various neurons and hope to provide anything revealing concerning the high-level mental phenomenon is hopeless. Just try understanding the physical workings of a laptop computer without making use of the conceptual resources of computer science! But it is possible to take one or another descriptive framework, typically combined with some mathematical or formal machinery, and make some headway.

It is here—the adoption of an explanatory framework—that contracting an iatrogenetic conceptual disease is a near certainty. There is a marked tendency to unreflectively identify a phenomenon of interest with the current favored set of tools for addressing that phenomenon.[22] If those tools aren't the right ones, they can end up doing more harm than good. New people entering the field can't help but think that the responsible thing to do is to get up to speed on the current accepted framework, but in doing so they run the risk of blinding themselves to the phenomenon of interest as understood apart from those tools. Crick was explicitly aware of the danger of just adopting whatever framework others seemed to be using.[23]

When it comes to psychological phenomena, a number of frameworks were in wide use when Crick was initiating his investigations. These include what has come to be known as good old-fashioned AI, as well as dynamical systems theory, and a handful of similar others. Crick managed to keep his eye on the ball and not be swayed by the dynamical systems theorists or the "computer metaphor" computationalists. He first articulated a plausible understanding of the problem the brain was solving, at least in terms of those operations that contribute to perceptual consciousness. And this was something like using the sensory inputs to construct a symbolic representation of the environment.[24] This understanding of the overall purpose of consciousness is not itself a framework. But having a description of the overall point of consciousness can help in the search for a suitable framework.

Extended meetings with researchers from a wide range of approaches were characteristic of Crick's method. He thrived on extended in-person discussions and was open to learning from just about anyone. In April of 1979 Crick arranged for David Marr and Tomaso Poggio to visit him in La Jolla. Both Marr and Poggio were using sophisticated mathematical tools for understanding neural phenomena. Crick immediately recognized that the mathematical framework that Marr and Poggio were using to model low-level neural functions—tools mainly known to and used by engineers and mathematicians—were suitable for his own investigations into consciousness. Crick called it "communication theory" but today in many circles it would more felicitously be referred to as *signal processing*. Crick collaborated with Marr and Poggio on an article that made the basic conceptual points.

This, finally, is where Crick's vision is manifest. Though he was interested in a high-level phenomenon—consciousness—he wasn't blinded by the frameworks that were being used by cognitive neuroscience at the time. Rather, as soon as Marr and Poggio (who were focused on much lower-level phenomena in visual processing) described the tools they used, he recognized that they held the key for approaching the higher-level phenomena.[25]

Before continuing, I should at this point mention Christof Koch. It is undeniable that Koch was central to Crick's investigations into the neural basis of consciousness, certainly more so than anyone else after 1980. Their close friendship and dozens of coauthored works speak to that relationship. But I am leaving Koch out of the discussion on the specific topic of what makes Crick a visionary because the key moves in Crick's visionary approach were made before he met Koch. Indeed, it is because Crick had already realized that he would be exploring the tools from "communication theory" in his search for the biological basis of consciousness that, in 1980, he visited Poggio in Tübingen, where he met Koch. At the time, Koch was a student of Poggio and Braitenberg, working on modeling electrical features of dendritic spines. So although Koch had unparalleled influence on the specifics of Crick's work on consciousness, the fact that Crick worked with Koch was an *effect* of, rather than a *cause* of, what I am identifying as Crick's visionary insight.

Crick's key visionary move—his ability to not be seduced by the frameworks that were currently being employed by most experts in the field and to recognize that a different framework would be more helpful—is more difficult than it might sound. It always sounds easy in retrospect. But if everyone could recognize that the current framework for understanding a phenomenon was a dead end and go on to find the correct framework, then we'd all have Nobel Prizes. Not only is it just a psychologically difficult thing to do, it comes at a high practical cost. One immediately loses the moorings that are provided by the dominant approach. But for that very reason, it puts one in a position to create new moorings.

It has taken a couple of decades, but the sorts of modeling tools and conceptual resources that Crick was drawn to and championed are now commonplace in the field. While his interest in consciousness may not have been trend-bucking, as is commonly supposed, and the record is mixed and unclear on the specifics of his proposals concerning the neural bases of various elements of consciousness, his recognition of what would be the more useful framework was visionary and helped to make a significant and lasting impact in the field by moving it in the right direction.

FURTHER READING

Crick, Francis. "Thinking about the Brain." *Scientific American* 241, no. 3 (1979): 219–33.

Crick, Francis, David Marr, and Tomaso Poggio. "An Information Processing Approach to Understanding the Visual System." MIT AI Laboratory Memo #557, April 1980.

Crick, Francis, and Christof Koch. "Towards a Neurobiological Theory of Consciousness." *Seminars in the Neurosciences* 2 (1990): 263–75. ISSN 1044-5765.

Dennett, Daniel. *The Intentional Stance*. Cambridge, MA: MIT Press, 1989.

Eccles, John. *Facing Reality: Philosophical Adventures by a Brain Scientist*. Berlin: Springer, 1970.

Fredericks, J. A. M. "Consciousness." In *Disorders of Higher Nervous Activity*, ed. P. J. Vinken and G. W. Bruyn, 48–61. Vol. 3 of *Handbook of Clinical Neurology*. Amsterdam: Elsevier, 1969.

Olby, Robert. *Francis Crick: Hunter of Life's Secrets*. Cold Spring Harbor, NY: Cold Spring Harbor Press, 2009.

Schrödinger, Erwin. *What Is Life?* New York: Macmillan, 1944.

NOTES

1. Robert Olby, *Francis Crick: Hunter of Life's Secrets* (Cold Spring Harbor, NY: Cold Spring Harbor Press, 2009).

2. Francis H. C. Crick, letter to Jonas Salk, Jacques Monod, Mel Cohn, and Ed Lennox, 1962, Francis Crick Papers, University of California, San Diego.

3. Francis H. C. Crick, "Thinking about the Brain," *Scientific American* 241, no. 3 (1979): 219.

4. Francis H. C. Crick, *The Astonishing Hypothesis: The Scientific Search for the Soul* (New York: Scribner, 1995).

5. Daniel Dennett, *The Intentional Stance* (Cambridge, MA: MIT Press, 1989).

6. See William Bechtel and Robert C. Richardson, "Vitalism," in *Routledge Encyclopedia of Philosophy*, ed. E. Craig (London: Routledge, 1998).

7. Dennett, *Intentional Stance*. See also his "Who's on First? Heterophenomenology Explained," *Journal of Consciousness Studies* 10 (2003): 19–30.

8. John C. Eccles, *Facing Reality: Philosophical Adventures by a Brain Scientist* (Berlin: Springer-Verlag, 1970).

9. See Erwin Schrödinger, *What Is Life?* (New York: Macmillan, 1944).

10. Francis H. C. Crick and Christof Koch, "Towards a Neurobiological Theory of Consciousness," *Seminars in the Neurosciences* 2 (1990): 263. See also Crick, *Astonishing Hypothesis*, at the beginning of which Crick includes as an epigraph the following quotation from John Searle: "As recently as a few years ago, if one raised the subject of consciousness in cognitive science discussions, it was generally regarded as a form of bad taste, and graduate students, who are always attuned to the social mores of their disciplines, would roll their eyes at the ceiling and assume expressions of mild disgust" (vii).

11. Eric R. Kandel, *Cellular Basis of Behavior: An Introduction to Behavioral Neurobiology* (San Francisco: W. H. Freeman, 1976).

12. Calvin S. Hall, *Psychology: An Introductory Textbook* (Cleveland, OH: H. Allen, 1960), 56. Hall opens his discussion of consciousness this way: "In order to approach the question of consciousness with some clarity, it will be helpful to make one simple assumption—that being conscious is itself a form of behavior" (55).

13. Robert E. Silverman, *Psychology* (Upper Saddle River, NJ: Prentice-Hall, 1978), 261.

14. See, for example, J. A. M. Fredericks, "Consciousness," in *The Handbook of Clinical Neurology*, vol. 3, *Disorders of Higher Nervous Activity*, ed. P. Vinken and G. Bruyn (Amsterdam: Elsevier, 1969).

15. See Crick, *Astonishing Hypothesis*, 13: "The majority of modern psychologists omit any mention of [consciousness], although much of what they study enters into consciousness. Most modern neuroscientists ignore it."

16. Crick, *Astonishing Hypothesis*, 15.

17. Crick, "Thinking about the Brain," 219.

18. Hall, *Psychology: An Introductory Textbook*.

19. My own view is that this is an interesting issue, but one that can be safely ignored for present purposes. The fact is that Crick was clear about what he meant by *consciousness* in his endeavor to understand its neural basis. And if we are in the position of reflecting on the contributions Crick made, we can do that while understanding that those contributions were intended to be of a certain, well-defined sort. The issue of whether or not the analysis of consciousness that he provided was the best one, or the right one, is a somewhat orthogonal issue.

20. Jeff Minckler, *Introduction to Neuroscience* (St. Louis: C. V. Mosby, 1972), 350.

21. Fredericks, "Consciousness," 49.

22. For instance, I've known a good many linguists who simply and pre-reflectively *equate* the study of language structure with tools from the generative tradition, to the point of not even being able to see anything that doesn't employ those tools as concerning the subject matter. It is clear from their approach and behavior that according to them, language structure *just is* (=) some member of the family of generative syntax theories.

23. Crick and Koch, "Neurobiological Theory of Consciousness," 264–65.

24. Crick, *Astonishing Hypothesis*, chap. 3.

25. See Francis H. C. Crick, Davis Marr, and Tomaso Poggio, "An Information Processing Approach to Understanding the Visual Cortex," Massachusetts Institute of Technology Artificial Intelligence Laboratory, A. I. Memo No. 557 (April 1980): esp. 9.

MARK E. BORRELLO

DAVID SLOAN WILSON
VISIONARY, IDEALIST, IDEOLOGUE

*I see things differently. For me, science is a medium for
listening and reflecting on the human condition, much
like religion and literature.*
—David Sloan Wilson, The Neighborhood Project, *2011*

David Sloan Wilson dreams of a Darwinian city. A professor
of evolutionary biology and anthropology at the University of Binghamton in
New York for the past three decades, he has undertaken to apply the Darwinian paradigm to reimagine his city and to engage in a reformation based on
these principles. Wilson, born the son of the novelist Sloan Wilson, author of
the mid-century classic novel *The Man in the Gray Flannel Suit*, has spent the
last decade stretching the bounds of evolutionary theory. He calls himself an
evolutionist, a theorist and practitioner, not bound by the traditional limits of
evolutionary biology, which, for many practitioners, focuses primarily on non-
human organisms, analyzing specific elements of particular traits. Instead,
Wilson is drawn to a much broader image of this science. The evolutionary
paradigm invites Wilson and his collaborators to explore the complexities
of human psychology and social relations in an attempt to use evolutionary
principles to guide us to better outcomes. As he describes it in the introduction to *The Neighborhood Project*, his recent book on this subject, "Science and
evolutionary theory can clarify what it means to have a soul in addition to
a body. Bodies and souls can transcend the skins of single individuals, making us part of something larger than ourselves. A city can have a body and a
soul, for example. . . . And then, if evolutionary theory can be used to *understand* the human condition, it can also be used to *improve* it."[1]

DREAMS IN THEORY
One could argue that Wilson's dream, *The Neighborhood Project*, was
a long time coming, and was born of his fascination with theory. After graduating with high honors from the University of Rochester in 1971, Wilson began graduate studies in evolutionary biology at Michigan State University
and completed his dissertation under the guidance of Don Hall in 1975. The
dissertation is famous among graduate students for being one of the shortest

Figure 15.1. David Sloan Wilson, 2006.

ever submitted to the university, just twelve pages long. According to legend, when the dissertation was returned to Wilson by the university as too short for binding, his advisor asked, "How many blank pages should he send over so that it could be bound?"[2]

Wilson's paper was subsequently published in the *Proceedings of the National Academy of Sciences* as "A Theory of Group Selection."[3] This paper represents the first element of the working out of Wilson's dream. Wilson had been interested in a question that had challenged Darwin himself—one that the great Victorian had left somewhat open. The question was, At what level does the mechanism of natural selection work? For most of the nineteenth century, the focus of evolutionary thinking was on the organism (and less often, the trait). With the rediscovery of Mendel's work at the turn of the century and the development of genetics and then population genetics in the 1920s and 1930s, the focus had shifted to genes and the frequency of genes in populations. Essentially, the measure of the effects of evolution was taken at the genetic level, and the fitness of individuals was the coin of the realm. That is, the value of a trait was assessed in terms of how it benefited the individual that possessed it, be it big horns or gaudy plumage. The alternative to this position was later to be known as group selection, but already in the *Origin of Species* and later in *The Descent of Man*, Darwin had written of processes that he called "community selection." He invoked this idea to address the challenge of neuter castes of social insects and extended the idea to include the evolution of moral sentiments in humans. In *Descent*, for example, Darwin famously argued,

> It must not be forgotten that although a high standard of morality gives but a slight or no advantage to each individual man and his children over the other men of the same tribe, yet that an advancement in the standard of morality and an increase in the number of well-endowed men will certainly give an immense advantage to one tribe over another. There can be no doubt that a tribe including many members who, from possessing in a high degree the spirit of patriotism, fidelity, obedience, courage, and sympathy, were always ready to give aid to each other and to sacrifice themselves for the common good, would be victorious over most other tribes; and this would be natural selection.[4]

Wilson's graduate paper provided a mathematical model that demonstrated that "the traditional concepts of group and individual selection appear to be two extremes of a continuum, with systems in nature operating in the interval in between."[5] This apparently unaffected conclusion belies a dramatic debate in evolutionary biology at the forefront of which Wilson had now placed himself. From this vantage point in the theoretical vanguard, Wilson would develop a career that would culminate in dreams of a Darwinian city in a Darwinian world. For Wilson, given his theoretical bent and his development of group selection theory in particular, this world would not be the nasty, brutish, and short version famously described by Thomas Hobbes, nor the hypercompetitive, laissez-faire world of the Victorian philosopher Herbert Spencer. The Wilsonian world is rather the world of the entangled bank described in Darwin's *Origin*: "It is interesting to contemplate an entangled bank clothed with many plants of many kinds, with birds singing on the bushes, with various insects flitting about, and with worms crawling through the damp earth, and to reflect that these elaborately constructed forms, so different from each other, and dependent on each other in so complex a manner, have been produced by laws acting around us."[6] This is a world where the factors that contribute to human culture, be they economic, sociological, religious, or political, can be put into the Darwinian machinery, subjected to the analysis of the Darwinian thinker, and out comes the Darwinian product: a harmonious city with minimal poverty, dynamic politics, and a peaceful balance between all members of the polity.

Wilson's dream is not brand new. Herbert Spencer envisioned a world where "survival of the fittest" would lead to a cosmic equilibrium that was warmly embraced by American oligarchs, such as Andrew Carnegie and John D. Rockefeller at the turn of the twentieth century and eugenicists, such as Francis Galton (Darwin's first cousin) and the American Charles Davenport. Nor does Wilson's dream present a radical alternative to the extant Darwinian paradigm. Rather, his dream is to extend the paradigm that Darwin described and use it to unify what Wilson calls the "archipelago of academic islands" under the banner of evolutionary theory. From his perspective, the twenty-first century presents an opportunity that was not available to his system-building predecessors, from Herbert Spencer to Edward O. Wilson. Now, given the newly achieved, multilevel perspective and the access to huge amounts of data via the Internet, it has become possible to weave together diverse human experiences catalogued across the various disciplines into a coherent evolutionary approach to human social organization. This is Wilson's dream and vision.

BECOMING A SCIENTIST

By his own account, Wilson had decided from a young age that a career as a scientist, "whatever that was," would be the means to escape his father's shadow. "I had to fill my father's shoes, but the prospect of equaling him at what he did so well was too daunting. I needed to do something that he would admire but couldn't do himself, preferably something that he couldn't even evaluate. I would become a scientist."[7] By the time he had graduated from boarding school in the Northeast, Wilson had decided that the life of an ecologist was the perfect means to achieve not only his goal of escaping his father's shadow, but to add to the rationalist project of increasing the stores of human knowledge. During his undergraduate studies at the University of Rochester, Wilson would build his own equipment to sample the zooplankton of various lakes and streams of upstate New York. By the end of his undergraduate experience, he had committed completely to the idea of becoming a scientist and had settled on the ecology of zooplankton as his specialty. But it wasn't long before Wilson was moving beyond his own conventional idea of the biologist who picks a particular problem within a particular system and works that problem from multiple perspectives. For Wilson, the introduction to evolutionary theory and his realization of its all-encompassing nature would transform both his life and his philosophy. Beginning with that very first graduate paper, he would remain at the center of the swirling debate about group selection. From this position, Wilson would not only advocate for a multilevel approach to evolutionary questions within the standard biological realm, but he would seek to expand the Darwinian framework to include the domains of human social evolution that had been the purview of social scientists. In his estimation, students of human culture had been unaware of the power of the evolutionary lens.

Introduced in the mid-1960s by the British evolutionary theorist, William Donald Hamilton, kin selection theory had come to dominate the explanatory landscape. In a pair of papers in the *Journal of Theoretical Biology* in 1964, Hamilton had argued that seemingly altruistic behaviors could be understood as the result of what he called "inclusive fitness." "Species following the model," he explained in the abstract, "should tend to evolve behaviour such that each organism appears to be attempting to maximize its inclusive fitness. This implies a limited restraint on selfish competitive behaviour and [the] possibility of limited self-sacrifices."[8] What Hamilton had created was essentially a model that described how social behavior evolved based on degrees of relatedness between individuals. "Clearly from a gene's point of view," he concluded, "it is worthwhile to deprive a large number of distant relatives in order to extract a small reproductive advantage."[9] It wasn't exactly a ringing endorsement for altruism.

In the coming years, Hamilton's formulation of evolutionary thinking was canonized by George Williams in his *Adaptation and Natural Selection* (1966) and apotheosized by Richard Dawkins in *The Selfish Gene*. Both men argued that too many biologists had been wooed by sloppy "good of the species" arguments that flew in the face of Darwinian logic. What came to be known as "Hamilton's rule" had reestablished the basic Darwinian logic of natural selection. The mild-mannered but exceptionally tenacious Wilson now took this as his challenge.

For the first fifteen years of his professional career, based initially at the University of California–Davis, and then at the Kellogg Biological Station at Michigan State University, Wilson worked tirelessly to counter the increasingly gene-centered approach to evolutionary biology. In the wake of Edward O. Wilson's hugely successful text *Sociobiology*, published in 1975, and Richard Dawkins's best seller *The Selfish Gene*, which came out the following year, scientific journals and the popular press alike were teeming with evolutionary explanations of human social behaviors. Most of these behaviors, especially those that weren't obviously the result of individual competition within species, were explained in terms of kin selection theory. Sloan Wilson and his support for group selection had become a beleaguered minority.

SWIMMING AGAINST THE TIDE

Throughout the seventies and eighties, as sole author or often with graduate students and select collaborators, Wilson continued to develop his theory of group selection and to meld this work with what came to be known as multilevel selection theory. This was the idea that natural selection worked simultaneously at many levels of biological organization, from the gene to the trait, the organism, groups of organisms, the species level, and perhaps beyond that. The theoretical work put him in conversation with a number of philosophers of science who had become intrigued by the nature of this scientific debate. Indeed, much of the work that established the philosophy of biology as a field of its own in the 1970s and 1980s revolved around questions of reductionism in biology and what came to be known as the levels-of-selection debate.[10] These philosophers were concerned that while genic selection could provide accurate accounting of evolutionary processes, it might not be giving the most informative causal account of how the process achieved these outcomes. Wilson engaged in both the theoretical and the philosophical conversations and began collaborating with the philosopher Elliot Sober, who had spent time in Richard Lewontin's lab at the Museum of Comparative Zoology at Harvard. During this period, Wilson continued his focus on theory but also expanded his view, working to place group selection and the levels-of-selection approach into a broader historical and philosophical framework.[11] It was also during this time that Wilson became increasingly interested in using his evolutionary toolkit to analyze and understand human behavior and social organization. In a paper from 1989, he argued that "levels-of-selection theory keeps the factors [the units (i.e., genes, organisms, and groups) that selection was acting on] separate, defining behaviors as self-interested when they increase relative utility within single groups, and group-interested when they increase the average utility of groups, relative to other groups. This provides a framework in which rational (utility-maximizing) humans need not be self-interested by definition."[12] Wilson, like Darwin before him, wanted a theory that could account for a human moral sense. He was convinced that Hamilton's idea of inclusive fitness did not tell the whole story. Selection at the level of the group could explain prosocial behavior between non-kin, as Wilson put it—a world in which "humans need not be self-interested by definition."

Throughout his career as a theorist, Wilson maintained a close connection to empirical work. In the field and in the lab, he published papers on organisms that crossed distant taxonomic categories, from plants to animals, microbes to megafauna. Much of this work was in the service of fleshing out and testing elements of group selection and multilevel selection theory. Nevertheless, Wilson had long been interested in the application of Darwinian theory to human populations, and during the 1990s, humans increasingly became the focus of his efforts. Looking back on his 2007 book, *Evolution for Everyone: How Darwin's Theory Can Change the Way We Think about Our Lives*, he wrote, "One reason that I became so passionate about group selection was because it so clearly related to the human condition, in addition to the rest of life. My professors and peers regarded themselves as *evolutionary biologists*. They respected the academic convention that studying humans is somehow not biology, as if we were set apart from the rest of nature. I had become an *evolutionist*, perhaps because I am the son of a novelist. For me, it was an unexpected homecoming."[13] Wilson was approaching the realization of his dream, moving toward the goal of using his experience and knowledge as an evolutionist to analyze, understand, and ultimately improve humanity. "I had been paddling away from my father, but now I had returned to ponder our own species, just like him, except through the lens of evolutionary theory rather than the lens of fictional narrative."[14]

EVOLUTION AND THE HUMAN CONDITION

This homecoming and the beginning of the realization of his dream of a Darwinian city, was manifested in his coauthored book with the philosopher Elliott Sober, *Unto Others: The Evolution and Psychology of Unselfish Be-*

havior (1998). This book grew out of a 1989 paper in the *Journal of Theoretical Biology*, "Reviving the Superorganism," in which the authors wrote,

> In one sense, levels-of-selection theory is a radical departure from the individualistic theories that have dominated evolutionary biology for the last twenty years. Most evolutionists have been taught, and many still teach their students, that higher levels of selection are so unlikely that they can be safely ignored. As a result, virtually all adaptations are explained in terms of benefits to individuals (or genes) and consequences for groups and communities are considered irrelevant. At the extreme, the entire process of natural selection is characterized by a metaphor of selfishness embodied in the concept of "selfish genes."[15]

This factionalized state of affairs, where the majority of biologists focused on the selection of genes in contrast to Wilson and the multilevel selectionists, who argued for their broader perspective, did not seem to be the kind of dispute that would be advanced by further clarification of theory. Sober had published a very well-received analytical treatment of evolutionary theory that pointed out the shortcomings of the gene-centered view and carefully outlined the benefits of a multilevel perspective. This approach expanded the domain of evolutionary explanations, accounting for previously poorly understood phenomena like deferred maturity and population restraint. Wilson had continued to expand and refine his mathematical models and add to the list of empirical examples in support of the multilevel approach to evolutionary processes, but he didn't seem to be winning over many hearts and minds. Nevertheless, in a somewhat ironic twist, the continued communal intransigence to the idea of group selection began to have a freeing effect on him. He would carry on. He would expand. He would demonstrate the power of Darwinian thinking on the most significant species on planet Earth: humans. Extending his father's novelistic efforts to describe and understand the human condition, Wilson would apply his evolutionary toolkit to more exactly describe, mechanistically understand, and ultimately improve the human condition.

It is perhaps important to remind ourselves here that Wilson was by no means alone nor original in this pursuit. Though Darwin was initially coy in the *Origin*, stating simply, "Light will be thrown on the origin of man and his history,"[16] many of his followers were certainly not, and, indeed, Darwin himself followed up with *The Descent of Man* (1871) and *The Expression of Emotion in Man and Animals* (1872). Darwin's most ardent contemporary supporters, the Englishman Thomas Henry Huxley and the German Ernst Haeckel, both published hugely popular works on human evolution in the late nineteenth century. Desmond Morris's *The Naked Ape* and Robert Ardrey's trilogy, *The Social Contract*, *The Territorial Imperative*, and *African Genesis* captured huge audiences for their evolutionary analyses of humans in the 1960s and 1970s. Wilson was, by the turn of the twenty-first century, ready to fully embrace the totalizing approach that had been pursued with varying degrees of success and ignominy by any number of predecessors.

While Wilson is aware of the negative connotations associated with the social Darwinism of the late nineteenth and early twentieth centuries, he remains undaunted. He has argued throughout his recent works that many of these historical figures, including Herbert Spencer, William Graham Sumner, and Andrew Carnegie were not in fact Darwinians at all. They invoked a theory they didn't completely understand. In some sense, this critique applies to some of the more sophisticated social Darwinists and eugenicists of the early twentieth century, including the British population geneticist Ronald Aylmer Fisher and the American eugenicist Charles Davenport. In the former's case, Wilson argues that it wasn't that Fisher was insufficiently Darwinian; it was his lack of a multilevel perspective. Davenport, for his part, not only had a naive view of genetics but was plagued by an overreliance on the science to solve complex social problems. In Wilson's view from the twenty-first century, we have the correct evolutionary theory and need only to build connections between the currently disparate academic disciplines to realize the dream of improving human society.

Through the first decade of the twenty-first century, Wilson would concentrate his efforts on building broader infrastructure (pedagogical and political) for his Darwinian dream. His first book of the new millennium, *Darwin's Cathedral: Evolution, Religion and the Nature of Society*, was described by one reviewer as a "model of how to pursue an evolutionary social science."[17] This was music to Wilson's ears. Indeed, *Darwin's Cathedral* was in many ways the culmination of his theoretical work. The multilevel framework and his emphasis on the power of group selection enabled an interpretation and analysis of human religiosity as a group-selected adaptive trait. Wilson's treatment ranged across boundaries of geography and time to explain the prosocial effects of religious belief systems in human social evolution and laid the groundwork for his expanding network of evolutionary programs. In *Darwin's Cathedral*, Wilson extended the view he had developed with Sober, thinking about human societies as organisms. He argued that if we analyze the varied human religious groups as organisms, we see that they function as single units rather than aggregations of individuals. We come to understand that morality and religion are biological and cultural adaptations that allow humans to achieve through collective action what could never be done alone. Scholars have long been intrigued by the role of religion in human culture. Wilson's argument for the evolution of religious belief and community emphasizes its adaptive function. From the multilevel perspective, religion functions in the human social organism as an adaptive trait that increases the interdependence of the members and thereby provides an evolutionary advantage over less integrated and cooperative groups.

INSTITUTIONALIZING THE DREAM

Wilson's next step toward realizing the dream of a Darwinian city was the establishment of the Evolutionary Studies (EvoS) program at Binghamton University. Wilson developed the EvoS program to facilitate the understanding of the "core principles of evolution and extend them in all directions from the biological sciences to all aspects of humanity."[18] Wilson intended this program to train students in the integrated application of evolution across disciplines. Further, and perhaps more importantly, the program would bring together faculty from the far reaches of the "ivory archipelago" (a metaphor that Wilson frequently uses to describe the disciplinary isolation of the contemporary university) to a common, unifying ground based on a shared evolutionary perspective. Wilson describes the EvoS program in his 2007 book *Evolution for Everyone* "as a new island in the ivory archipelago, a tropical paradise so to speak."[19] The EvoS program at Binghamton includes course work from eleven different departments, including some obvious ones like biology and anthropology, and a number of less obvious ones like English, and industrial and systems engineering.

As the program developed at Binghamton, Wilson embarked on another element of his dream. In 2007 in collaboration with Jerry Lieberman, then the president of the Humanists of Florida Association, he co-founded the Evolution Institute, a think tank that takes as its mission "to accomplish for the world of public policy what EvoS attempts to accomplish for the world of higher education."[20] The Evolution Institute, according to Wilson, is perfectly positioned at the beginning of the twenty-first century to "provide a direct connection between current evolutionary research and real-world applications." In Wilson's view, the institute will bring together ideas and problems from across the spectrum of human experience and then analyze and assess them using evolutionary approaches in order to develop policies and institutions to better serve humanity. In a rare and glancing nod to history, Wilson and Lieberman assure their readers and potential clients and donors that we ought not to be scared off by negative ideological notions associated with social Darwinism. They acknowledge that "it is common for political ideologies to claim the support of any authoritative idea, religious, scientific, or otherwise." Nevertheless, they continue, "The nature of ideological thinking, exploitation and cooperation within groups, and exploitation and cooperation among groups are all subjects that urgently need to be understood from a genetic and cultural evolutionary perspective, leading to knowledge that can be used to formulate human social policies agreed upon by consensus."

The expression of this sentiment, while laudable, is a bit pat. One might think of Google's corporate motto: "Don't be evil." Wilson maintains that the self-correcting nature of the scientific process, when properly instantiated, will avoid the kinds of policies and prescriptions of the social Darwinists of the past. Wilson maintains a positivist perspective on science that is unaffected by the last fifty years of scholarship in the history of science. He seems not to recognize the fact that science can never be free from its social and historical context, and that political and economic concerns often trump scientific consensus. One need only look at current debates regarding global climate change to see clear evidence of this.

BINGHAMTON NEIGHBORHOOD PROJECT

In a 2011 *Nature* article, "Darwin's City," science writer Emma Marris asks, "Has David Sloan Wilson fallen in love with his field site?"[21] This is an intriguing question for a historian of science, as we've long been fascinated by the relationship of researchers to their organisms, their labs, and their theories. In Wilson's case, it is clear that Binghamton itself is not the object of his affection. Rather, he sees it as an opportunity to make manifest his dream. There is nothing special about the city; it has a history similar to other upstate New York cities, but it happens to be where Wilson has lived for the past thirty years. In his book about the project, Wilson exclaims, "What I have to offer is the evolutionary paradigm. . . . If we don't use the tools of evolutionary theory to reflect on a Darwinian world, then *fuhgeddaboudit*; . . . this should be looked upon as one of the most positive developments in the history of intellectual thought. Think of all that we are on the verge of understanding."[22] "This" in the passage above, refers to Wilson's practical application of evolutionary theory to his city of Binghamton. Using maps that indicate areas of high and low prosocial behavior (that is, areas of more or less cooperation and community-focused effort), Wilson has begun projects with the Binghamton public schools, the Parks Department, various neighborhood associations, and local businesses to implement new public policies and projects that have resulted from his evolutionary analysis. In her article in *Nature*, Marris recalls Wilson describing his project: " 'I really wanted to see a map of altruism,' he says. 'I saw it in my mind.' And with a frisson of excitement, he realized that his models and experiments offered clues about how to intervene, how to structure real-world groups to favour prosociality. 'Now is the implementation phase.' Evolutionary theory, Wilson decided, will improve life in Binghamton."[23]

In the first of these projects to reach the implementation stage, Wilson has collaborated with the city of Binghamton and the Broome County United Way to establish a set of community-planned parks in various neighborhoods that had been assessed by Wilson's research as the most needy and the most likely to benefit. Given the maps of prosociality generated by Wilson's team, potential sites for parks were identified and community members were invited to participate in the design and naming of the parks. The goal of the parks was to improve the environment of the selected neighborhoods thereby increasing the social capital and the level of prosocial behavior. While this may not seem particularly evolutionary to many of the observers and even some of the participants in this program, Wilson insists that without an evolutionary lens guiding the process, the likelihood of success would be minimal. In a pair of articles in the *American Journal of Play*, he and his coauthors describe how the information collected by their group is analyzed from an evolutionary perspective to identify the areas where their intervention would have the greatest impact.[24] They argue that the three components of evolutionary processes—variation, selection, and inheritance—are a part of this program. Indeed, they claim that the Design Your Own Park project is "explicitly designed as a managed process of cultural evolution." They continue, "There will surely be variation in the plans that groups submit, which can be selected according to carefully designed judging criteria. Implementing the best plans and making them available to all groups for future competitions counts as inheritance, the cultural equivalent of genetic inheritance mechanisms. A second round of selection occurs when plans that are implemented succeed in varying degrees."[25] While this process is inherently evolutionary from the perspective of Wilson and his team, it is not particularly difficult to see how, for many of the participants, the evolutionary elements of the process look quite a bit like managed trial and error.

In another of the projects currently underway, Wilson's team is studying the role of religion in quality of life. The questions guiding the research include such matters as how the organization and function of social groups influence the well-being of its members and whether secular and religious communities differ in terms of community functioning and member well-being. This research is being conducted in collaboration with the Norwegian Institute for Public Health and aims to improve quality of life for the citizens of Binghamton. Here again, the evolutionary elements of this project are not readily apparent. Yet, Wilson insists that the evolutionary lens is crucial to realizing the goal of a human city as organized and efficient as a colony of integrated and industrious honeybees.

Most fascinating about Wilson as a dreamer is his ability to maintain the enthusiasm of the neophyte. He enters each new project, whether on the nature of religious belief or the analysis of pedagogy in public schools, with the bright-eyed excitement of a first-year graduate student. While this enthusiasm is immensely inspiring and contagious to many of Wilson's collaborators, I find this confidence in the power of evolutionary thinking essentially ahistorical. In all his enthusiasm, Wilson is unable to acknowledge how the dreams of scientists (among other social saviors) have so often gone wrong. He remains convinced that balancing the effects of selection acting at multiple levels will avoid the kind of simplified, top-down social engineering that undermined previous forms of social Darwinism. This democratic, data-driven, consensus-based, hive-mind approach, according to Wilson's dream, needs only an opportunity to demonstrate its effectiveness, and Binghamton will become the model organism.

REALIZING OR IDEALIZING THE DREAM?

Writing about a new project to assess the belief in the afterlife and its potential evolutionary significance, Wilson asserts, "I'm optimistic about reaching a consensus on the subject of religion, because the major evolutionary hypotheses make testable predictions, and we are overflowing with information about religions around the world."[26] He goes on to describe the respectful interactions and collaborations he has developed with religious scholars and theologians, and likens his evolutionary approach to religion to Darwin's original effort.

> The more I became involved in the study of religion, the more I regarded myself in a situation comparable to Darwin's. The natural historians of his day had accumulated enormous amounts of information about plants and animals around the world. . . . But it wasn't organized. Darwin's great achievement was to organize all of that information with his theory of evolution. . . . My need to consult with scholars of religion was precisely like Darwin's need to consult with natural historians. . . . By cultivating a respectful relationship it might be possible to organize the voluminous information on religion in the same way that Darwin organized natural history information. . . . After a large number of cases are examined, we'll be closer to having a fully rounded understanding of religion from an evolutionary perspective, which can be applied on a small spatial scale, such

as my city of Binghamton, in addition to around the world and throughout history.[27]

The ease with which Wilson toggles between the disciplines of natural history and religious studies, the geographical scale of the city of Binghamton and the globe, and the temporal scale from contemporary to "throughout history" is breathtaking. It also belies a deep and abiding commitment to the power of the process of science in general and evolutionary theory in particular. It is a commitment, however, that to this historian's eye, lacks historical awareness. This is particularly ironic, given Wilson's career of swimming against historical tides in his own advocacy for group selection. Wilson's positivist belief that the truth will out is not consistent with the history of science or even his own experience. While the ideal of the scientific process is envisioned as some inexorable march toward truth, the history of science makes clear that the path is meandering, often lost, sometimes blocked by varied obstacles, be they theoretical, technological, political, or economic. The revolution in human health promised at the completion of the Human Genome Project more than a decade ago remains an increasingly complex and distant goal. Ongoing disputes regarding the reality of global climate change and the appropriate response give the lie to the notion that science, properly applied, can solve complex human problems.

Wilson is a dreamer who wants his science to change the world. He first realized as a graduate student that theory and mathematical models could change the way we understand the evolutionary dynamics underlying altruism. Throughout his career, ranging across all the levels of nature and all its groups, he continued to look to evolution for the answers to deep questions about human social organization. In some sense Wilson has achieved his dream. His city is his laboratory, and his neighbors are his subjects. He works frenetically to realize new projects and continues to collect data and subject that data to evolutionary analysis. Will Binghamton be improved? Will Wilson's dream be realized? We'll have to wait and see.

FURTHER READING

Borrello, Mark. *Evolutionary Restraints: The Contentious History of Group Selection.* Chicago: University of Chicago Press, 2010.

Comfort, Nathaniel. *The Science of Human Perfection: How Genes Became the Heart of American Medicine.* New Haven, CT: Yale University Press, 2014.

Oreskes, Naomi, and Erik Conway. *Merchants of Doubt: How a Handful of Scientists Obscured the Truth on Issues from Tobacco Smoke to Global Warming.* New York: Bloomsbury Press, 2011.

Ruse, Michael. *The Evolution Wars: A Guide to the Debates*. New Brunswick, NJ: Rutgers University Press, 2002.

Wilson, David Sloan. *Darwin's Cathedral: Evolution, Religion and the Nature of Society*. Chicago: University of Chicago Press, 2002.

———. *Evolution for Everyone: How Darwin's Theory Can Change the Way We Think about Our Lives*. New York: Delacorte Press, 2007.

———. *The Neighborhood Project: Using Evolution to Improve My City One Block at a Time*. New York: Little, Brown, 2011.

———. *Does Altruism Exist? Culture, Genes, and the Welfare of Others*. New Haven, CT: Yale University Press, 2015.

NOTES

1. David Sloan Wilson, *The Neighborhood Project: Using Evolution to Improve My City One Block at a Time* (New York: Little, Brown, 2011), 7.

2. "Dynamic Ecology," last modified May 21, 2014, online at https://dynamicecology .wordpress.com/2014/05/21/what-are-the-greatest-ecology-evolution-dissertations -ever/.

3. David Sloan Wilson, "A Theory of Group Selection," *Proceedings of the National Academy of Sciences* 72, no. 1 (1975): 143–46.

4. Charles Darwin, *The Descent of Man and Selection in Relation to Sex* (London: John Murray, 1871), 166.

5. Wilson, "Theory of Group Selection," 145.

6. Charles Darwin, *On the Origin of Species; or, The Preservation of Favoured Races by Means of Natural Selection* (London: John Murray, 1859), 489.

7. David Sloan Wilson, *Evolution for Everyone: How Darwin's Theory Can Change the Way We Think about Our Lives* (New York: Random House, 2007), 326.

8. William D. Hamilton, "The Genetical Evolution of Social Behavior," *Journal of Theoretical Biology* 7 (1964): 1.

9. Hamilton, "Genetical Evolution of Social Behavior," 16.

10. See, for example, the work of philosophers Elisabeth Lloyd, David Hull, Bill Wimsatt, and Elliot Sober, among many others.

11. David Sloan Wilson, "The Group Selection Controversy: History and Current Status," *Annual Review of Ecology and Systematics* 14 (1983): 159–87; and "Levels of Selection: An Alternative to Individualism in Biology and the Human Sciences," *Social Networks* 11 (1989): 257–72.

12. Wilson, "Levels of Selection," 269.

13. Wilson, *Evolution for Everyone*, 342.

14. Wilson, *Evolution for Everyone*, 342.

15. David Sloan Wilson and Elliott Sober, "Reviving the Superorganism," *Journal of Theoretical Biology* 136 (1989): 352.

16. Darwin, *Origin of Species*, 488.

17. Peter Corning, "Unmasking *Darwin's Cathedral*: It's Not Just about Religion" (re-

view of *Darwin's Cathedral*, by David Sloan Wilson), *Skeptic Magazine*, 2003, online at http://www.skeptic.com/reading_room/unmasking-darwins-cathedral/.

18. See "EvoS," online at http://evolution.binghamton.edu/evos/.

19. Wilson, *Evolution for Everyone*, 9.

20. "Why a Think Tank?," online at https://evolution-institute.org/about/why-a-think-tank/.

21. Emma Marris, "Darwin's City," *Nature* 474 (2011): 146–49.

22. Wilson, *Neighborhood Project*, 161.

23. Marris, "Darwin's City," 146.

24. David Sloan Wilson, "The Design Your Own Park Competition: Empowering Neighborhoods and Restoring Outdoor Play on a Citywide Scale," *American Journal of Play* 3 (2011): 545–46; D. S. Wilson, D. Marshall, and H. Iserhott, "Empowering Groups That Enable Play," *American Journal of Play* 3 (2011): 523–38.

25. Wilson, "Design Your Own Park," 548.

26. Wilson, *Neighborhood Project*, 310.

27. Wilson, *Neighborhood Project*, 311.

Part VI: The Systematizers

Figure 16.1. D'Arcy Thompson. Courtesy of the University of St. Andrews Library. Image ms48534-ph-2-16.

TIM HORDER

D'ARCY THOMPSON
ARCHETYPICAL VISIONARY

16

INTRODUCTION

It is difficult to imagine a biologist more worthy of the term *visionary* than D'Arcy Wentworth Thompson (1860–1948). In his day he stood out distinctively from mainline science with his idiosyncratic but penetrating and highly original approach to the fundamentals of biology, in particular to the explanation of biological forms. His scientific contribution is contained in his extraordinary book, *On Growth and Form* (*G&F*), which is still in print—in paperback, abridged, and with an introduction by Stephen J. Gould—a testament in itself to the uniquely striking nature of his biological vision.[1] Thompson's originality of approach lies in his seeking a *causal* explanation for biological forms at a time when biologists were content simply to classify them. Thompson introduced a new method for explaining the morphological structures of animals and plants as part of a general consideration of the adequacy of physico-chemical forces as the basis for the formation of living organisms. At the time his book was published in 1917, many biologists—those favoring an alternative in vitalism—would have denied this possibility.

This chapter examines the origins of Thompson's perspective and the influence of his book at the time, along with its continuing appeal. It is important to bear in mind that the book was written at a formative period in the history of biology when many foundational notions we now take for granted were in flux. Most of the fundamental principles of biology were then controversial and poorly defined, including the role and nature of genes, the structure of cells, how cells collaborate and develop into tissues, and even the adequacy of Darwin's theory of evolution to explain the whole range of biological phenomena. Biochemistry was in its infancy, and molecular biology was unheard of. The limitations on biologists' understanding of their subject at that time and the cursory way in which Thompson deals with many of the key questions make interpretation somewhat difficult for present-day readers.

The structure of *G&F* can perhaps best be understood by taking seriously Thompson's own insistence that his intention was a limited one—namely, to explore the possibility of advancing biology by applying the methodological precision of mathematics and physics. The book largely consists of a series of carefully selected examples of biological structures potentially amenable to precise description using mathematical methods. It starts with single-celled organisms and progresses finally to vertebrates. Thompson uses his examples to speculate on how their shapes and sizes might be explicable on the basis of laws of physics already well understood at the time. The now-outdated term *Form* in the title of the book identifies a feature that was, up to Thompson's day, regarded as the key to characterizing the differences between types of organisms. Overall shape in itself was thought sufficient to capture the essential distinguishing characteristics of individual organisms, irrespective of their component parts.

LIFE HISTORY AND CHARACTER

Thompson fitted the image of a visionary perfectly (fig. 16.1). He was, in the words of his friend and obituarist Clifford Dobell, "a very Viking of a man"; tall, red-tinged hair, flowing beard, bright blue eyes, specially made, wide-brimmed hat, sometimes parading with his parrot on his shoulder.[2] A popular lecturer who loved teaching, he was also described by his daughter as sometimes "brusque, . . . volatile, impulsive, and individual."[3]

Thompson was born in Edinburgh, an only child. His maternal grandfather was a well-known veterinary surgeon, and his father was an outstanding classicist.[4] After two years of medical studies in Edinburgh (1878–1880), he moved to Trinity College, Cambridge, to study natural sciences (1880–1883). In 1884, aged twenty-four, he was appointed founding Professor of Biology in a small university college in remote and impoverished but up-and-coming Dundee, with all that entailed in terms of teaching and administration (including creating a museum).[5] Moreover, he took on the time-consuming role of UK representative on international fishing commissions, which involved traveling, administration, and extensive international correspondence: this alone he reckoned to be "a full-time post."[6] His paper *Morphology and Mathematics* (1915)—which would become chapter 17 in *G&F*—made an impression and may have helped to earn him his Royal Society Fellowship the next year. Even before *G&F* was published, he seems to have emerged from the position of a hard-working if undistinguished academic to be a figure of note. He became Professor of Natural History at St. Andrews in 1917. Knighted in 1948, he died at age 88.[7]

"D'Arcy was a true Victorian in many ways," wrote his daughter.[8] He was part of a distinguished generation of British scientists. Among his fellow undergraduates at Cambridge were Charles Scott Sherrington and Alfred North Whitehead, who, like Thompson, were much concerned with a "holistic" approach to fundamental scientific problems, both pursuing philosophical interests later in their careers.[9] Thompson's viewpoints showed remarkable parallels with those of another contemporary, William Bateson, Britain's leading Mendelian.[10] Both men's perspectives reflect the transition in biology from mid-nineteenth-century styles of investigation—descriptive and classificatory—to an increasingly experimental and analytical, laboratory-based science, led by physiology. Thompson's methodological recommendations are heavily influenced by his experience at Cambridge, where the brilliant achievements of physics were taken to be ideals of scientific advance, and mathematics offered the ultimate form of "proof," a route to precision in the face of the vast accumulation of complex data and the absence of firmly established explanatory concepts that set biology apart from the physical sciences at the time.[11]

Thompson's loyalties and interests were, unsurprisingly, split between biology and the classics. In 1929 he was president of the Classical Association of England. His scholarly and still-valued compendium, *A Glossary of Greek Birds* (in his words "the apple of my eye"), illustrates his encyclopedist's method of working. His idol was Aristotle, and he followed the teachings of Plato and Pythagoras, which he identified as the "message of Greek wisdom"—that all aspects of nature can be elucidated and expressed mathematically.[12] One cannot appreciate and fully understand Thompson's approach without taking into account the classics-influenced and sometimes opaque literary style of *G&F*, which so impressed early reviewers. In his allusions and untranslated quotations in many languages Thompson assumed readers as highly educated as himself. "To spin words and make pretty sentences is my one talent, and I must make the best of it," he wrote; "[it is] the one thing I am a bit proud and vain of—the one and only thing."[13]

His bibliography includes some three hundred immensely diverse items, ranging from book reviews, obituaries, and marginalia to learned studies on collecting, the classics, and anatomical or taxonomic topics—as many appeared in *Country Life* or the *Classical Review* as in *Nature*. His zoological work was in the British tradition of natural history study that he acquired as a schoolboy—he relished the work of the museum curator—and he does not seem to have attempted laboratory experimentation at any point. His lack of focus was seen as a difficulty early in his career; in numerous letters, his mentor, Michael Foster, founder of the Cambridge school of physiologists, offered stern advice: "I want to urge you to complete one or other of your pieces of work." (1885); later he warned that "time is getting on—and if you don't take care, the research will become difficult and in the end impossible" (1888).[14] As Dobell put it, "His basic philosophy was an extreme form of eclecticism."[15]

Dobell summed up Thompson's scientific contribution as follows: "His contributions to ordinary zoology . . . were nearly all published during the first thirty years of his working life. . . . I still find it hard to see any thread connecting them, or any link with his major works. They deal mainly with morphology and systematics, and reveal a wide field of interest—especially in out-of-the-way or problematic animals—but seem to exhibit no specialist's expertise. D'Arcy Thompson will not, I feel sure, be remembered for his workaday contributions to zoology, but for his far greater legacy to general biology and learning. . . . I know of nobody else, living or dead, who could have conceived and written *Growth and Form*."[16]

ON GROWTH AND FORM

Thompsons's reputation as a scientist is today solely based on the one scientific book that he published when he was fifty-seven, which is best seen as a reflection of his many-sided character. He was a true scholar-naturalist, a traditionalist, a man of contradictions and of a forthright, Scottish independence of mind. Prior to 1909 he had had no intention of writing such a book until encouraged to do so by colleagues.[17] Letters to him from renowned embryologists Wilhelm Roux and Richard Assheton show that writing was underway by October 1911 and it was completed by 1915.[18] His "little" 793-page book, as he put it,[19] was well reviewed, and by 1922 it needed reprinting. *G&F* was in effect Thompson's single and only coherent presentation of his vision and is therefore the focus for this chapter. So why did such an unusual book, of forbidding length and complexity, have an impact—and one that has endured?

In large part *G&F* follows the scholarly habits that were evident in his earlier book on ancient Greek lexicography. It is a compendium of illustrative examples of morphological forms throughout nature—also including data tabulations, evidence of Thompson's statistical interests, and many digressions. There is a clear underlying theme. It is "an easy introduction to the study of organic Form, by methods which are the common-places of physical sciences."[20] Further, he explains, it addresses "how . . . the forms of living things . . . can be explained by physical considerations; . . . no organic forms exist save such as are in conformity with ordinary physical laws."[21] The chapters of the book broadly progress in sequence from considering simple, unicellular organisms—where forces such as diffusion, viscosity, coherence, osmosis, gravity, and especially surface tension offer plausible causal explanations of shape—to consideration of complex and large-scale biological forms, such as human skulls or whole fish.[22] Throughout, Thompson uses illustrations to show striking analogies between living and nonliving structures.

In 1942 a second edition of *G&F* was published, enlarged from 793 to 1,116 pages. The changes essentially amount to the addition of new or more detailed examples; most of the original text is untouched. Key programmatic sections—including much of the prefatory note, chapter 1 and the epilogue—are retained word for word. Some errors remained uncorrected.[23] Most of the revisions were probably made in the 1930s, but printing was delayed, and references were added as late as 1940.[24] Thus the new edition was basically an expansion of the first, in which Thompson carried on—during his seventies—the method of accumulating illustrative examples started early in his career.[25] A notable feature is that no account is taken of many of the momentous scientific advances that had occurred since 1917, including the "evolutionary synthesis" of the 1930s, in which a modern understanding of genetics became integrated into a now generally accepted Darwinian evolution theory.[26] Interpretation of Thompson's position must therefore be based on the first edition and on the perspectives of 1917.

One thing is clear: the book is not intended as a systematic review of the relevant contemporary literature, and its references to many of the then-controversial foundational issues of biology are merely brief asides scattered across the text. As Thompson notes, themes "often deemed to be fundamental" are "spoken of by the way."[27] For this reason, it is no easy matter to characterize Thompson's views on key theoretical issues. In particular, his treatment of Mendelian genetics and Darwinian evolution—in a cursory and apparently questioning manner—has often puzzled commentators. The book contains no comprehensive coverage or taxonomically arranged survey of biological knowledge such as might have been found, for example, in a standard zoology textbook. The absence of any treatment of important theories and concepts in the biology of the time can be explained and understood if we recognize Thompson's objective. The entire book is devoted to promoting a particular methodological approach to biology based on mathematics and the physical sciences, a method that he hoped would bring precision and order to a subject that was still struggling to achieve clarity and consensus concerning its most basic principles. It must be judged with this in mind.

THOMPSON'S THEORETICAL CONCERNS

Thompson's British Association lecture in 1894 was entitled "Difficulties of Darwinism."[28] At that time it was commonly assumed that the Darwinian mechanism of natural selection explained the elimination of the unfit, but left inexplicable the evolutionary novelties that make phyletic "progress" possible.[29] Thus the teleological notion of "orthogenesis" was often introduced: this envisaged an inbuilt potential within living organisms toward greater perfection and adaptedness.[30] Thompson had taken this approach in 1884, but later came to emphasize an even greater difficulty with Darwinism.[31]

As he wrote in 1926, "It was Bateson, far more than any other, who showed us that there were difficulties in Darwinism, that the problems of Evolution were far from settled, that the origin of species was, in fact, an unsolved mystery. His study of Variation shattered the 'crude belief' that Natural Selection had, of itself, impressed form and symmetry on the organic world."[32] In 1894 Bateson had drawn attention to sudden discontinuities among the varieties of animal forms ("saltations"), which seemed incompatible with the gradualism that was implied by Darwinian natural selection as the origin of new species through successive small adaptations.[33] When Thompson points out that "the adequacy of natural selection to explain the whole of organic evolution has been assailed on many sides," it does not follow that he does not accept the basic concept of Darwinian evolution through natural selection; rather, he is probably referring—following a line typical of many in his generation of biologists—to these various open questions regarding the explanatory adequacy of the theory.[34] He noted that "it is very manifest that there is abroad on all sides a greater spirit of hesitation and caution than of old."[35]

Turning to the material basis of biological phenomena, Thompson avoided a simplistic reductionism based on elemental structures, while alternatively emphasizing mechanical forces to explain morphology. Influenced by the Cambridge physicists—especially by the biophysicist William Hardy—Thompson agreed with others at the time that cells should be viewed in dynamic terms rather than in the structural terms of the organelles described by microscopists.[36] As he put it, "Matter as such produces nothing, changes nothing, does nothing; [cells and their contents] can never act as matter alone, but only as seats of energy and as centres of force."[37] Thompson's approach to the physical nature of genetic factors paralleled Bateson's well-known struggle during the 1920s in coming to terms with T. H. Morgan's crucially important linking of Mendelian methods of tracing patterns of inheritance to a basis in the microscopic structure of chromosomes.[38] As Thompson wrote in a letter in 1923, "The chromosome people are having a good inning; but their theories are top-heavy, and will tumble down of their own weight. It is of little use, meanwhile, to argue with them."[39] Regarding Mendelism, he acknowledged that "there can still be no question whatsoever but that heredity is a vastly important as well as a mysterious thing."[40]

While writing his book, Thompson took the opportunity, as British Association section president, to survey wider biological perspectives, including "the hypothesis of a Vital Principle . . . [that] has come into men's mouths as a very real and urgent question, the greatest question for the biologist of all."[41] At that time it was tenable—and widely accepted by philosophers (e.g., Henri Bergson) as by scientists—that some version of vitalism was unavoidable, given the perceived inexplicability of the uniquely defining characteristics of living organisms, their integrity, adaptedness, and apparent teleological purposefulness, as seen for example in their goal-directed behavior.[42] Vitalism posited a superadded force specific to biology in order to explain the unique character of living organisms that differentiated them from the nonliving. *G&F* is an explicit attempt to avoid vitalism by promoting strictly physicochemical mechanisms.[43] As Thompson made clear, "There are no problems connected with Morphology that appeal so closely to my mind, or to my temperament, as those that are related to mechanical considerations, to mathematical laws, or to physical and chemical processes."[44]

Overall one can see that Thompson is seeking to define a rigorous methodology that avoids purely hypothetical notions, such as orthogenesis or vitalism, and that he is reacting critically against the speculative—and merely descriptive—construction of phylogenetic trees as indicators of the path of evolution typical of the earlier zoological tradition. "The study of form may be descriptive merely, or it may become analytical," he says.[45] Thus, in *G&F*, Thompson is offering just such an "analytical" methodology—one, moreover, that potentially provides causal explanations for the full complexities of whole organisms. The book largely consists of examples illustrating explanations of shapes (*form*), subsumed collectively into the all-embracing term *growth*.[46]

TRANSFORMS

Thompson's method of transforms, the subject of the final chapter, is the best known feature of *G&F*. His objective is to show how individual anatomical parts must be subject to overall structural constraints when considering larger organisms.[47] The coordinate transform method will, he argues, potentially "constitute a proof" that a "comprehensive 'law of growth' has pervaded the whole structure."[48] He demonstrates the method by means of numerous diagrams. In each case the patterns of deformation of superimposed grid lines serve to demonstrate a possible integrated pattern of growth that might explain the differing anatomies of, for example, types of fish. Thompson puts his thinking in typically vivid terms: "Deep-seated rhythms of growth . . . are the chief basis of morphological heredity, [that brings] about similarities of form."[49]

This "theory"—as it is described in the chapter title—has limitations in its applicability within biology which need to be understood if the method is to be properly evaluated.[50] The coordinate transforms do not describe or account for specific anatomies as such, but merely highlight similarities among readily comparable, relatively closely related anatomical types of organisms.

The nature of the "new system of forces" that is proposed—a potential explanation for evolutionary transitions between species—is not identified but merely described in mathematical form. Thompson acknowledges this when he says, "The *deformation* of a complicated figure may be a phenomenon easy of comprehension, though the figure itself have [*sic*] to be left unanalysed and undefined."[51] A more serious limitation of the method is that it cannot encompass types of organisms that are as widely different anatomically as, for example, fish and quadrupeds. In a section added in the second edition, Thompson now seems to assume that such fundamentally different groups of organisms evolve through saltations, as Bateson had proposed, rather than through gradual deformation.[52] He concludes that "a 'principle of discontinuity' . . . is inherent in all our classifications, whether mathematical, physical or biological."[53]

THOMPSON'S LEGACY

Thompson's theory regarding transforms has been seen as the single most original and readily applicable element of his general methodological approach, and a few younger scientists took it up enthusiastically, most notably the evolutionist Julian Huxley.[54] From 1916, Huxley analyzed growth experimentally and extended Thompson's method to describe differential growth rates of anatomical structures algebraically, a technique now known as *allometry*. In 1932 he published the important *Problems of Relative Growth*, dedicated to Thompson.[55] However, the notable diversity of approaches evident in the D'Arcy Thompson *Festschrift* of 1945 suggests uncertainty regarding the general applicability of Thompson's approach.[56] By 1950 some of the same contributors to a follow-up symposium were more openly expressing their doubts.[57]

Few people, one imagines, have read *G&F* from cover to cover, even in its successfully abbreviated version. Nonetheless, successive generations have been influenced by it in constantly changing ways. Structuralists, such as mathematician and biological theorist Brian Goodwin, have continued to repeat his arguments against reductionism. Allometric methods have since been usefully deployed by biologists such as Stephen J. Gould and John Tyler Bonner, the Princeton developmental biologist and evolutionist. Physicists Erwin Schrödinger and Alan Turing acknowledged the book's impact on them. The English Nobel Laureate biologist and essayist, Peter Medawar, was simultaneously admiring and yet deeply critical: he judged that *G&F* was "courageous (though inevitably often faulty)."[58] Accepted almost from the start as a classic, that status has proved to be self-confirming. Gould saw *G&F*

as a model of scientific writing, and his first book, *Ontogeny and Phylogeny*, is dedicated to Thompson.

Written with persuasive conviction, even today *G&F* is striking in the sheer range of biological material covered and in the hard evidence offered in pictorial form. The eye-catching and suggestive array of pictorial images—the second edition has 554 of them—draws attention to patterns and shapes in their own right, whether seen in engineering, geological or biological systems, whether artificial or natural. The appeal of the comparison of the iconic cantilever Forth Bridge with the skeletal struts of a dinosaur is immediate and powerful. Thompson created some of the most enduringly familiar images in the history of biology. From an early stage, *G&F* provided stimulus for admirers outside science: artists, architects, and engineers have often described how they have been inspired by Thompson.[59] A museum combining zoological material and art works has now been named in his honor in Dundee.

A SUMMATION

Suitably interpreted, Thompson's *magnum opus* can be seen as a vivid and searching portrayal of the concerns of biologists generally in 1917. The book reveals how a dauntingly knowledgeable contributor viewed the fundamentals of his subject. On account of "his heresies," as he called his propositions,[60] Thompson can be considered a significant participant in an important stage in the modernizing of biology.[61] Regarding the degree to which his overall hoped-for analytic approach to biology had been fully realized, Thompson was himself cautious. He insisted that *G&F* was "all preface."[62] In summary, he concludes, "How far . . . mathematics will suffice to describe, and physics to explain, the fabric of the body no man can foresee."[63] A primary objective for Thompson was to encourage mathematicians and biologists to extend his pioneering efforts.[64]

The scientific status of Thompson's contribution remains uncertain. *G&F* stands as a classic source on the importance of biomechanics. But his vision is difficult to square with today's molecular genetic insights, which explain so much about how organisms develop.

Thompson's influence should not simply be judged on the basis of the theory of transforms: this was in a sense little more than an exploratory exercise in mathematical methodology. The adaptedness and the coordination of complex anatomies were the defining characteristics of organisms that he most wanted to understand, and it is in these terms that his continued influence may perhaps be explained. Successive generations of biologists may have recognized in *G&F* a uniquely powerful expression of the mystery of the structural integrity of living organisms, the central Thompsonian problem that challenges our understanding even today, and for which Thompson offered such a beguiling and original answer. In this respect Thompson was not just a dreamer but truly a visionary.

FURTHER READING

Ball, Philip, and Matthew Jarron, eds. "D'Arcy Thompson and His Legacy." Special issue, *Interdisciplinary Science Reviews* 38, no. 1 (2013).

Esposito, Maurizio. *Romantic Biology, 1890–1945*. London: Routledge, 2013.

Gould, Stephen J. *The Structure of Evolutionary Theory*. Cambridge MA: Harvard University Press, 2002.

Thompson, Ruth D'Arcy. *D'Arcy Wentworth Thompson: The Scholar Naturalist, 1860–1948*. London: Oxford University Press, 1958.

Whyte, Lancelot Law. *Aspects of Form*. London: Lund Humphries,1951.

NOTES

I am grateful for helpful comments received from Richard Boyd, Oren Harman, Nick Hopwood, Matthew Jarron, Eddie Small, and Andrew Woodfield.

1. D. W. Thompson, *On Growth and Form* (Cambridge: Cambridge University Press, 1917; 2nd ed., 1942; abridged ed., edited by John Tyler Bonner, 1961; 2nd abridged ed., 1971). Quotations in this chapter are taken from the 1917 edition of *G&F*; square-bracketed page references are from the 1942 edition.

2. Clifford Dobell, "D'Arcy Wentworth Thompson," *Obit. Not. Fell. Roy. Soc.* 6 (1949): 603.

3. Ruth D'Arcy Thompson, *D'Arcy Wentworth Thompson: The Scholar Naturalist, 1860–1948* (London: Oxford University Press, 1958), 159.

4. His mother died following his birth, and he was brought up by her family in Edinburgh. He got to know his much-admired father—academic, schoolmaster, classicist—only later in Ireland. They shared the same names. "Father and son were both, mentally and physically, several sizes larger than the common run of mankind, and they were both fated to live in backwaters" (Dobell, "D'Arcy," 614). After thirty years of collaboration, they published a translation (with annotations) of Aristotle's *Historia Animalium* in 1910.

5. Eddie Small, *Mary Lily Walker: Forgotten Visionary of Dundee* (Dundee: Dundee University Press, 2013). Thompson's earliest publications included a translation of Müller's *The Fertilisation of Flowers*, with a preface by Darwin, in 1883, and *A Bibliography of Protozoa, Sponges, Coelenterata, and Worms, . . . (1861–1883)* in 1885.

6. Thompson, *D'Arcy*, 148. His involvements included the Behring Sea Commission (1896–1897), which "changed the course of his life" (ibid., 99); the Fisheries Board of Scotland; and the International Council for the Exploration of the Sea (1902–1947), as editor of its *Bulletin Statistique* (vols. 8–27, 1911–1927).

7. Married in 1901, he was much involved in charitable work in Cambridge (Thompson, *D'Arcy*, 62) and Dundee (Small, *Mary Lily Walker*). He "was deeply religious, but

he had no 'religion,' in a sectarian sense." (Dobell, "D'Arcy," 614). "He was at heart a lonely man, and sometimes felt his intellectual isolation acutely" (ibid., 613). "The disappointment of never becoming a Fellow of his beloved College was one that he never got over" (Thompson, *D'Arcy*, 64); it was "a bitter blow" (Dobell, "D'Arcy," 602). Marked by his failures to obtain prestigious jobs, "he felt his isolation in Dundee acutely" (Thompson, *D'Arcy*, 159), having spent "thirty years of his early life 'in the wilderness'" (ibid., 164).

8. Thompson, *D'Arcy*, 182.

9. Maurizio Esposito, *Romantic Biology, 1890–1945* (London: Routledge, 2013); Maurizio Esposito, "Problematic 'Idiosyncrasies': Rediscovering the Historical Context of D'Arcy Thompson's Science of Form," *Science in Context* 27 (2014): 79–107.

10. Bateson's 1894 book *Materials for the Study of Variation* (London: Macmillan, 1894) shares the compendious structure of *G&F*.

11. As a biology student, he was much influenced by Francis Balfour (Thompson, *D'Arcy*, 46–47, 52, 70–71); "personally Balfour's death [in 1882] was a crushing blow" (ibid., 55). The Greeks apart, other influences were Herbert Spencer and various Cambridge physicists; Clerk Maxwell, for example, is referred to many times in *G&F* (e.g., 9, 40, 44, [964]). See also Robert Olby, "Structural and Dynamical Explanations in the World of Neglected Dimensions," in *A History of Embryology*, ed. T. J. Horder, J. A. Witkowski, and C. C. Wylie (Cambridge: Cambridge University Press, 1985).

12. Dobell, "D'Arcy," 608; Thompson, *Growth and Form*, [1097]. Also see Thompson, *Growth and Form*, 10 [15], and 717 [1026]. Thompson wrote that "numerical precision is the very soul of science and its attainment affords the best, perhaps the only criterion of the truth of theories and the correctness of experiments" ([2]). He aims at "something of the use and beauty of mathematics" (ibid, 778–79 [1096–97]). Dobell writes that "he took infinite delight in mathematical reasoning" (Dobell, "D'Arcy," 610).

13. Dobell, "D'Arcy," 612.

14. Thompson, *D'Arcy*, 63 and 80. W. E. Le Gros Clark and P. B. Medawar, eds., *Essays on Growth and Form Presented to D'Arcy Wentworth Thompson* (Oxford: Clarendon Press,1945; includes a bibliography). In a chapter entitled "On Biological Transformations," Joseph Woodger pinpoints ways in which Thompson's new methods fail to confront problems of anatomical homology and the complexities of early embryogenesis.

15. Dobell, "D'Arcy," 611.

16. Dobell, "D'Arcy," 605–6.

17. Thompson, *D'Arcy*, 161.

18. Thompson, *D'Arcy*, 161.

19. John Whitfield, *In the Beat of the Heart: Life, Energy and the Unity of Nature* (Washington, DC: Joseph Henry Press, 2006), 10. *G&F* was originally intended to be 144 pages long.

20. Thompson, *Growth and Form*, "Prefatory Note," v.

21. Thompson, *Growth and Form*, 10 [15]. Concerning his aims in *G&F*, see chapter 1, 486, 497, 673, 711–28, and the epilogue.

22. Topics are covered as follows: chapters 2–3, the general principles of scaling,

size, weight and volume; chaps. 4–6, cell structure; chaps. 7–8, forms of cell aggregates; chap. 9, chemistry and mechanics of calcareous skeletons (spicules), as in diatoms, radiolarians, foraminifera and sponges; chaps. 10–11, spirals; chap. 12, constraints and variability among varieties of foraminifera; chap. 13, horns, teeth, and tusks; chap. 14, plants and Fibonacci formula, phyllotaxis, leaf form; chap. 15, eggs and hollow structures; chap. 16, adaptation in vertebrate bone structures; chap. 17, coordinate transforms. For a good summary of *G&F*, see Stephen. J. Gould, *The Structure of Evolutionary Theory* (Cambridge MA: Harvard University Press, 2002).

23. Bonner, in introducing his abbreviated version of *G&F*, refers to errors and the neglect of contemporary literature. See Thompson, *Growth and Form*, abridged edition, ed. John Tyler Bonner (Cambridge: Cambridge University Press, 1961).

24. Thompson, *D'Arcy*, 163; see, for example, Thompson, *Growth and Form*, [35].

25. As he wrote in a letter in October 1889, "I have taken to Mathematics, and believe I have discovered some unsuspected wonders in regard to the Spirals of the Foraminifera"(Thompson, *D'Arcy*, 89). Early concerns included ligaments (1884) and the shapes of eggs (1908). He referred to "laws of growth" in his 1894 British Association lecture (*Nature* 50 [1894]: 435), further discussed in letters to a doubtful Foster (June–Oct. 1894; Thompson, *D'Arcy*, 89–90). The transform method probably originated some years before 1915 (Thompson, *Growth and Form*, 757). *G&F* appeared "nearly thirty years after D'Arcy had first begun to meditate upon its problems" (Thompson, *D'Arcy*, 162).

26. Ernst Mayr and William. B. Provine, *The Evolutionary Synthesis* (Cambridge, MA: Harvard University Press, 1980).

27. Thompson, *Growth and Form*, 778 [1096].

28. *Nature* 50 (1894): 435.

29. Thompson, *Growth and Form*, 137–38 [269–70].

30. Thompson, *Growth and Form*, 549 [807].

31. D. W. Thompson, "The Regeneration of Lost Parts in Animals," *Mind* 9 (1884): 415–20. Written as a riposte to J. S. Haldane, he here invokes holism, orthogenesis, and recapitulation. Haldane, born the same day and place as Thompson, was a distinguished physiologist, but a leading vitalist. Thompson frequently opposed him in debates.

32. D. W. Thompson, preface to *Nomogenesis; or, Evolution Determined by Law*, by Leo S. Berg (Cambridge, MA: MIT Press, 1926), xiii–xiv. In the 1969 edition Dobzhansky provides historical context for Berg's latter-day vitalistic theory.

33. Bateson, *Materials*.

34. D. W. Thompson, "*Magnalia Naturae*; or, the Greater Problems of Biology," *Nature* 87 (1911): 325. In this eloquent and revealing lecture, he distinguishes between the experimental, reductionist approach of "physiologists" and the traditional, descriptive perspectives of "zoologists" as morphologists (see also Thompson, *Growth and Form*, 2). For background see P. J. Bowler, *The Eclipse of Darwinism: Anti-Darwinian Evolutionary Theories in the Decades around 1900* (Baltimore: Johns Hopkins University Press, 1983). Although not spelled out (Thompson, *Growth and Form*, 716), Thompson's

position implied Lamarckism (the inheritance of acquired characteristics), which was widely accepted at the time.

35. Thompson, *Magnalia*, 326.

36. Thompson, *Growth and Form*, 172 [303–4]. Thus, "Cell and tissue, shell and bone, leaf and flower, are so many portions of matter, and it is in obedience to the laws of physics that their particles have been moved, moulded and conformed" (10).

37. Ibid., 14–15 [20]; see also 157 [287], [333–35], 194–200 [341–5], 286 [457]. In the United States, C. O. Whitman was promoting a similar cell theory: see "The Inadequacy of the Cell-Theory of Development," *Journal of Morphology* 8, no. 3 (1893): 639–58.

38. T. H. Morgan, A. H. Sturtevant, H. J. Muller, et al., *The Mechanism of Mendelian Heredity* (New York: Henry Holt, 1915). Bateson only came to accept Morgan's position in the mid-1920s (see W. Coleman, "Bateson and Chromosomes: Conservative Thought in Science," *Centaurus* 15 [1970]: 228–314). Bateson and Thompson rejected Weismann's earlier particulate theory of heredity.

39. Whitfield, *Beat of the Heart*, 19–20.

40. Thompson, *Growth and Form*, 715 [1023]. "With the 'characters' of Mendelian genetics there is no fault to be found. . . . But when the morphologist compares one animal with another, . . . character by character, these are too often the mere outcome of artificial dissection and analysis. Rather is the living body one integral and indivisible whole, in which we cannot find . . . any strict dividing line even between the head and the body, the muscle and the tendon, the sinew and the bone" (726 [1036–37]).

41. Thompson, *Magnalia*, 325.

42. Regeneration was also seen as evidence for teleological forces (see note 31). Thompson expresses the problem as follows: "It has been by way of the 'final cause,' by the teleological concept of 'end,' of 'purpose,' or of 'design' . . . that men have been chiefly wont to explain the phenomena of the living world." (Thompson, *Growth and Form*, 3 [4]). Against this he argues, "To seek not for ends, but for 'antecedents' is the way of the physicist, who finds 'causes' in what he has learned to recognise as fundamental properties . . . of matter and energy" (5 [6]).

43. Thompson, *Magnalia*, 326: "We keep an open mind on this matter of Vitalism."

44. Thompson, *Magnalia*, 328.

45. Thompson, *Growth and Form*, 719 [1026].

46. At that time, the term *growth* embraced the whole range of developmental phenomena (Thompson, *Growth and Form*, 10). Chapter 3 in *G&F* shows growth to be multifactorial: "The form of organisms is a phenomenon to be referred in part to the direct action of molecular forces, in part to a more complex and slower process, indirectly resulting from chemical, osmotic and other forces, by which material is introduced into the organism and transferred from one part of it to another. It is the latter complex phenomenon which we usually speak of as 'growth'" (53). *Form* (11) concerns *morphology* (see Geddes, "Morphology," in *Encyclopaedia Britannica*, 9th ed., 1878), as understood by Plato or Goethe. See also Stefan Helmreich and Sophia Roosth, "Life: Forms: A Keyword Entry," *Representations* 112, no. 1 (2010): 27–53.

47. The penultimate chapter shows how the structure of bones illustrates their direct adaptation to requirements of mechanical forces (in response to environmental—for example, gravitational—and contextual influences), and its last section (Thompson, *Growth and Form*, 712–18), entitled "The Problem of Phylogeny," introduces the theme that is to be addressed by the transform method.

48. Thompson, *Growth and Form*, 727 [1037].

49. Thompson, *Growth and Form*, 717–78 [1025].

50. Thompson, *Growth and Form*, 774–77 [1087–90]. Thompson recognized that structures may grow non-uniformly (776–77) and that his transforms had not addressed the third dimension (774–75). Some anatomies defied transform analysis (772 [1085]), and others required revisions (750 [1064]). Like the Scottish marine biologist T. W. Fulton before him, he recognized limits to mathematical analysis (98–99). Chapters 11, 14, and 17 provide the most direct applications of mathematics to complex structural forms. In chapter 13 he abandons formal mathematical treatment in favor of classifying "configurations" (612). Thompson seems to have been unconcerned about priority regarding the originality of his program. His methods were anticipated by Descartes, Galileo, Dürer, Camper [742], and Herbert Spencer (18), among others (777), especially Theodore Cook (*Spirals in Nature and Art*, 1903) and Arthur Church (639–41). Thompson's examples and illustrations were often supplied by colleagues (see "Prefatory Note," 768). He himself lacked artistic skills (Dobell, "D'Arcy," 612).

51. Thompson, *Growth and Form*, 723 [1032].

52. Thompson, *Growth and Form*, [1092–95].

53. Thompson, *Growth and Form*, [1094]. See chapter 12 for key evidence of saltation.

54. On Huxley, see Thompson, *D'Arcy*, 229–31. Huxley's allometric method explains orthogenesis. Thompson queried Huxley's contribution (Thompson, *Growth and Form*, [193, 205–12]).

55. T. J. Horder, "A History of Evo-Devo in Britain," *Annals of the History and Philosophy of Biology* 13 (2008): 101–74. The large Thompson archive housed at the University of St. Andrews includes an extensive correspondence with Huxley.

56. Le Gros Clark and Medawar, *Essays*. Thompson's noticeable—and surprising—disregard of embryological examples or evidence—he focused on unicellular organisms or post-embryonic (larval) stages—is perhaps explained by his skepticism concerning the speculative Haeckelian concept of recapitulation (see Thompson, *Growth and Form*, 3, 51, 155, 196–97, 608–9) and Balfour's approach to it (57). He was conscious of the earlier rejection of His's mechanical theories (55–56 [84–85]).

57. S. Zuckerman, ed., "A Discussion on the Measurement of Growth and Form," *Proceedings of the Royal Society* B137 (1950): 433–523.

58. Thompson, *D'Arcy*, 227.

59. Philip Ball and Matthew Jarron, eds., "D'Arcy Thompson and His Legacy," special issue of *Interdisciplinary Science Reviews* 38, no. 1 (2013). Many of those influenced by Thompson are listed.

60. Dobell, "D'Arcy," 610.

61. At the time, experimental methods were, for example, being applied to the problem of embryonic development, and Thompson corresponded intensively with many of the leading figures involved (e.g., Wilhelm Roux, Hans Driesch).

62. Thompson, *Growth and Form*, "Prefatory Note" and "Epilogue."

63. Thompson, *Growth and Form*, 8 [13]. Thompson studied mathematics in his first Cambridge year, but said, "I pretend to no mathematical skill." (v).

64. Thompson, *Growth and Form*, "Prefatory Note," 778 [1096].

SÉBASTIEN DUTREUIL

17

JAMES LOVELOCK'S
GAIA HYPOTHESIS
"A NEW LOOK AT LIFE ON EARTH" . . .
FOR THE LIFE AND THE EARTH SCIENCES

James Lovelock (b. 1919) was described by the curators of an exhibition at London's Science Museum in 2014 as a "scientist, inventor and maverick."[1] He was clearly an eclectic inventor from the very beginning of his career as a research engineer in the 1940s. He was an accomplished scientist before formulating the Gaia hypothesis in the 1970s, having done pioneering work in analytical chemistry, biochemistry, and cryobiology. He was perhaps a maverick, quitting academia in 1964, at age forty-five to settle as an "independent scientist."

But the Gaia hypothesis, his major accomplishment, is that of a dreamer. As he described it, Gaia was "for those who like to walk or simply stand and stare, to wonder about the Earth and the life it bears, and to speculate about the consequences of our own presence here."[2] From the very beginning, the Gaia hypothesis was perceived as something big, challenging the paradigmatic views prevailing in the earth and life sciences, redefining the boundaries and questions of these disciplines, providing a new conception of life, nature, and Earth.

But for many biologists, the Gaia hypothesis was just the dream of a "romantic and new-ager,"[3] wandering astray from science. Here is what the microbiologist John Postgate says about it: "Gaia—the great mother Earth! The planetary organism! Am I the only biologist to suffer a nasty twitch, a feeling of unreality, when the media invite me yet again to take her seriously?"[4]

Was Gaia just a misty dream? Was it only an evocative *metaphor* comparing Earth with an organism, as it has often been called? By postulating a new entity emerging from the interconnection of life and geological processes, the Gaia hypothesis not only had a revolutionary influence on the constitution of a new scientific field of the Earth sciences, but on the way we collectively think about nature.

Figure 17.1. Lovelock and his daughter collecting air samples in the summer of 1970, County Cork. These measures of air composition were but some of many he carried out in the early 1970s in connection with the Gaia hypothesis. Courtesy of the *Irish Examiner*.

DREAMING ABOUT LIFE IN THE SOLAR SYSTEM . . . AND BACK TO EARTH

Lovelock received, on October 19, 1961, a letter of invitation from NASA to work at the *Jet Propulsion Laboratory* (JPL), Pasadena, as a consultant engineer on instruments related to space exploration—a chromatograph for the 1964 Mariner B mission to Mars. As a science fiction reader, Lovelock was thrilled, and he accepted.

As he commonly tells the story, the Gaia hypothesis traces back to a very practical problem he encountered while working at JPL: how would one detect life on a distant planet, such as Mars or Venus? In 1965, Lovelock turned his back on the *biochemical* approaches to life that prevailed in exobiology, to focus on physical, thermodynamical approaches.[5] He pointed out that Earth's atmosphere is in great thermodynamic disequilibrium: for example, methane and oxygen coexist in proportions that are orders of magnitude out of what thermodynamic equilibrium would predict. This, he underlined, is the consequence of living beings' influence on their planetary environment (constantly producing methane and oxygen). Important thermodynamic disequilibrium would thus be a sign for the presence of life.

This proposal challenged and redirected research in the nascent field of exobiology. Back on Earth, the crucial recognition that life massively influences its planetary environment soon led Lovelock to the development of the Gaia hypothesis. In 1968, at a NASA meeting on the origins of life, Lovelock met Lynn Margulis (1938—2011), at that time a young microbiologist. They collaborated to develop the Gaia hypothesis in a series of coauthored papers from 1973 through 1978, before the publication of Lovelock's book.[6]

The Gaia hypothesis was meant to account for the long-term stability of Earth's environment, which had kept the Earth habitable by life for billions of years in spite of external perturbations, such as the increase of solar luminosity. The Gaia hypothesis accounted for this stability by positing that "the ensemble of living organisms which constitute the biosphere might act as a single entity to regulate chemical composition, surface pH, and possibly also climate."[7] These supposed regulatory properties led Lovelock to compare Gaia, the entity composed of the living beings and the geological environment with which they interact, to a living, homeostatic entity.[8]

After 1978, Lovelock and Margulis published separately. They came from two very different intellectual backgrounds—Margulis clung explicitly to a nineteenth-century romantic and naturalist tradition, foreign to Lovelock's chemical and cybernetic background—and had different views about Gaia.[9] After 1979, Gaia became Lovelock's major and central scientific concern. For Margulis, Gaia was an instance of a more general notion, that of symbiosis, encompassing cellular and microbial associations as much as relationships between living beings at a global scale. For this and other reasons, Lovelock is often presented as *the* author of the Gaia hypothesis. Yet Margulis had a decisive role in the 1970s: she drew Lovelock's attention to the major ecological role of microbes; she brought Gaia into evolutionary biology; and she had an important role in the diffusion of the Gaia hypothesis in American counterculture.[10]

THE STANDARD ACCOUNT: EVOLUTIONARY
BIOLOGY RIDICULED GAIA AS PSEUDOSCIENCE
The standard account of the reception of the Gaia hypothesis, the most popular in the life sciences, claims that it was rejected by scientists after famous critiques made by evolutionary biologists Ford Doolittle (see chapter 7) and Richard Dawkins in the early 1980's.[11]

The supposition that living organisms may act *in order* to regulate a larger whole seemed to reintroduce agency in the natural world at the wrong level of the biological hierarchy, and reminded evolutionary biologists of the heated debate of the 1960s and 1970s over the explanation of biological altruism, where benefit to the whole emerged from the altruistic actions of the individuals.[12]

These early critiques paved the way for the denunciation of Gaia as *pseudoscience*, barely good for neo-pagan worshipers of mother Earth. The idea that Earth or the cosmos is (like) a living creature finds roots in stoic philosophy; it was dismissed with the rise of modern and mechanistic science and partly revived in the *Naturphilosophie* of nineteenth-century German romanticism. Gaia also reminded contemporary scientists, especially evolutionary biologists, of the idea, linked with natural theology, that there is a "balance of nature." The Gaia hypothesis was altogether considered to be an extreme form of holism and of naively benevolent views of nature: a metaphor at best, and pseudoscientific mysticism at worst.[13]

What is particularly remarkable about Gaia's reception in evolutionary biology is its homogeneity in the entire field: Gaia has been dismissed by Richard Dawkins but also, and sometimes on the ground of strikingly similar arguments, by people who usually disagree about every other issue with Dawkins, such as Stephen Jay Gould and Richard Lewontin.[14] The fame of these biologists' critiques ultimately contributed to the diffusion of this standard account about Gaia.

In spite of its popularity, nothing of this standard account of Lovelock's hypothesis sounds right. It gives the impression that Lovelock was some kind of guru of a neo-pagan community, or a retired romantic in the countryside writing green poetry about a re-enchanted nature, or an evolutionary biologist with odd ideas. As we shall see, the reality of Lovelock's approach could not be more different.

LOVELOCK: AN INDEPENDENT AND PRACTICAL SCIENTIST, A CHEMIST, AN ENGINEER

Lovelock was born in Hertfordshire on July 26th, 1919, and grew up in London. His parents owned a small painting shop in Brixton. But Lovelock was more moved by his early and frequent visits to the Science Museum than by the artistic environment that surrounded him. He graduated in chemistry at the University of Manchester in 1941. He started working at the National Institute for Medical Research at Hampstead and obtained a PhD in medicine in the London School of Hygiene and Tropical Medicine. For twenty years, he worked on biochemical and engineering issues related to medical problems.

He published an important number of pioneering articles in *Nature*, some of them cited hundreds of times, on various issues: transmission of infections, the effect of heat on biological tissues and blood coagulation, cryobiology, and the resurrection of frozen hamsters.

He excelled in the invention of small-scale instruments, usually made to detect chemical substances. His most renowned invention, the one to which he owes his invitation from NASA, and (in 1997) the *Blue Planet Prize*, remains the Electron Capture Detector (ECD) which he invented in 1957. This small device enables scientists to detect minute quantities of chemical compounds (with a precision orders of magnitude above what was attainable beforehand).

So before the 1960s, Lovelock was already an accomplished scientist and a gifted engineer. In 1964 he quit academia to "bury [himself] in the country village of Bowerchalke,"[15] in the southwest of England. Reflecting an important romantic theme, he constantly presents himself with the ethos of a solitary and creative thinker, doing his best when working alone, free of every institutional constraint and the bureaucracy of contemporary science, and finding his inspiration in the walks he took in the countryside.

But make no mistake about what his "independent scientist" status meant concretely. The first thing he purchased was a Hewlett-Packard 9800 to solve differential equations for his ECD. In Bowerchalke, he did not settle into a library with rare alchemical books, but a homemade laboratory in the garage with chromatographs and electronic circuits. He quickly obtained a formal attachment to the University of Reading, in the department of Cybernetics, after one of his papers was refused because he only had a private address, and no institutional one. To pay for his living expenses and his own research expenses, like other "scientific entrepreneurs" of the 1960s and 1970s, he worked as a consultant engineer in big industries: Shell and Hewlett-Packard mostly—not the typical places where you expect to find bearded wizards making naked incantations to mother nature.[16]

If Lovelock had a disciplinary home, it was chemistry. He was trained early on as a practical chemist while a laboratory assistant, a job he obtained after high school where, he says, he learned "to regard accuracy in measurements as almost sacred."[17] The 1960s marked, for Lovelock, the transition from the study of the chemistry of living bodies (biochemistry) to the chemistry of Earth's surface (geochemistry), a field foreign to him before the late 1960s.

Aside from chemistry, the other intellectual matrix that played a prominent role in shaping Lovelock's research was cybernetics. When forced to represent what Gaia is, to draw what his vision of Gaia is, Lovelock does not call for the graphic artists that Cameron hired for the movie *Avatar* to picture a network of interrelated animated entities; he designs an electronic circuit (fig. 17.2). Electronic circuits were central components in most of the small-scale devices he invented as an engineer. And first-order cybernetics, the science of thermostats, systems, and feedbacks, occupies a central place in Gaian publications when it comes to proposing a mechanism that would maintain Earth's stability.

Various moving passages of Lovelock's autobiography, as well as scientific papers he published, reveal his style of science in the 1960s and 1970s. As he recalls, "It was a family ritual at Bowerchalke to measure the haze density using a sun photometer"[18] (see fig. 17.1). Troubled by an unusual presence of fog, persuaded that it was of anthropic origin, he settled to measure and trace the presence of chlorofluorocarbons (CFCs), a compound only produced by artificial means, to show that the atmospheric masses had been polluted. In the early 1970s, Lovelock embarked in the marine vessel Shackleton to measure the CFCs over the Atlantic, thanks to his ECD.[19] These first global measures of CFCs were decisive for the imputation of these compounds as the causal agents of ozone destruction by Mario Molina and Franklin Rowland,[20] winners of the Nobel Prize for this discovery. Intrigued by certain predictions of Molina and Rowland's theory, Lovelock did not propose an alternative theory: he found a meteorological plane flying at stratospheric heights which would enable him to make the chemical measurements he needed. On board of the marine vessel Shackleton, Lovelock also measured the quantities of dimethyl suphide (DMS). He had earlier realized that this sulfur compound was produced in great quantities by algae as he walked along his cottage grounds in Ireland, identifying algae and measuring their emissions with a chromatograph. These DMS measures were, for him, linked with the Gaia hypothesis. In 1973, in a famous paper published in *Nature*, he suggested that the important emissions of DMS by algae were essential for the closing of the global sulfur cycle.[21]

The idea that Gaia traces back to Lovelock's thinking about life detection has been prominently put to the fore by Lovelock. Yet after the mid-1960s and up to the early 1980s, Lovelock's central scientific activities were not focused on this issue but on global pollution, a problem rendered central to Lovelock's thinking and preoccupations through his work as a consultant for the greatest chemical and petroleum industries. And these thoughts were at that time central to the elaboration of Gaia.

This quick overview enables us to acknowledge that Lovelock was not a philosopher or a poet trying to resurrect a romantic view of nature. He was not even a theoretician, but a chemist and engineer, with the hard-core ethos of a practical scientist. He found his problems and arguments not in books but in the chemical compounds he smelled and measured.[22] In the early 1970s,

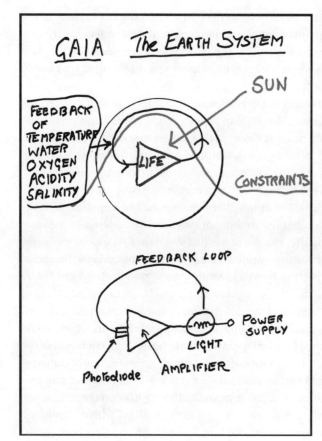

Figure 17.2. Lovelock's illustration of Gaia, showing the importance of cybernetics in his thinking and underlining the peculiar and central role he attributed to life in Earth's system. Copyright *Science Museum*. Permission courtesy of the Science & Society Picture Library.

the foundational decade of the Gaia hypothesis, Lovelock was not creating a mathematical model of the genetics of a population, sorting *Drosophila* in the lab, making ethological observations of chimpanzees in central Africa, dancing and making invocations with druids: he was measuring chemical compounds across the Atlantic, in the stratosphere, and in the English and Irish countryside.

THE INSTITUTIONAL CONTEXT IN WHICH GAIA
DEVELOPED AND SPREAD: EARTH SCIENCES
AND ENVIRONMENTAL COUNTERCULTURE

The community to which Gaia was explicitly addressed was originally that of Earth scientists, mostly geochemists. At NASA's meetings in the 1960s, Lovelock had not only met Lynn Margulis but also Lars Sillén (1916—1970), a Swedish chemist who had a decisive influence on oceanography and geochemistry, and Heinrich "Dick" Holland (1927—2012), a top-notch contemporary geochemist of Earth's atmosphere and oceans. Dawkins can nag that Gaia reminds him of naive views of a "balance of nature," but he entirely missed that Lovelock had no interest whatsoever in *plant and animal* demography—the epicenter of the balance-of-nature tradition ever since Linnaeus.[23] If Lovelock was interested in demography, it was that of atmospheric gases. And, back in the early 1970s, the unraveling of the long-term chemical *history* of Earth's atmosphere and oceans was in its infancy. Three central and pioneering figures of these geochemical researches, Robert Garrels, Abraham Lerman, and Fred Mackenzie, in a 1976 famous paper entitled "Controls of Atmospheric O_2 and CO_2: Past, Present, and Future,"[24] presented a conclusion "in agreement with Lovelock and Margulis's 'Gaia' hypothesis (1974)." This conclusion is that of a paper that used box models to describe geochemical processes to estimate the long-term evolution of atmospheric O_2 and CO_2—not of two pages of abstract argument over group selection and altruism, as was Dawkins's argument. This is not to say that all geochemists embraced Gaia enthusiastically. In the 1970s, Gaian papers were not cited that much.

Yet in the 1980s, Gaia became a major topic, not only because Lovelock changed his audience and published a book for the general public, but also, in part, thanks to the crucial editorial work of the famous climatologist Stephen Schneider. Though wary and cautious about the meaning of the Gaia hypothesis, Schneider wanted Gaia to be subjected to *scientific* debate.[25] In 1988, he organized, with Penelope Boston, the first international scientific conference on Gaia as a Chapman Conference of the American Geophysical Union (AGU). In 2000, he co-organized the second such conference. In the early 2000s, he managed a space for the discussion of Gaia in his journal *Climatic Change*.

If Schneider carried out this work, it is because he did not want to leave Gaia to American and environmental counterculture. In the 1970s, and early 1980s, papers by Lovelock and Margulis, but also Doolittle's famous critique, were published in Stewart Brand's journal, *CoEvolution Quarterly*. This journal was the successor of the famous *Whole Earth Catalog*, which sold millions of copies and enabled Brand to become a central figure of American counterculture, in which contemporary cyberculture finds its roots.[26] Gaian systemic and cybernetic views of Earth deeply resonated with the spirit of *CoEvolution Quarterly*, and this journal had a very important role in the diffusion of Gaian papers. In addition, Lovelock (and Margulis) also published on Gaia in the two major environmentalist journals in the United Kingdom: *Resurgence* and *The Ecologist*, the last one founded by Lovelock's friend and sponsor, Edward Goldsmith.

THE DEVELOPMENT OF THE HYPOTHESIS AND THEORY

Initially many critics complained about the paucity of mechanisms that could account for global homeostasis. In the 1980s, Lovelock and others offered such mechanisms in three major papers.

In 1983, Andrew Watson and Lovelock published the computational model Daisyworld, developed specifically to address biologists' critiques and to show that the regulation of a global environmental variable could emerge from the influence of living organisms on their environment. The model depicted a fictive planet in which temperature is regulated by the proportion of populations of black and white daisies influencing the climate through their albedo. This Daisyworld model sparked more than one hundred of papers proposing variants of the original model.[27] The two other major papers of the 1980s were not about abstract mathematical models but about specific empirical mechanisms. The first one, published in *Nature* by Lovelock and Michael Whitfield in 1982, proposed a mechanism by which life, through its influence on rock weathering, may have counteracted the long-term increase of solar luminosity over Earth's long history and thus maintained Earth habitable.[28] This paper was an important stepping stone in the study of Earth's chemical and climatic history. Finally, in 1987, the paper exposing what would soon be known as the CLAW hypothesis was published in *Nature*.[29] The hypothesis suggested that algae may regulate the climate through a negative feedback loop involving the emission of DMS, which has a crucial role in cloud formation. This paper has been cited more than three thousand times, and certain atmospheric scientists have devoted their careers to the influence of DMS on contemporary climate.

Lovelock considered these three mechanisms (rock weathering, DMS, and Daisyworld) as "arguments" in favor of the Gaia hypothesis: the empirical examples were for geochemists and climatologists; Daisyworld was for evolutionary biologists. Clearly, for Lovelock, "Gaia" was the name of a hypothesis (and then a theory)—that is, a general proposition which can make predictions and can be *tested* against empirical facts. Lovelock placed a particular emphasis on Gaia as a theory or hypothesis in the early 1990s, after the famous critiques made by the geomorphologist James Kirchner in his Popperian lecture.[30] Kirchner distinguished four or five different formulations of the Gaia hypothesis. He then argued that the weaker versions were trivial, and the stronger ones were not testable.

After the 1990s, the development of the Gaia theory continued, focused on the elaboration of new versions of Daisyworld. Timothy Lenton, now a climate and Earth system scientist, who did his PhD under the supervision of

Andrew Watson (and Lovelock), pursued this development with Watson and a team of students in England.[31]

Overall, scientists were very willing to discuss life's influence on rock weathering, the influence of DMS on climate, and Daisyworld modeling in different specialized journals of (bio)geochemistry, climatic sciences, and biological theory. Yet Gaia, as a more general hypothesis or theory, was *not* discussed in specialized journals, as we expect every run-of-the-mill scientific hypothesis to be. The contexts of publications on Gaia per se were always *exceptional*: one to three international conferences per decade, a general paper in *Nature* and *Science* on some occasions, special issues of *Climatic Change*, books, but not regular discussion in *specialized* journals such as, say, *Geochemistry, Geophysics, Geosystems*. Scientists normally don't write books to discuss and test run-of-the-mill scientific hypotheses. But it is thanks to books that Lovelock was mostly known for Gaia and through which Gaia was criticized.[32]

GAIA, THE GREAT RESEARCH PROGRAM

For Lovelock, Margulis, and other Gaia supporters, Gaia was not *only* the name of a hypothesis to be tested with empirical arguments and elaborated with mathematical models, but that of a research program or a paradigm, a new discipline, a new way of doing science, of viewing and studying the world. Rather than talking about "Gaian science," they tried several terms such as "geophysiology" or "geognosy." Gaian science or geophysiology were often compared, contrasted, or opposed to other names of established scientific *disciplines*: geochemistry, biogeochemistry, chemical oceanography, microbial ecology, ecology, environmental science.

This research program made explicit methodological claims. The first claim was that the influence of living beings on their geological environment must be recognized and taken into account. In the 1970s, this claim was addressed to Earth scientists. They were accused of only considering rocks and chemistry and neglecting the pervasive influence of living beings. Then, beginning in 1983, Lovelock started to criticize the notion of adaptation. He argued that the fit that exists between life and its environment could be the result of life's influence on its geological environment, rather than the result of organisms adapting to their environment.

The second methodological claim was that Earth should be studied "as a system," "as a whole," or with a holistic and not a reductionist perspective. Lovelock and Margulis lamented over the separation between life and Earth sciences, as well as the splitting of the Earth sciences into atmospheric chemistry, climatology, fluid dynamics, study of rocks, and so forth. On many occasions, they used Gaia as the name of a revolutionary way of doing science that resolved the two centuries of divorce between biology and geology. All these entities of Earth's surface, such as rocks, bacteria, soils, chemical compounds of the oceans and atmospheres, were to be studied as aspects of one unified science.

These methodological claims rapidly turned into historiographical claims: for its authors, the emergent "Earth system science" was Gaia's research program with another name. Earth system science emerged in the 1990s and 2000s with new departments, institutes, centers, chairs, and textbooks of Earth system science out of the institutional work carried out by NASA and then the International Geosphere-Biosphere Program (IGBP) in the 1980s. Earth system science was not so much conceived as a new scientific discipline, but as an entirely new and revolutionary way of looking at Earth and of organizing the Earth sciences in an interdisciplinary fashion. Its central aim was "to describe and understand the interactive physical, chemical, and biological processes that regulate the total Earth system, the unique environment that it provides for life, the changes that are occurring in this system, and the manner in which they are influenced by human actions."[33] Interestingly, even those skeptical about Gaia as a hypothesis, or wary of some of Lovelock's environmental and political claims, such as Stephen Schneider and Ann Henderson-Sellers, credited Lovelock with having brought climatologists' attention to the pervasive influence of life on Earth's history. Ecosystem ecologists, such as Eugene Odum, who shared a cybernetic and systemic framework with Gaia theory, also pointed to Gaia's important role in emphasizing life's influence on planetary chemical cycles and in contributing to the emergence of global ecology.[34] And so did the founders of a new field, "geobiology," studying the interactions between the history of life and its environment, a research agenda departing from the one of the paleobiology of the 1970s.[35]

Most importantly, major actors in the field of Earth system science acknowledged Gaia's decisive role in calling scientists' attention to the existence of a new object of study: the "Earth system."[36]

GAIA, PHILOSOPHY OF NATURE, AND ENVIRONMENTAL PRESCRIPTIONS

Finally, Gaia was also for Lovelock the name of a philosophy of nature: "[Gaia] is an alternative to that pessimistic view which sees nature as a primitive force to be subdued and conquered. It is also an alternative to that equally depressing picture of our planet as a demented spaceship, forever travelling, driverless and purposeless, around an inner circle of the sun."[37]

The first sentence opposes our modern view of nature inherited from Bacon and Descartes.[38] The second opposes the metaphor of "spaceship Earth," very popular in the 1970s, conceiving of Earth as a vessel, in which environmental problems ought to be managed by experts.[39] Though, again, Lovelock is not anchored in a literary philosophical tradition, there is no doubt that Gaia was to propose a reconception of important and related concepts which were: life, nature, the environment. And certainly his philosophy of nature is not to be found in his explicit reflections over the categories of life and nature, but in the tacit assumptions he made while building models, in what needs to be taken for granted to engage in the research questions he envisaged.

To address Gaia as a philosophy of nature, the relevant attitude is not to "test" it against empirical facts. It is not to argue about the priority of such-and-such scientific issues or about the way scientific institutions should be organized, as it was for Gaia as a research program. It is to embrace and elaborate Gaia's categories and worldviews or to reject and dismiss them.

Another attitude one can adopt is to make explicit Lovelock's metaphysics, ontology, or categories and to contrast them with other ways to think about nature, life, and the world. And here again, Gaia found important echoes. The anthropologist and philosopher Bruno Latour sketched a symmetry between the way Galileo contributed to the overthrow of the Aristotelian conception of the cosmos to make us consider alike the physical properties of terrestrial bodies and of celestial bodies, and the way Lovelock has, conversely, managed to render Earth so peculiar and local in the solar system, influenced as it is by living entities.[40]

For conservative environmentalists, such as Edward Goldsmith, Gaia was truly to be thought of as an organismic ordered whole: not to follow the "natural rules" of this ordered whole should be seen as something deeply wrong.

But for Lovelock also, ever since the 1970s, Gaia has not only been a grandiose view of life on Earth but also a framework to think about the very concept of pollution, from which he derived many practical and concrete environmental and political prescriptions. In the 1970s, against the (green) current, he opposed the ban on the CFCs responsible for ozone's hole; he has long been actively militating for nuclear energy; and in the recent years, he has been criticized for taking radical positions ranging from the suspension of democracy and human rights to the proposition of geoengineering techniques and to a more or less voluntary reduction of the world's population.

Interestingly, in the three emblematic critiques made *by scientists* to the Gaia hypothesis—those of Doolittle and Kirchner in the 1980s, and of Tyrrell in his 2013 book—Lovelock's environmental and political prescriptions made in Gaia's name were mentioned as *the* central issues and as the reason why it mattered whether Gaia hypothesis was "right or wrong," "true or false." And many supporters of the Gaia hypothesis or research program, and even as a way of thinking about nature, found it difficult to follow Lovelock's "Gaian" environmental prescriptions.

CONCLUSION

The Gaian literature is vast, with enthusiasts and critics talking about rich and complex issues in very different domains. Gaia was not initially addressed to evolutionary biologists, but to geochemists. The very ambiguity of what Gaia meant had an important role in Gaia's pervasive diffusion and is what makes Gaia so rich and interesting. Lovelock, in the same papers and books, used the word "Gaia" to refer to very different things: a hypothesis, about which you can argue with empirical arguments and mathematical models, dealing with the peculiar influence that life may have had during Earth's history on its geological environment; a research program guiding and imposing the way Earth and life sciences should study Earth's chemistry, climate, and living beings; a philosophy of nature challenging our modern conceptions of life and nature. While dismissed as a problematic hypothesis, Gaia has been credited by Earth scientists for its role in fostering new research programs, such as those in Earth system science.

Certainly the most decisive and revolutionary contribution of the Gaia hypothesis was an ontological one. Gaia has called for the recognition of a new entity: the system composed of the entirety of life and the geological environment with which it interacts. The recognition of this new entity laid the ground for a new research program in the Earth sciences and offered a new framework within which to think about nature. Gaia has been central to our contemporary accepted view of Earth as a planetary system of interrelated entities, teeming with life, but also as a planet with stable states that can be overthrown, as is now dramatically pictured by the anthropocene discourse.

FURTHER READING

Dick, Steven, and James Strick. *The Living Universe: NASA and the Development of Astrobiology*. New Brunswick, NJ: Rutgers University Press, 2004.

Latour, Bruno. *Facing Gaia: Eight Lectures on the New Climatic Regime*. Cambridge: Polity, 2017.

Lenton, Timothy M., and Andrew Watson. *Revolutions That Made the Earth*. Oxford: Oxford University Press, 2011.

Ruse, Michael. *The Gaia Hypothesis: Science on a Pagan Planet*. Chicago: University of Chicago Press, 2013.

Tyrrell, Toby. *On Gaia: A Critical Investigation of the Relationship between Life and Earth.* Princeton, NJ: Princeton University Press, 2013.

NOTES

1. Exhibition, "Unlocking Lovelock: Scientist, Inventor, Maverick," Science Museum, London, 2014.

2. James Lovelock, *Gaia: A New Look at Life on Earth* (Oxford: Oxford University Press, 1979), 11.

3. Stephen Jay Gould, *The Structure of Evolutionary Theory* (Cambridge, MA: Harvard University Press, 2002), 612.

4. John Postgate, "Gaia Gets Too Big for Her Boots," *New Scientist*, 7 April 1988, 60.

5. James Lovelock, "A Physical Basis for Life Detection Experiments," *Nature* 207 (1965): 568. The most detailed historical account of the constitution of exobiology and of Lovelock's place in this adventure is provided by Steven Dick and James Strick, *The Living Universe: NASA and the Development of Astrobiology* (New Brunswick, NJ: Rutgers University Press, 2004).

6. Lovelock, *Gaia*.

7. James Lovelock and Lynn Margulis, "Atmospheric Homeostasis by and for the Biosphere: The Gaia Hypothesis," *Tellus* 26, nos. 1–2 (1974): 3.

8. Lovelock, *Gaia*.

9. See Michael Ruse, *The Gaia Hypothesis—Science on a Pagan Planet* (Chicago: University of Chicago Press, 2013), with the important following caveat: Ruse presents Lovelock as a typical Cartesian, yet cybernetics and systems thinking—Lovelock's disciplinary matrix—have often been considered at odds with typical Cartesian science.

10. Margulis introduced Gaia to Doolittle and encouraged him to publish his review of Lovelock's book in *CoEvolution Quarterly*, which Doolittle did not know (Doolittle, personal communication). Stewart Brand heard of Gaia through Margulis, thanks to her ex-husband, Carl Sagan (Brand, personal communication).

11. Ford W. Doolittle, "Is Nature Really Motherly," *CoEvolution Quarterly* 29 (Spring 1981): 58; Richard Dawkins, *The Extended Phenotype: The Gene as the Unit of Selection* (Oxford: Oxford University Press, 1982).

12. Oren Harman, *The Price of Altruism: George Price and the Search for the Origins of Kindness* (London: Vintage books, 2011).

13. In his recent book, *The Gaia Hypothesis*, Michael Ruse shows in detail how violent the reaction to Gaia was. But by focusing on Gaia's reception *in evolutionary biology*, he neglects the scientific disciplines to which Gaia was meant to contribute: geochemistry and earth sciences. See Sébastien Dutreuil, "Review of Michael Ruse, *The Gaia Hypothesis*," *History and Philosophy of Life Sciences* 36, no. 1 (2014): 149.

14. See Gould, *Structure*, 612; and Richard C. Lewontin, *Biology as Ideology: The Doctrine of DNA* (Ontario: Anansi Press, 1995), 18.

15. James Lovelock, *Homage to Gaia: The Life of an Independent Scientist* (Oxford: Oxford University Press, 2000), 2.

16. On scientific entrepreneurs, see Steven Shapin, *The Scientific Life: A Moral History of a Late Modern Vocation* (Chicago: University of Chicago Press, 2008).

17. Lovelock, *Homage to Gaia*, 38.

18. Lovelock, *Homage to Gaia*, 192.

19. James Lovelock, R. J. Maggs, and R. J. Wade, "Halogenated Hydrocarbons in and over the Atlantic," *Nature* 241 (1973): 194.

20. Mario J. Molina and Frank S. Rowland, "Stratospheric Sink for Chlorofluoromethanes: Chlorine Atom-Catalysed Destruction of Ozone," *Nature* 249 (1974): 810.

21. James Lovelock, R. J. Maggs, and R. A. Rasmussen, "Atmospheric Dimethyl Sulphide and the Natural Sulphur Cycle," *Nature* 237 (1972): 452.

22. In a letter to Arnold Kotler, he confesses to scarcely read besides fiction, and most of his intellectual debts go to oral discussion rather than to written materials.

23. Frank N. Egerton, "Changing Concepts of the Balance of Nature," *Quarterly Review of Biology* 48 no. 2 (1973): 322.

24. Robert M. Garrels, Abraham Lerman, and Fred T. Mackenzie, "Controls of Atmospheric O_2 and CO_2: Past, Present, and Future," *American Scientist* 64 (1976): 306.

25. Stephen H. Schneider, "A Goddess of the Earth: The Debate on the Gaia Hypothesis," *Climatic Change* 8 no. 1 (1986): 1.

26. William Bryant, "Whole System, Whole Earth: The Convergence of Technology and Ecology in Twentieth-Century American Culture" (PhD diss., University of Iowa, 2006); Fred Turner, *From Counterculture to Cyberculture: Stewart Brand, the Whole Earth Network, and the Rise of Digital Utopianism* (Chicago: University of Chicago Press, 2010).

27. For a review, see Andrew J. Wood, Graeme J. Ackland, James Dyke, et al., "Daisyworld: A Review," *Reviews of Geophysics* 46 no. 1 (2008). For an epistemological discussion, see Sébastien Dutreuil, "What Good Are Abstract and What-If Models? Lessons from the Gaïa Hypothesis," *History and Philosophy of the Life Sciences* 36 (2014): 16.

28. James Lovelock and M. Whitfield, "Life Span of the Biosphere," *Nature* 296 (1982): 561.

29. Robert Charlson, James Lovelock, Meinrat Andreae, et al., "Oceanic Phytoplankton, Atmospheric Sulphur, Cloud Albedo and Climate," *Nature* 326 (1987): 655.

30. James Lovelock, "Hands Up for the Gaia Hypothesis," *Nature* 344 (1990): 100.

31. Timothy M. Lenton and Andrew Watson, *Revolutions That Made the Earth* (Oxford: Oxford University Press, 2011).

32. Toby Tyrrell, *On Gaia: A Critical Investigation of the Relationship between Life and Earth* (Princeton, NJ: Princeton University Press, 2013).

33. This is an iconic statement of the first report of the IGBP, *The International Geosphere-Biosphere Programme: A Study of Global Change. Final Report of the Ad Hoc Planning Group, ICSU 21st General Assembly, Berne, Switzerland, 14–19 September, 1986* (p. 3), chaired by the famous Swedish meteorologist Bert Bolin, editor-in-chief of the journal *Tellus*, which published the iconic Lovelock and Margulis article, "Atmospheric Homeostasis."

34. Eugene P. Odum, "Great Ideas in Ecology for the 1990s," *BioScience* 42 (1992): 542.

35. For a philosophical and historical overview of paleontology and paleobiol-

ogy, see David Sepkoski and Michael Ruse, *The Paleobiological Revolution: Essays on the Growth of Modern Paleontology* (Chicago: University of Chicago Press, 2009); Derek Turner, *Paleontology: A Philosophical Introduction* (Cambridge: Cambridge University Press, 2011). No equivalent study exists for contemporary geobiology. On geobiology's debt to Gaia, see Andrew H. Knoll, Donald E. Canfield, and Kurt O. Konhauser, *Fundamentals of Geobiology* (Oxford: Wiley, 2012).

36. The details of the historical relations between Gaia and Earth system science cannot be fully sketched here—for details see Sébastien Dutreuil, *Gaïa* (PhD diss., Université Paris 1 Panthéon-Sorbonne, Institut d'Histoire et de Philosophie des Sciences et des Techniques, 2016). It would require a detailed history of various separate disciplines in which Earth system science finds its roots from 1950s to the 1970s: ecosystem ecology, climatology, biogeochemistry, and systems theory, but also the geochemistry and Earth history of Robert Garrels and Heinrich "Dick" Holland. On NASA, Earth system science, and Gaia, see Erik M. Conway, *Atmospheric Science at NASA: A History* (Baltimore: Johns Hopkins University Press, 2008). On Earth system science and IGBP, see Chunglin Kwa, "Local Ecologies and Global Science Discourses and Strategies of the International Geosphere-Biosphere Programme," *Social Studies of Science* 35 (2005): 923; Ola Uhrqvist, "Seeing and Knowing the Earth as a System: An Effective History of Global Environmental Change Research as Scientific and Political Practice" (PhD diss., University of Linköping, 2014).

37. Lovelock, *Gaia Hypothesis*, 11.

38. Carolyn Merchant, *The Death of Nature: Women, Ecology and the Scientific Revolution* (New York: Harper, 1980).

39. Sebastian Grevsmühl, *La Terre vue d'en haut: L'invention de l'environnement global* (Paris: Seuil, 2014); Robert Poole, *Earthrise: How Man First Saw the Earth* (New Haven, CT: Yale University Press, 2008).

40. Bruno Latour, *Face à Gaïa: Huit conférences sur le nouveau régime climatique* (Paris: La Découverte, 2015), 105.

EHUD LAMM

BIG DREAMS FOR
SMALL CREATURES
ILANA AND EUGENE ROSENBERG'S
PATH TO THE HOLOGENOME THEORY

INTRODUCTION

"How did the idea originate? My bugs were not doing what they were supposed to." That's Eugene Rosenberg, eighty years old, an idiosyncratic, cigar-smoking, English-speaking, American-born Israeli microbiologist hailing from Tel Aviv, explaining what prompted him and his wife Ilana to rethink evolution, ecology, and what it means to be an individual. What started as an empirical investigation into how corals respond to changing sea temperature eventually led the Rosenbergs to the claim that bacteria are integral parts of evolving individuals, far from their typical portrayal as pathogens. Indeed, they and their colleagues have further suggested that by altering organisms' mate choices, symbiotic bacteria are an important factor in the origin of new species, and that their mechanisms for cell-to-cell adhesion and signaling render them strong candidates for having paved the way to multicellular organisms. Symbiosis with bacteria, so it would seem, is crucial to understanding both evolution and the self.[1]

Along the way, so I will claim, the Rosenbergs are nudging us beyond seeing the individual organism—each of us—as something with clear boundaries. Rather, individuals are similar to clouds, sharing and exchanging water droplets with nearby billows and with the environment, the droplets being bacteria. Individuals in this picture are multispecies consortia. If this pulsating picture of the living world is true, the boundary lines between evolutionary change and ecological change will have to be redrawn.

Eugene Rosenberg is a biochemist. But he has spent his career as a microbiologist, eventually working on microbial marine life and collaborating with marine biologists without becoming one. He has ended up thinking about evolution, not having spent much time on evolutionary questions during most of his career. But throughout his career, he always had specific bacteria in mind. The systematic new picture of the living world that he and his intellectual and life partner Ilana promote has its origins not in interaction with previous ideas or from mere theorizing. It came from a firm belief that you cannot understand fundamental things about the living world without an

Figure 18.1. Eugene Rosenberg and Ilana Zilber-Rosenberg.

intimate feeling for what it means to be a bacterium. The Rosenbergs, above all else, have been involved microbial dreamers. And their story begins with Eugene Rosenberg trying to understand why corals lose their color.

WHY DO CORALS LOSE THEIR COLOR?

Corals are poster children for symbiosis. The coral itself is an animal, an immobile, stony, distant relative of jellyfish and hydra (of phylum Cnidaria, kingdom Animalia). Rosenberg was studying *Oculina patagonica*, a rather pedestrian, off-white coral. Corals consist of polyps, vase-like structures with a mouth surrounded by tentacles. These polyps harbor algae, unicellular plankton. Algae are eukaryotes; each single cell has a nucleus, unlike bacteria, which lack nuclei. This symbiotic system of multicellular coral and unicellular algae harbors vital bacteria. Coral reefs are the foundation of rich ecosystems of other fish and marine animals that make up colorful underwater dream worlds of symbiotic relationships. When environmental conditions change, corals often lose their algae, bleach, and die, taking with them the entire ecosystem. Eugene Rosenberg's work was concerned with the role of "bugs" in this story.

When Eugene Rosenberg talks of bugs, he means bacteria. He and his students determined that *Vibrio shiloi*, a rod-shaped bacterium of 2–4 square micrometers, was the cause of Mediterranean Sea coral bleaching. They discovered that bleaching is the result of *shiloi* turning pathogenic when water temperatures rise. The bug then kills the pigmented zooxanthellae algae, which reside inside the coral and in the mucus it is covered in, and inhibits photosynthesis, the main source of energy for the coral. In other words, bleaching was a disease. The chemist and oceanographer Robert Buddemeier made an alternative suggestion earlier. He argued that bleaching allows the coral to replace one zooxanthellae symbiotic partner with another, one that is more tolerant to the new temperature. This idea was termed the adaptive bleaching hypothesis, and suggested that the coral system, composed of all partners, can react in adaptive ways, just like a single organism. While Rosenberg had a different perspective on bleaching, this idea took hold in his mind.[2]

Understanding that *Oculina* bleaching was a disease and that it was caused by *Vibrio shiloi* were important discoveries. But the story we are interested in begins after a decade of working with corals. Trying to understand how bleaching occurred and how the bacteria were released back to the environment, Rosenberg and his students tried to infect corals with *shiloi* to get them to bleach. But they could no longer infect the corals. The experimental system stopped working. Recall Rosenberg explaining that the bugs were not doing what they were supposed to do. Ariel Kushmaro, the graduate student who did the original experiments in which infection with *shiloi* caused bleaching was even rushed over to make sure the infection experiments were done in exactly the same way as before—and he was by now no longer a student! Still no luck.[3] For many of us, the reaction in such circumstances would be to leave well enough alone. But for Eugene Rosenberg, that was not an option. Perhaps because of his devotion to playing sports in his early years, be it basketball or baseball, he loves a challenge and often repeats a Buckminster Fuller quip to the effect that there are no "failed experiments," only "unexpected outcomes."

GERMS PROTECTING FROM OTHER GERMS

Why were the pathogenic bacteria not infecting anymore? Since corals do not have an adaptive immune system that learns to recognize harmful bacteria, as do humans, what could explain the acquired resistance to infection? Or rather *who*? Rosenberg, who says about himself that he thinks "like a bug," and who, through a gift for mentoring, infected his graduate students with a love for them, ended up suggesting that it was other bugs, other bacteria, that protected the coral.[4] Not only were the bacteria not necessarily bad, but the bacterial population in the coral was almost like an organ, a muscle that could be flexed when the need arose.

A first clue came from studying the healthy, colorless corals growing inside the grottoes of Rosh HaNikra in the north of Israel. This involved diving, breaking a piece of coral, quickly bagging it, taking it to the lab, and trying to get the bacteria in your sample to grow in a selective agar called thiosulfate citrate bile salts sucrose (TCBS), that turns yellow when *Vibrio* is present. The *Vibrio* colonies were then sequenced to determine the species. No *Vibrio shiloi* were found.[5] This indicated that the bacterial population in the coral is affected by the absence of the colorful algae found in corals infected with *Vibrio shiloi*. Another clue was the idea coming from the adaptive bleaching hypothesis that the coral and its symbionts form a single, adaptive entity. If so, why couldn't the bacterial population be its immune system?

Was Rosenberg's bug-thinking alone responsible for this novel suggestion? The close-knit organismic unit that was Rosenberg's flip-flop-wearing team, considered rather odd birds by the rest of the department, was by this time affected by an unexpected outside influence.[6]

Enter Ilana Zilber-Rosenberg—who many years earlier had been a PhD student in Eugene's lab and was now, after a lifetime of adventures, Eugene's life partner. Following her PhD work and postdoctoral research, Ilana became a clinical nutritionist, and worked with institutionalized mental patients, trying to help them with the digestive problems caused by psychiatric drugs. An outsider to the daily activities of the small, family-like group in the lab, which included regularly sharing lunch (during which no talk about science was allowed), Ilana's influence on the trajectory of Eugene's work steadily grew.[7] Her background in nutrition exposed her to the idea of probiotics—live microorganisms in food that promote health—an idea made increasingly famous by nutritionists and yogurt manufacturers since the mid-1970s. Remarkably, a germ-free guinea pig can be killed by fewer than ten *Salmonellae* bacteria, while a billion are needed to kill an animal with normal gut-bacterial composition.[8] The Rosenbergs extended the nutritionists' idea to include acquisition of beneficial bacteria from the environment, and called their suggestion about the source of coral immunity the "coral probiotic hypothesis."[9]

Curiosity about matters beyond biological research and clinical work had previously led Ilana to study philosophy and sociology. In retrospect, sociological reflection about the relation between individuals and society may have been in the background of her early thoughts about individuals as consortia. It wasn't altogether surprising, therefore, that she should propose a probiotic, cooperative explanation for the breakdown of Eugene's experimental system.

So, who first came up with the suggestion, wife or husband? The Rosenbergs do not really say, nor is it clear that they know or that it matters.[10] As in the symbiotic systems they study, in which a metabolic process may involve metabolites passing between several partners, breaking down the system in order to study it may be as misleading as it can be enlightening. A recent article about symbiotic systems like corals put it succinctly: "Remove one [part] to study it reductionist style and you learn not what that part does but how the now changed [system] adapts."[11] Probably at some point during their evening walks together, the idea first appeared, and the next morning Eugene carried it with him to the lab and let his students loose with it. "The Coral Probiotic Hypothesis" was published in 2006. It "posits that a dynamic relationship between the symbiotic microorganisms and the environmental conditions" allows corals to adapt to changing conditions more quickly than if they had to rely on accumulating advantageous mutations via natural selection, the "accepted" evolutionary mechanism of producing complex adaptation to the environment.[12]

If we skip ahead in the story to the present day, we find that the collection of microbes associated with organisms, called the microbiome, and its effects on health, has become an area of intensive study. The ability to identify and measure the microbiome in animals and plants improved significantly from the early 2000s and is constantly improving. It has become routine to use genome sequencing to identify bacteria that cannot be grown in culture, techniques that were on the cutting edge when Rosenberg used them to study corals. Alas, most experiments so far have been concerned with human development and health, and not with multi-generational evolutionary dynamics. Corals, like humans, are far from being an ideal experimental system: the life cycle is too long, and it is hard to reproduce results,[13] which makes them especially inappropriate for experimental studies of evolutionary processes.

In the first decade of the century, with a difficult model system for studying evolution, with the required technology only emerging, and with the traditional conceptual toolkit of microbiology insufficient for the questions at hand—the Rosenbergs' work stood a chance to avoid a dead end by becoming more programmatic. And thus, sitting at a restaurant in Vienna in 2007, the night before presenting their ideas to microbiologists interested in coral health, the Rosenbergs came up with the next step in the story. Should the symbiotic microorganisms and host be considered separate biological entities, or are they a single "superorganism," facing natural selection as one tightly knit entity? They would argue that it was clearly the latter. They referred to the combined entity as a *holobiont*, reusing a term introduced by Lynn Margulis in 1981 in describing her groundbreaking work on the role of symbiosis of bacteria in the evolution of the eukaryotic cell.[14] They called their theory the *hologenome* theory of evolution (the hologenome being the combined genomes of the holobiont). The idea that organisms coexist and interact with multiple microorganisms that may change throughout life wasn't new.[15] Nor was the use of the term *holobiont* to describe corals.[16] But the Rosenbergs pushed the superorganismic view further than ever before—indeed introducing a conceptual novelty—by taking it to be central for understanding evolution.[17] At least one person was intrigued, an editor from *Nature Reviews Microbiology*, who attended the meeting and commissioned the Rosenbergs to write up the proposal for publication in his journal. Not having worked on evolution, the Rosenbergs began poring over textbooks.

A WORLD OF SUPERORGANISMS

On the face of it, the hologenome idea falls squarely within a longstanding tradition. Thinkers considering systems as diverse as ant colonies and termite mounds, as well as human societies, have long suggested that they are superorganisms. Looking at human society as analogous to a single superorganism, with different classes performing different tasks, all of which serve the whole, goes back at least to Plato. The frontispiece of Thomas Hobbes's *Leviathan* (1651) is a famous illustration of this idea. It depicts the state as a huge, human-like entity in the shape of the king, whose body, upon closer examination, is made up of thousands of tiny individuals.

For modern biology, the more immediate influence was Herbert Spencer, Darwin's contemporary and coiner of the phrase *survival of the fittest*. Spencer saw human society as an organism in which the parts depended on the whole and vice versa. He thought all systems evolved toward increasing heterogeneity of mutually dependent parts and differentiation from the surrounding medium, in this way achieving greater degrees of independence and coherent individuality.[18] Spencerian thinking influenced early twentieth-century entomologists, at Harvard and Chicago in particular, who were quick to repurpose the metaphor and apply it first and foremost to ant and termite colonies. These "collectivist" views were common between the two world wars, but as World War II approached, more pessimistic views that cautioned against human society evolving toward mindless individuals with mob mentality, akin to social insects, took the stage.[19]

During the Cold War, the prevailing attitude was individualistic, often termed "Darwinian." In world affairs this manifested itself in the application of game-theoretical models, such as the Prisoner's Dilemma, to problems of strategic planning, such as nuclear deterrence, and was epitomized by approaches such as Mutual Assured Destruction, or MAD. In such scenarios, selfish behavior is kept in check by playing a game reiteratively, so that players need to take into consideration the long-term effects of their behavior. The attitude in biology during the period—in particular, views about social insects—was also mostly individualistic. Especially influential were the views of the Harvard entomologist E. O. Wilson, author of the agenda-setting book *Sociobiology*. His explanations of social behavior appealed to natural selection working on individuals who share evolutionary interests due to family relatedness, rendering selfish behavior suboptimal. The theoretical notions of kin selection and selfish genes formalized this reasoning. This rollercoaster of optimism and pessimism about human society as an integrated, well-functioning, superorganism, and by analogy the idea of mutualistic superorganisms in nature, took an upturn after the Cold War. Today, less individualistic options are being explored than in past decades. In fact, E. O. Wilson himself made an about-face and began expounding an explanation of social behavior as originating in group living and selection between colonies. In remarkable synchronicity, Wilson published his revised view about ant nests as superorganisms at roughly the same time that the Rosenbergs began expounding the holobiont perspective in 2007.[20]

And yet the Rosenbergs seemed to be drawing on a tradition other than the one described above. The Spencerian tradition is concerned with the division of labor in the superorganism and with its integrated development and growth. It is deeply sensitive to what makes something an individual. But while Spencer thought that systems always evolved toward greater complexity, he made it clear that one could enumerate the parts of the system and identify what is not part of it at each time point. This flies in the face of the idea of the holobiont.

A source of new insight into such matters, and a backdrop to the Rosenbergs' proposal, is the study of dynamic systems and systems biology. This work got its start in military research during World War II.[21] The illustrious MIT mathematician Norbert Wiener worked on anti-aircraft weapons and later translated his discoveries into hypotheses about how biological systems can act in a goal-directed fashion by employing feedback loops.[22] Cybernetics, the science he evangelized, was supposed to be a science of control, unifying both living and nonliving systems. The focus on dynamics echoes a much older tradition, going back centuries, about the balance of nature. In the 1920s, Walter Cannon in physiology and Charles Elton in ecology, respectively, continued this tradition and argued that organisms and ecosystems are able to maintain stable internal conditions (homeostasis). This dynamic view seems to fit much more naturally with the holobiont perspective than traditional examples of superorganisms, such as ant colonies. The holobiont individual, rather than having a protective border, seems to be a constantly fluctuating entity with microbes coming in from the environment, going out, and competing within the host.

MICROBIAL OPTIMISM

Surveying a large amount of data, the Rosenbergs proposed extending the probiotic hypothesis to other multicellular organisms and their symbiotic microorganisms. The key to their generalization was that all animals and plants have symbiotic relations with microorganisms. Symbiosis is the rule rather than a curious exception, as it is too often portrayed when microbes are neglected. Moreover, as the coral example showed so clearly, the relationship can be either beneficial or harmful, and this can change dynamically. In this sense the relationship is what is primarily significant; it is what makes the holobiont one unit, while the nature of the relationship comes second and can change over time. Microbes can enter the holobiont from the sea water, can go out, and may then come back in again. There is no simple threshold beyond which the system becomes one individual, as other researchers argue.[23] Indeed, the picture we have of bacteria, even in the most well-studied system, the human gut, is incomplete. A species of bacteria may exist in relatively small numbers—below 0.01% of the total bacteria in the organism that we can currently detect—and have no easily noticeable effect, until some change causes the bacteria to proliferate.[24] The idea that individuals are loosely coupled consortia seems uniquely appropriate for an age of Internet-based social networks, based on voluntary and transient connections described using generic terms such as "friending" or "following"— dyadic connections from which communities with varying degrees of stability sometimes emerge.

The exciting implications of the consortium of animal and microorganisms derive from the unique properties that bacteria have. The microbial community or ecology can change, for good or bad, as a result of environmental challenges (recall what presumably happened to make the coral immune to infections by *Vibrio shiloi*). This can involve changes in the relative number of each species of bacteria or the acquisition of new bacteria from the environment. The symbiotic bacteria can also change due to natural selection among themselves. This evolutionary change can change the holobiont much more quickly than evolution of the multicellular host, since bacteria have much shorter life cycles and their populations are huge compared to the host population. By relying on the bacteria, the consortium is supposedly able to respond quickly and adaptively to the environment.

Working from these pathbreaking ideas, the Rosenbergs ended up making three claims that are hard for mainstream Darwinians to swallow. First, symbiosis, rather than simply a relationship forged by natural selection, is a dynamic developmental process. Second, the primary unit of evolution is not individual organisms but these symbiotic ensembles; and, third, these evolve via the dreaded Lamarckian inheritance of the adapted microbiome. These three challenges are not simply incremental suggestions about separate issues; they stem from a vision achieved through an intimate feeling for one specific biological system and a resolute demand that this vision be applied to life everywhere. This is not simply an incremental advance in evolutionary description, but a true theoretical novelty. Let's see how it plays out.

ROCKING THE BOAT

By pointing out that the ensemble can adapt to challenges before natural selection operates on the holobiont, the hologenome theory provides a negative answer to a perennial question puzzling evolutionary thinkers: does natural selection explain all adaptations? A major process, long argued to be crucial for understanding evolution, is the ability of organisms to react sensibly to changing environments, modifying their organization or behavior. Such so-called developmental, or behavioral, plasticity might seem like an evolutionary dead end, since developmental changes are not inherited, but there are nonetheless two main approaches for explaining how plasticity can play a role in evolution. One is taken by those who have argued through the years that acquired changes are, at least in some cases, hereditary. These approaches are typically called Lamarkcian, after the much maligned pre-Darwinian evolutionist, Jean-Baptiste Lamarck, who famously thought that this was possible. Alternatively, or in addition, it has been noted that short-term changes can affect the direction of natural selection—for example, by causing organisms to migrate to new, more hospitable surroundings or by favoring mutations that work well in tandem with the direction in which the organism adjusts to environmental challenges. In this way, it has been argued, genes may be "followers" in evolution, often merely cementing changes that were originally the result of developmental plasticity. Two seminal books arguing for such notions appeared at the beginning of the millennium: Mary Jane West-Eberhard's book *Developmental Plasticity and Evolution*, and Eva Jablonka and Marion Lamb's *Evolution in Four Dimensions*.[25]

The idea that a coral can adapt via the ecology and evolution of its microbiome is exciting and challenging, and extends the notion of development to encompass the symbionts as part of the developmental apparatus of the host. Symbiosis, from this perspective, is a dynamic relationship, in which the identity of the microbial partners changes throughout the life of the host, depending, for example, on the season. The symbiotic system is an ecosystem and yet may function as a regulated developmental system. The adaptive response may surely take time.

However, proving that changes in the microbiome are a result of an adaptive process and not random fluctuations of self-interested individuals is no simple task. Critics were quick to point out that changing climate tends to produce holobionts that are less well adapted than before—for example, by being more prone to disease.[26] Readily rising to the challenge, the Rosenbergs went even further. The adaptive changes, they argued, are indeed hereditary, since the microbiome passes "vertically" from parent to offspring. It was an overtly Lamarckian claim. Crucially, however, it hinged on a further claim, one that harked back to an old debate among evolutionists. The holobiont, claimed the Rosenbergs, is a unit of selection, possibly even *the* primary unit of selection in evolution. To many evolutionary thinkers over the years, the claim that there are multiple kinds of units of selection seemed outrageous. They demanded units of selection to be cohesive entities, reproducing with high fidelity and existing in large multitudes to produce selection pressures. Indeed, for the most part, mainstream thinking rejected the idea of multiple kinds of units in favor of the primacy of the gene as the unit of selection, a view epitomized by Richard Dawkins in his 1976 classic book *The Selfish Gene*.

Most of the debate about the unit of selection was concerned with whether groups of organisms of a single species were units of selection. If they were, that could explain behaviors that benefited the group but were detrimental to the individual performing them. This is at the core of the puzzle about the evolution of altruism. One proposal, made prominent in the 1960s, suggested that competition between groups explained prosocial behavior; however, for many years, group selection was considered both theoretically and empirically problematic.[27] One of the problems was that typical groups did not reproduce and transmit their group properties to daughter groups. Yet if group selection was not the cause of group beneficial behavior, what was? One approach was kin selection; another was evolutionary game theory; a third was reciprocal altruism, the scientific version of "You scratch my back, I'll scratch yours." The latter two ideas apply to multispecies symbiosis. The heated discussions on the evolution of altruism obscured the option of mutualistic relationships in which both partners benefit. Once again, after the turn of the millennium, possibly affected by loosening geopolitical fault lines, as well as by developments in theory and rigorous experiments, both multilevel selection and mutualism became less verboten, if still suspect. Since the hologenome theory emphasized the common interest of host and microorganisms and among the microorganisms themselves, evolutionists were bound to be suspicious.[28] The bleaching story, however, demonstrates that this rests on a misunderstanding: symbionts can be beneficial or become pathogens, depending on circumstances.

So, the holobiont perspective offered at least three foundational arguments: Symbiosis is developmentally plastic; the primary unit of evolution transcends the individual organism; and Lamarckian inheritance, via the inheritance of the adapted microbiome, is alive and kicking. Taken together they offered a dramatic challenge to core beliefs and a novel perspective on evolution.

REACTIONS

It took about five years for biologists to begin to respond to the hologenome idea.[29] The holobiont idea preceded the hologenome theory, but whereas the Rosenbergs were committed to seeing the holobiont as an integrated unit of selection in evolution, other proponents of holobionts favored a more ecological perspective. They argued that the relationships between host and symbionts were much more fluid than the hologenome theory predicts, and much more determined by the environment.[30] This controversy can be approached empirically by studying multiple holobionts and host species, and seeing whether the microbial community is associated with conditions, with the species of the host, or with neither. This is an indirect way to assess the extent to which the microbiome is hereditary.[31] According to a recent study, most of the symbionts are moving in and out the holobiont, while a small core of stable partners persists.[32] The researchers argue that if the main source of symbionts is the environment rather than the parent holobiont, then selection most likely operates on each member individually rather than on holobionts as wholes. The symbionts and host affect each other's evolution, and consequently they may coevolve, but selection operates on each member individually.

Another critique came from the prominent symbiosis researcher Nancy Moran and her colleague Daniel Sloan, in a provocatively entitled article: "The Hologenome Concept: Helpful or Hollow?"[33] They point out that microorganisms can have intimate and long-term relationships with hosts that are beneficial to both, without having evolved to benefit the host. Close association does not imply evolution at the level of the holobiont nor even coevolution (the host, for example, may have adapted to make use of typically available symbionts). They conclude that holobionts are units of selection in some cases, but that this is far from being the general case. They are not thrilled with the idea of a primary unit of selection, but they think that if there is one, it is unlikely to be the holobiont, because of the inherently divergent interests of the parties. In response, the Rosenbergs and thirteen colleagues state that there are many units of selection, and the holobiont is one such unit (nevertheless, they insist that the holobiont approach is distinct from group selection). Which unit is most important depends on the trait in question.[34] The objections raised by Moran and Sloan seem to echo the historical unease of Darwinians with superorganismic views that stressed balance, equilibrium, and integration.[35] The philosopher Michael Ruse wrote about such "organicists" that they see "an integrative aspect in nature that operates outside of or beyond selection. If . . . they are forced to put things in evolutionary terms, then group selection comes into play, but it is secondary to the basic way of nature." He concludes that, for them, "there is something wholesome about nature that the hard-line Darwinian misses."[36] The Rosenbergs' view of individuals as ecologically fluid, yet nonetheless as evolutionary individuals, is incongruent with the standard approach, which remains committed to stable individuals. Yet, paradoxically, objections from top researchers, based on sophisticated versions of the dominant views, are what can help refine the Rosenberg's novel ideas.

A more fundamental objection is that the hologenome theory is genecentric.[37] The Rosenbergs defined the hologenome as the sum of the genetic information in the holobiont, arguing that selection shapes the sum total of genetic information by selecting among holobionts. The holobiont is analogous to an organism, and the hologenome is analogous to the organism's genome. The gene-based understanding of evolution underlying this analogy is not without its detractors. Recall the idea that developmental plasticity can affect evolutionary change, highlighted by the "genes-as-followers" approach. If natural selection can operate on sources of variation between individuals that are not genetic but that affect development just as powerfully, a focus on genes may miss what is really going on.[38] Selection can operate on the constitution of the microbiome community if it is inherited vertically from parent to offspring, as in the cases the Rosenbergs emphasize. Previously, we have seen the Rosenbergs integrate emerging ideas, still at the cutting edge; the focus on gene change as what is important in evolution is more traditional. Far from being a curious aspect of their work, gene-centricity may undercut the role of nongenetic evolutionary history in explaining the constitution of holobionts and strengthen various arguments against the theory. In a recent reply to criticism, the Rosenbergs and their colleagues maintain the claim that variation in the hologenome yields variation in holobiont phenotypes rather than allowing that evolution can operate directly on microbiome variation.[39]

BUGS ARE EVERYWHERE

Eugene Rosenberg found himself espousing a view of life, something professional biologists are not encouraged to do. Nearing the end of his university career, he barged into fields like evolution and marine biology armed with an outsider's perspective. At the same time, he rode in with a big reputation in microbiology. This is a useful combination. It also helps to have a reliable instigator, sounding board, or partner in crime, and this role was played brilliantly by his wife Ilana. At the same time, if you have novel ideas, inspired by other disciplines, and a penchant to push them to the limit, like Ilana, it helps to have a partner in crime with solid empirical chops. That the hologenome theory was a generalization of empirical work that Eugene and his students had done was critical. Moving to more theoretical pursuits allows Eugene to continue working in his eighties with Ilana, after having to shut down his lab. The mark of success would be the extent to which their ideas ignite new, empirically testable questions and lead to new insights into concrete living systems. The debates that are now starting about what the view actually demands are a necessary step for making it rigorously testable, not just evocative—a dream, perhaps, but not a fantasy.

There are several ways in which one can make general claims—as dreamers are wont to do—in biology, a science notorious for being obsessed with special cases. One way is to invoke generic concepts, like the unit of selection, which may turn out to be applicable to various kinds of biological systems. Another is to look for analogous processes that, while significantly different form each other, may have similar consequences. This is arguably how some Lamarckians find evidence for inheritance of acquired characters in very diverse domains, such as the copying of molecular marks attached to DNA (epigenetics) and human social learning through language. Neither of these routes to generality is what is going on here with the hologenome theory. What makes it of general significance, rather, is that bugs are everywhere. Like the air we breathe, only more so. This is not a necessary result of the logic of natural selection. It is an empirical fact. It is similar to observing the general fact that, on this planet, hereditary material consists of nucleic acids. As Eugene likes to point out, bacteria were already around when eukaryotes began to evolve, and more than two billion years of prior evolution taught them how to do many important things. Reminding us of his original training in biochemistry, he notes that microbes are "the world's best biochemists."

The debates outlined above, significant as they are for the hologenome theory as it stands, may obscure what is ultimately at stake. The perspective urged on us by the Rosenbergs *begins* by rejecting the standard notion of individuality. It does not simply change the location of the boundaries of individuals or make them a matter of degree. Rather, the very idea of protective borders, or shared genetic destiny, that define the individual is replaced with dynamic processes and networks of interaction.[40] But this is not all: what happens "inside" the holobiont may be natural selection, while the evolution at the level of the holobiont may involve developmental acquisition and Lamarckian inheritance, and will depend on ecological factors. Thus, ecological changes and evolutionary changes, where and how they happen—and to what entities—are all realigned.

■

A unique amalgam of a playful attitude toward received ideas, bug-thinking, work on nutrition, a study of sociology, and a leap of imagination had all played their role. A difficult experimental system, corals, may have left more room for imagination, tempering failure with intimate knowledge to produce an audacious new idea. Indeed, the scope of the holobiont idea is general, a systematic, new approach to evolution, but it is always the "bugs" that play the pivotal role. It is very different from other superorganism theories that apply to multiple levels in the hierarchy of life and from still older views that posited a social impulse in all living things. Indeed, comparing human societies to ant colonies is a far cry from saying that ants are everywhere. In the final analysis, it is perhaps not surprising that Eugene and Ilana Rosenberg are not too interested in symbiosis as such—only when microbes are involved.[41] Born from the point of view of a bug, as if in a dream, the hologenome theory is Gaia for the little guy.

FURTHER READING

Gilbert, Scott F., and David Epel. *Ecological Developmental Biology*. Sunderland, MA: Sinauer Associates, 2009.

Harman, Oren. *The Price of Altruism: George Price and the Search for the Origins of Kindness*. New York: W. W. Norton, 2010.

Margulis, Lynn. *The Symbiotic Planet: A New Look at Evolution*. New York: Basic Books, 1998.

Rosenberg, Eugene, and Ilana Zilber-Rosenberg. *The Hologenome Concept: Human, Animal and Plant Microbiota*. Berlin: Springer Science and Business Media, 2014.

Ruse, Michael. *The Gaia Hypothesis: Science on a Pagan Planet*. Chicago: University of Chicago Press, 2013.

Sapp, Jan. *Evolution by Association: A History of Symbiosis*. New York: Oxford University Press, 1994.

NOTES

I thank Ilana and Eugene Rosenberg, Omry Koren and Gil Sharon for generously answering my questions and sharing their thoughts.

1. Eugene Rosenberg, Gil Sharon, Ilil Atad, et al., "The Evolution of Animals and Plants via Symbiosis with Microorganisms," *Environmental Microbiology Reports* 2, no. 4 (2010): 500–506; Gil Sharon, Daniel Segal, John M. Ringo, et al., "Commensal Bacteria Play a Role in Mating Preference of Drosophila Melanogaster," *Proceedings of the National Academy of Sciences* 107, no. 46 (2010): 20051–56.

2. Robert W. Buddemeier and Daphne G. Fautin, "Coral Bleaching as an Adaptive Mechanism," *BioScience* 43, no. 5 (1993): 320–26, doi:10.2307/1312064; author interview with Ilana and Eugene Rosenberg, 9 February 2016.

3. Author interview with Omry Koren, 15 March 2016.

4. Author interview with Omry Koren, 15 March 2016.

5. Author interview with Omry Koren, 15 March 2016.

6. Author interview with Omry Koren, 15 March 2016.

7. Author interview with Gil Sharon, 13 March 2016.

8. Michael de Vrese and J. Schrezenmeir, "Probiotics, Prebiotics, and Synbiotics," in *Food Biotechnology*, ed. Ulf Stahl, Ute E. B. Donalies, and Elke Nevoigt, Advances in Biochemical Engineering/Biotechnology 111 (Berlin: Springer , 2008), 1–66, online at http://link.springer.com/chapter/10.1007/10_2008_097.

9. Author interview with Eugene and Ilana Rosenberg, 11 April 2016.

10. Author interview with Ilana and Eugene Rosenberg, 9 February 2016.

11. J. Gordon, Nancy Knowlton, David A. Relman, et al., "Superorganisms and Holobionts," *Microbe Magazine* 8, no, 4, 2013, 152–153; online at http://www.asmscience.org/content/journal/microbe/10.1128/microbe.8.152.1.

12. Leah Reshef, Omry Koren, Yossi Loya, et al., "The Coral Probiotic Hypothesis," *Environmental Microbiology* 8, no. 12 (2006): 2068–73, doi:10.1111/j.1462-2920.2006.01148.x.

13. Author interview with Gil Sharon, 13 March 2016.

14. Lynn Margulis, *Symbiosis in Cell Evolution: Microbial Communities in the Archean and Proterozoic Eons* (New York: W. H. Freeman, 1993).

15. Forest Rohwer, Victor Seguritan, Farooq Azam, et al., "Diversity and Distribution of Coral-Associated Bacteria," *Marine Ecology Progress Series* 243 (November 13, 2002): 1–10, doi:10.3354/meps243001.

16. Rohwer, Seguritan, Azam, et al., "Diversity and Distribution."

17. Rohwer, Seguritan, Azam, et al., "Diversity and Distribution."

18. Herbert Spencer, "Transcendental Physiology," first published as "The Ultimate Laws of Physiology," in *National Review*, October 1857; Michael Ruse, *The Gaia Hypothesis: Science on a Pagan Planet* (Chicago: University of Chicago Press, 2013), 102.

19. Ruse, *Gaia Hypothesis*, 107; Oren Harman, *The Price of Altruism: George Price and the Search for the Origins of Kindness* (New York: W. W. Norton, 2010), chap. 5.

20. Ruse, *Gaia Hypothesis*, 116.

21. Ruse, *Gaia Hypothesis*, 113.

22. Ehud Lamm, "Theoreticians as Professional Outsiders: The Modeling Strategies of John von Neumann and Norbert Wiener," in *Biology Outside the Box: Boundary Crossers and Innovation in Biology*, ed. Oren Harman and Michael R. Dietrich (Chicago: University of Chicago Press).

23. For example, on the notion of an *aboluta iunctio* [*sic*] (absolute linkage) raised in Eric R. Hester, Katie L. Barott, Jim Nulton, et al., "Stable and Sporadic Symbiotic Communities of Coral and Algal Holobionts," *ISME Journal*, 10 November 2015, doi:10.1038/ismej.2015.190.

24. Author interview with Ilana and Eugene Rosenberg, 9 February 2016.

25. M. J. West-Eberhard, *Developmental Plasticity and Evolution* (New York: Oxford University Press, 2003); E. Jablonka and M. J. Lamb, *Evolution in Four Dimensions: Genetic, Epigenetic, Behavioral, and Symbolic Variation in the History of Life* (Cambridge, MA: MIT Press, 2005).

26. William Leggat, Tracy Ainsworth, John Bythell, et al., "The Hologenome Theory Disregards the Coral Holobiont," *Nature Reviews Microbiology* 5, no. 10 (2007), doi:10.1038/nrmicro1635-c1.

27. E. Sober and D. S. Wilson, *Unto Others: The Evolution and Psychology of Unselfish Behavior* (Cambridge, MA: Harvard University Press, 1998); Mark E. Borrello, *Evolutionary Restraints: The Contentious History of Group Selection* (Chicago: University of Chicago Press, 2010); David Sloan Wilson, *Does Altruism Exist? Culture, Genes, and the Welfare of Others* (New Haven, CT: Yale University Press, 2015).

28. The claim that the theory was rendered uninteresting to evolutionists by the belief that there is no conflict is made in Hester, Barott, Nulton, et al., "Stable and Sporadic Symbiotic Communities."

29. Author interview with Eugene and Ilana Rosenberg, 11 April 2016.

30. Author interview with Eugene and Ilana Rosenberg, 11 April 2016.

31. See also Edward J. van Opstal and Seth R. Bordenstein, "Rethinking Heritability of the Microbiome," *Science* 349, no. 6253 (2015): 1172–73, doi:10.1126/science.aab3958.

32. Hester, Barott, Nulton, et al., "Stable and Sporadic Symbiotic Communities."

33. Nancy A. Moran and Daniel B. Sloan, "The Hologenome Concept: Helpful or Hollow?," *PLoS Biology* 13, no. 12 (2015): e1002311, doi:10.1371/journal.pbio.1002311.

34. Kevin R. Theis, Nolwenn M. Dheilly, Jonathan L. Klassen, et al., "Getting the Hologenome Concept Right: An Eco-Evolutionary Framework for Hosts and Their Microbiomes," *bioRxiv* 2 (February 2 2016), 038596, doi:10.1101/038596.

35. See Ruse, *Gaia Hypothesis*, 100.

36. Ruse, *Gaia Hypothesis*, 118.

37. Hester, Barott, Nulton, et al., "Stable and Sporadic Symbiotic Communities."

38. For influential development of this line of thought, see Susan Oyama, Paul E. Griffiths, and Russell D. Gray, *Cycles of Contingency: Developmental Systems and Evolution* (Cambridge, MA: MIT Press, 2003); and Jablonka and Lamb, *Evolution in Four Dimensions*.

39. Theis, Dheilly, Klassen, et al., "Getting the Hologenome Concept Right."

40. For discussion of related issues, see Ehud Lamm, "Conceptual and Methodological Biases in Network Models," *Annals of the New York Academy of Sciences* 1178, no. 1 (2009): 291–304, doi:10.1111/j.1749-6632.2009.05009.x.

41. Author interview with Ilana and Eugene Rosenberg, 9 February 2016.

JOAN ROUGHGARDEN

EPILOGUE THE SCIENTIST DREAMER

What makes for a scientist dreamer, independent of any blessings of success bestowed late in life or posthumously? The profiles assembled here show scientists with a variety of talents and motivations. Some might be described simply as independent thinkers, others as following a deeply held passion, others as ambitious entrepreneurs, still others as philanthropists. Among all of these profiles, do any represent what might uniquely be considered as a "scientist dreamer"? That is, not just any old dreamer, but a special kind of dreamer that is unique to the pursuit of science?

Indeed, one might ask whether, in the end, dreaming is reserved for artists and musicians—or for dilettantes. Creativity would seem antithetical to science. Scientists shouldn't compose facts, sculpt evidence, or fuse autobiography with theory. Yet surely creativity has a role in science. How else does a scientist—really just a human being after all—imagine where a thesis might lead, or a new experiment to conduct, or a model to build? The state of science at any particular time doesn't dictate what steps to take next. If it did, science could be outsourced to robots programmed to read out what tomorrow's experiment or model should be. Still, a scientist doesn't enjoy the creative freedom that a novelist has when conjuring a plot or its dialogue. The story a scientist tells must potentially be a true story. A scientist's dream can't be intended as fantasy. It must express a vision with a chance of being at least partially accurate.

This brief essay presents what I suggest are seven distinctive features of scientific dreaming, along with pieces of advice to prospective scientific dreamers. Among the preceding profiles, scientist dreamers who might fit all seven features may include Lamarck, Kropotkin, and Lovelock, among others.

1. Above all, scientific dreamers sense that something is wrong, dreadfully wrong, with contemporary science. They discern that something big is missing or being disgracefully misrepresented. A scientific dream is conceived from a visceral dissatisfaction, even anger, with the scientific status quo. Anything less than reform results in tailoring existing science to accommodate new information. Instead, scientific dreams require a qualitative departure from existing understandings.

2. Contemporaries invariably question the dreamer's sanity, wondering if

dreamers inhabit an alternative reality. To contemporaries, a radically new account simply isn't needed—why entertain a theory that the sky is green when it's obviously blue?

3. The dilemma faced by scientific dreamers is to avoid self-deception, to avoid living in a bubble. After all, the odds are that a contrarian alternative to existing science is itself incorrect. The sky is blue, not green. The contemporary consensus is probably correct, and contrarian alternatives are probably mistaken. But not always. Professional ethics requires that dreamers be scrupulous in subjecting their dreams to rigorous self-criticism.

4. Scientific dreamers must respect disagreement and not disparage skeptics as being too dim-witted to appreciate the shining, crystalline truth placed before them. Even if the main points of a dream turn out to be true, not all of them will. Humility is needed. Beware of invoking the specter of Galileo on one's own behalf and of heroically casting oneself as speaking truth to power while denigrating opponents as slaves to political correctness.

5. Dreamers are radioactive. Inhabitants of an alternative reality rarely make good cocktail party guests. Dreamers should not expect to occupy a central node in the scientific citation network. Nonetheless, dreamers must seek out scientific interaction. A scientist dreamer cannot be a hermit. The input from engaging with other scientists, however awkward and unpleasant, is necessary, though not sufficient, to ensure that a scientific dream comes true.

6. Dreamers self-fund. Scientist dreamers, unlike creatives in the arts, need more than a living wage. They need funds for travel, equipment, and assistants. Hence, they must appeal to government agencies or foundations for support. But once they do, scientist dreamers encounter peer review. Peer review ensures that agency or foundation funds are responsibly dispensed— peer review will never go away. Funding personnel often declare their intention to take risks, to support projects with little chance of success but with enormous promise. Risky projects though, come in two varieties—extensional and disruptive. Everyone wins when an extensional project succeeds. Reviewers feel validated, and program officers brag to Congress or donors that their dollars have been well spent. In contrast, the scientific dreamer's project promises to be disruptive, not extensional, with few winners and lots of losers. A dreamer's project invariably crashes and burns while ascending the cliff of peer review. How can one expect peer reviewers to act against their self-interest by advocating a project that undercuts their own teaching and research? And how can a project officer tell Congress or their donors, "Remember what we told you when we thanked you for your past support? Well, um, that might be all wrong, and um, we'd like some more dollars to see if we

were mistaken." Not going to happen. So dreamers become starving scientists, like starving artists.

7. Are dreamers born or made? Maybe some are born. Sometimes little children are recognized as especially inquisitive, independent, perhaps unruly, and possibly destined to become scientist dreamers. Other scientist dreamers are surely made, having survived a life-changing experience that sets them on a new path.

If you're a scientist dreamer, let it happen; don't fight it. Really, you have no choice. It's what you are. In the end, you might be correct, after all. And remember, be a disciplined dreamer. The scientist dreamer must aim to tell a true story, not a fantasy.

If you're not a scientific dreamer yourself, but know someone who is, be kind to them.

CONTRIBUTORS

MARK E. BORRELLO
Department of Ecology, Evolution and
Behavior
University of Minnesota
St. Paul, MN 55108
USA

JANET BROWNE
History of Science Department
Harvard University
Cambridge, MA 02138
USA

RICHARD W. BURKHARDT JR.
Department of History
University of Illinois–Urbana-
Champaign
Champaign, IL 61820
USA

LUIS CAMPOS
Department of History
University of New Mexico
Albuquerque, NM 87131
USA

MICHAEL R. DIETRICH
History and Philosophy of Science
Department
University of Pittsburgh
Pittsburg, PA 15217
USA

SÉBASTIEN DUTREUIL
Institut d'Histoire et de Philosophie
des Sciences et des Techniques
Université Paris 1 Panthéon-Sorbonne
75006 Paris
France

KIRSTEN E. GARDNER
Department of History
University of Texas at San Antonio
San Antonio, TX 78249
USA

RICK GRUSH
Department of Philosophy
University of California, San Diego
La Jolla, CA 92093
USA

OREN HARMAN
Program in Science, Technology, and
Society
Bar-Ilan University
Ramat Gan 52900
Israel

TIM HORDER
Department of Physiology, Anatomy
and Genetics
Oxford University
Oxford OX1 3DW
England

PHILIPPE HUNEMAN
Institut d'Histoire et de Philosophie
des Sciences et des Techniques
Centre National de la Recherche
Scientifique
Université Paris 1 Sorbonne
75006 Paris
France

CHARLOTTE DECROES JACOBS
Department of Medicine, Division of
Oncology
Stanford University
Stanford, CA 94305
USA

EHUD LAMM
The Cohn Institute for the History and
Philosophy of Science and Ideas
Tel Aviv University
Ramat Aviv, Tel Aviv 69978
Israel

LAURA L. LOVETT
Department of History
University of Massachusetts, Amherst
Amherst, MA 01003
USA

MAUREEN A. O'MALLEY
LaBRI
Université de Bordeaux
33076 Bordeaux
France;
HPS
University of Sydney
NSW 2006
Australia

DALE PETERSON
USA

ANYA PLUTYNSKI
Department of Philosophy
Washington University
St. Louis, MO 63130
USA

ROBERT J. RICHARDS
Department of History
University of Chicago
Chicago, IL 60637
USA

JOAN ROUGHGARDEN
Department of Biology
Stanford University
Stanford, CA 94305
USA

BRUNO J. STRASSER
University of Geneva
1211 Geneva
Switzerland

INDEX

The letter *f* following a page number denotes a figure.

Binghamton neighborhood project, 248–50

biodiversity: dispersal-assembly views, 188; laboratory approaches to, 132–34; niche-assembly views, 188; patterns of, 176, 178–79; randomness of, 184

biogeography, ecology and, 178

bioinformatics, Dayhoff and, 6, 128–42

Biologics License Applications, 94

biology, historians of, 2

biology, philosophy of, 243–44

biomechanics, 265

bioremediation, living machines and, 163

biospheres, organism-like, 117–18

birds, pesticides and, 202

Birth Control Federation of America (BCFA), 73

Bishop, J. Michael, 96

Bissell, Mina, 96–109, 97f; approach used by, 13, 14; background of, 98–101; "The Differentiated State of Normal and Malignant Cells in Culture," 104; "How Does the Extracellular Matrix Direct Gene Expression?," 106–7; impetus to dream, 11–12; response to the work of, 106; selection for this book, 5–6

Bleek, Wilhelm, 46

Blue Planet Prize, 276

Bolin, Bert, 286n33

bones, structure of, 270n47

Bonner, John Tyler, 264, 268n23

Boston, Penelope, 279

Brand, Stewart, 279, 285n10; *CoEvolution Quarterly*, 279; *Whole Earth Catalog*, 163, 279

breast tissue, remodeling of, 105

Bronn, Georg, 39

Buddemeier, Robert, 290

Burbank, Luther, 145, 155

Bureau of Fisheries (subsequently US Fish and Wildlife Service), 198–99, 201

Burroughs, Edgar Rice, 211; *Tarzan of the Apes*, 211

Calcaronea, Haeckel's study of, 48n3

calcium ions, 99–100

cancer: causation, 96–109; cell reversion to normal phenotype, 105; DDT exposure and, 204; glucose metabolism of cells, 103; hallmarks of, 96; Mary Lasker's research, 77; microenvironment of, 98; research into, 75

Cancer Institute Council, 76

Cancer Research Institute, 77

Cannon, Walter, 295

Carnegie, Andrew, 241, 246

Carson, Rachel, 196–208, 197f; background of, 196, 198; at Bureau of Fisheries, 198–99; *The Edge of the Sea*, 200; "A Fable for Tomorrow," 205; and gender issues, 207; "Help Your Child to Wonder," 201; impetus to dream, 13; on pesticides, 6; religious influences on, 198; *The Sea Around Us*, 200; selection for this book, 6; *Silent Spring*, 196; *Under the Sea Wind*, 199

Carter, Jimmy, 92

Caswell, H., 191

cell cultures, conditions for, 100–103

cell nuclei, hereditary material in, 35

cell-to-cell adhesion, 288

Cellular Basis of Behavior (Kandel), 231

cellular behaviors, determination of, 106

Chaikovsky Circle, 58

Challenger expedition, 44

chance, ecological, 64n3

characteristics, acquired, 49n18

chimpanzees: diet of, 217; ethnological observations of, 278

chlorofluorocarbons (CFCs), 277, 283

chloroplasts, origin of, 115

Christianity, beliefs on nature in, 55

Church, Arthur, 270n50

Church, George, 143–60, 144f; approach used by, 14; background of, 145–46; impacts of, 9; influences of, 10; *Regenesis*, 150, 152; selection for this book, 6; sensationalism of, 154–55

Citizens Committee for the Conquest of Cancer (CCCC), 78

CLAW hypothesis, 280

Davenport, Charles, 241, 246

Dawkins, Richard: on the Gaia hypothesis, 118, 274–75, 279; *The Selfish Gene*, 242–43, 297

Dayhoff, Margaret, 128–42, 129*f*; *Atlas of Protein Sequence and Structure*, 128–34; impetus to dream, 12–13; selection for this book, 6; vision of, 129–30

DDT (dichloro-diphenyl-trichloroethane): food chains and, 203; impacts of, 199; problems with, 164; shortcomings of, 201; use of, 203–6

deferred maturity phenomena, 245

de Kruif, Paul, 72, 81n3

Dennett, Daniel, 229, 230

Department of Agriculture, US (USDA), 202–3

Descent of Man, The (Darwin), 66n30, 240, 245

Developmental Plasticity and Evolution (West-Eberhard), 297

DeVita, Vincent, 79

De Vore, Irven, 216–17

Diamond, Jared, 183

dieldrin, 202, 206

"Differentiated State of Normal and Malignant Cells in Culture, The" (Bissell), 104

dimethyl sulfide (DMS), 277, 280, 281

disagreements, respect for, 306

Discomedusa Desmonema Annasethe, 46, 47*f*

DNA: storage safety, 158n39; structure of, 227; synthesis on microchips, 147; synthetic, data storage using, 150

Dobell, Clifford, 258, 259–60

Dolittle, John, 118, 119*f*, 211

dominance-diversity curves, 180*f*

Doolittle, W. Ford, 112*f*, 113–27; approach used by, 13, 14; background of, 113–14; early career of, 114–15; ENCODE debate and, 117; Eric Bapteste and, 120–23; on the Gaia hypothesis, 274–75, 283; influences on, 116; "Phylogenetic Classification and the Universal Tree," 121*f*; selection for this book, 6; tree of life cartoon, 121*f*

Douglas, Mary Stoneman, 202; *The Everglades*, 202

dreamers: categorization of, 4; communication skills of, 15; community relationships of, 3; controversies and, 143, 145; financial support, 306; impetus to dream, 11–13; innovations of, 3; questioning of, 305–6; scientists as, 305–7; traits common to, 14; transformations and, 8–11

dreams, analysis of, 8–11

Driesch, Hans, 35, 45

D. T. Watson Home for Crippled Children, 89

dualist approaches, 229

Dubois, Eugene, 36

Dutch elm disease, 204

dynamical systems theory, 234

dynamic reciprocity hypothesis, 98

Earth atmospheric disequilibrium, 273–74

earth sciences: life and, 281–82; roots of, 287n36

Eccles, John, 230

ecological dependence: Darwin's concept of, 64n3; interrelatedness and, 202; pesticide use and, 202

ecological disturbances: cycles of, 193n16; human interference and, 196

ecological drift, 185

ecological engineering: description of, 164–65; John Todd and, 162–75

Ecological Engineering Associates (EEA), 167, 170

ecological equivalence, 193n17; neutral theory of ecology and, 184–87

ecology: changes and evolution, 288; concept of, 35; cycles of disturbances, 193n16; definitions of, 178; island biogeography and, 189; neutral theory of, 176; species abundance distributions, 180*f*

ecomachines, 163; Harwich system, 170, 171*f*; Sugarbush experiment and, 169

Edge of the Sea, The (Carson), 200

Galton, Francis, 241
Garrels, Robert, 279
Gegenbaur, Carl, 38, 40
GenBank, 134
gender issues: cultural ideas about, 151; family responsibilities and, 138–39; professional trajectories and, 138–39; Rachel Carson and, 207; women in science, 199
gene drives, 153
Generelle Morphologie der Organismen (Haeckel), 42
genes: frequencies of, 240; selfish, 294; social behaviors and, 242
genetic diversity, Neanderthal genome and, 149
genic selection, 243
genomics revolution, 139, 143
geobiology, 282, 286–87n35
germline editing, human, 153
Gilbert, Walter, 116
Global Polio Eradication Initiative, 93
global warming, 148
Glossary of Greek Birds, A (Thompson), 259
glycolysis, process of, 101–3
Godfrey-Smith, Peter, 118
Goethe on God and nature, 42
Goldsmith, Edward, 279–80, 283
Gombe Stream Chimpanzee Reserve, 219, 220–24
Goodall, Jane, 210f, 211–26; approach used by, 13, 222–23; background of, 211–12; Gombe Stream Reserve and, 220–24; impacts of, 8, 9; impetus to dream, 12; selection for this book, 6
Goodwin, Brian, 264
gorillas, mountain, 218
Gorini, Luigi, 99
Gould, Stephen Jay, 264; on the Gaia hypothesis, 275; influence on Doolittle, 116; *Ontogeny and Phylogeny*, 264–65; recapitulation hypothesis and, 49n19
Graur, Dan, 117
Gray, Michael, 115
group selection, 240, 243–44, 298
growth: development and, 269n46; form

and, 260–61; transform method and, 268n25

habits, environmental influences on, 25
Haeckel, Ernst, 35–50, 36f; and Anna Sethe, 37–38, 40–41; approach used by, 13, 14; *Challenger* expedition, 44; devotion to Darwinian theory, 41; *Generelle Morphologie der Organismen*, 42; impetus to dream, 13; *Kunstformen der Natur*, 46, 47f; mind-brain relationship theory, 37; *Der Monismus als Band zwischen Religion und Wissenschaft*, 45–46; *Natürliche Schöpfungsgeschichte* (Natural history of creation), 35; *Die Radiolarien*, 39; recapitulation hypothesis and, 36–37; religious views of, 45–46; selection for this book, 5; support for Darwin, 245; *Das System der Medusen*, 41, 44, 46; travels by, 38–39, 43–44; *Die Welträtsel* (Riddles of the world), 35
Haldane, J. S., 268n31
Hall, Don, 238
"Hallmarks of Cancer" (Hanahan and Weinberg), 107
Hamilton, William Donald, 242, 244; "Hamilton's rule," 242, 243
"Hamilton's rule," 242, 243
Hanahan, Douglas, 107; "Hallmarks of Cancer," 107
Handbook of Clinical Neurology, 232
Hanski, Ilkka, 185
Hardy, William, 262
Harlow, Harry, 215–16
Harwich system, 170, 171f
health policy, political discourse on, 74
Heart Institute Council, 76
Helmholtz, Hermann von, 44
"Help Your Child to Wonder" (Carson), 201
Henderson-Sellers, Ann, 282
heptachlor, 202
herbicides, 203
herd immunity, 86
Hertwig, Oskar, 35, 44
Hertwig, Richard, 35, 44

Kimura, Motoo, 185, 187, 191, 195n49
King's Garden (Jardin du Roi), 22. *See also* National Museum of Natural History, France
kin selection theory, 242, 298
Kirchner, James, 280, 283
Koch, Christof, 231, 235
Koonin, Eugene, 116
Kortlandt, Adriaan, 220, 221–22
Kropotkin, Peter, 51–67, 52*f*; approach used by, 13; impetus to dream, 13; influences of, 10; life of, 55–56; *Mutual Aid*, 60–61; selection for this book, 5; Victorian support for, 67n44
Kropotkin, Pyotr Alexeyevich. *See* Kropotkin, Peter
Kuhn, Thomas, 2, 17n11
Kunstformen der Natur (Haeckel), 46, 47*f*
Kushmaro, Ariel, 290–91

Lamarck, Jean-Baptiste, 20*f*, 21–34, 30*f*; Academy of Sciences and, 21–34; career trajectory of, 31; conchology and, 29–30; *Considerations relative to the natural history of animals*, 24–25; course on invertebrate zoology, 23–24; Georges Cuvier, on, 21, 28–29; *Hydrogéologie*, 24; impacts of, 9; impetus to dream, 11; on inheritance, 297; issues with orthodoxy, 25–26; issues with religious orthodoxy, 25–26; issues with scientific orthodoxy, 25–26; at King's Garden (Jardin du Roi), 22; National Museum of Natural History, France, and, 22; *Natural History of the Invertebrates*, 24; *Researches on the Organization of Living Bodies*, 24, 28; selection for this book, 5; theories on inheritance, 27; theory of organic evolution, 21–22; theory of organic mutability, 23; *Zoological Philosophy*, 24, 25–28
Lamarckism, 268–69n34
Lamb, Marion, 297; *Evolution in Four Dimensions*, 297
Lamb, Ruth Deforest, 201
Lamm, Ehud, 7, 10

language: social learning and, 301; study of, 237n22
langurs, 217
Lasker, Albert Davis, 73; donations by, 81n6; Florence Mahoney and, 82n13; Lord & Thomas and, 81–82n7
Lasker, Mary Woodard Reinhardt, 70*f*, 71–82; background of, 70–82; Florence Mahoney and, 76; impacts of, 9; impetus to dream, 12; selection for this book, 5, 6
Lasker Foundation "fact books," 77
lateral gene transfer (LGT), 119–20
Latour, Bruno, 283–84
Latreille, Pierre, 29
leachfields, 168
Leakey, Louis Seymour Bazett, 12, 215, 218–20, 221
Lederberg, Joshua, 7, 115
Ledley, Robert S., 12, 130–32
Lee, Eva, 103, 106
Lenfant, Claude, 76
Lenton, Timothy, 280–81
Lerman, Abraham, 279
Leviathan (Hobbes), 293
Levins, Richard, 178, 185
Lewontin, Richard, 178, 243; on the Gaia hypothesis, 275; influence on Doolittle, 116
Lieberman, Jerry, 247
life-history traits, trade-offs, 187
life sciences, novelty in, 2–3
limiting similarity, competition and, 179
linguists, 237n22
Linnaeus, 29, 148
living machines, 163, 173n2
log-normal curves, 192n7
Loon, Hendrik Willem Van, 199
Lovelock, James, 273*f*; approach used by, 13; background of, 275–78; Gaia hypothesis and, 117–18, 272–87; influences of, 10; reading by, 286n23; selection for this book, 7
Loving v. Virginia, 152
Lyell, Charles, 32

National Institutes of Health (NIH): competition for grants, 94; expansion of, 77–78, 79; funding by, 136, 137

National Institutes of Health and Cancer Research, 72

National Museum of Natural History, France, 22. *See also* King's Garden (Jardin du Roi)

National Science Foundation (NSF), 136, 166

natural history, molecular biology and, 138

Natural History of the Invertebrates (Lamarck), 24, 32

natural sciences: internal enemies of, 122–23; motives of, 122–23

natural selection: adaptation and, 176; altruism and, 65n25; competition and, 61; D'Arcy Thompson on, 262; Gaia hypothesis and, 118; group selection and, 240; group vs. individual, 240–41; multilevel, 243; neutral theory of ecology and, 176; social behavior and, 294; sources of variation and, 300. *See also* selection

natural theory, Kimura's, 185

nature: non-morality of, 55; philosophy of, 282–84; science of, 59–60

Nature Conservancy, 202

Natürliche Schöpfungsgeschichte (Natural history of creation) (Haeckel), 35

Naturphilosophie, 275

Neanderthal children, 143, 149; cloning of, 158n36; surrogates for, 149, 151, 158n35

Neanderthal genome, 149

Neanderthals in space programs, 150–51

Neighborhood Project, The (D. S. Wilson), 238

neutral theory of ecology, 176, 178; dynamics of, 186; ecological equivalence and, 184–87; Hubbell's explanation of, 181–84; unification in, 188

New Alchemy Institute, 164–67; Arks, 165; founding of, 163, 164; funding for, 166, 168; Todd test farm, 165

niches: competition and, 179; differences, 190

Nissen, Henry W., 217

Nixon, Richard, 78, 79

Nordenskiöld, Erik, 35

novelty: definitions of, 10; in the life sciences, 15

null hypotheses, 191

null models, 183

Ocean Arks International, 167

O'Connor, Basil, 87–88, 89, 90–91, 93

Oculina patagonica, 290

Odoesky, Vladimir, 58

Odum, Eugene, 164, 188, 282

Odum, Howard, 164–65; *Environment, Power, and Society*, 164–65

Office of Scientific Research and Development (OSRD), 74–75

oncogene paradigm, 96

100,000,000 Guinea Pigs (Kallett and Schlink), 201

On Growth and Form (*G&F*) (Thompson), 257–58, 259, 260–61, 263–66, 267–68n22

ontogeny: concept of, 35; phylogeny and, 36–37

Ontogeny and Phylogeny (Gould), 264–65

organic change: gaps in, 28, 118; theory of, 25, 28–31

organic evolution, theory of, 21–23

organicists, 299

organic mutability, 23

organisms: development of, 296–97; forms of, 269n46; structural integrity, 266. *See also* superorganisms

Orgel, Leslie, 117

originality, definitions of, 10

Origin of Species (Darwin), 44–45, 241; illustrations in, 35; Russian evolutionists and, 57–59

orthodoxy: Doolittle and, 119–20; Lamarck's issues with, 25–26; questioning of, 14

orthogenesis, 261

Osborn, Rosalie, 218

Ospovat, Dov, 49n19

ozone destruction, 277, 283

Rowland, Franklin, 277

Rubin, Harry, 96, 98, 100–101

Ruse, Michael, 285n5, 285n13, 299

Russell, E. S., 49n19

Russian Nights (Odoesky), 58

Rutherford, Ernest, 128

Ryan, Francis, 115

Ryther, John, 164

Sabin, Albert, 88, 89, 91, 92, 93

Sagan, Carl, 132, 285n10

Salk, Jonas, 83–95, 84*f*; approach used by, 13, 14; backlash against, 90–92; career of, 85; impacts of, 93–94; impetus to dream, 13; influences of, 10; the media and, 15, 91–93; selection for this book, 5; Thomas Francis and, 86–87

Salk vaccine, fate of, 92–93

Sanger, Frederick, 128

Sanger, Margaret, 81n5

Sapienza, Carmen, 117

Sarnoff, David, 81n5

Schaller, George, 220, 221, 222–23

Schlink, Frederick, 201; *100,000,000 Guinea Pigs*, 201

Schneider, Stephen, 279, 282

Schneirla, Theodore, 182

Schrödinger, Erwin, 231, 264; *What Is Life?*, 231

science: authorship and credit, 135; communities within, 134–38; creativity and, 305; orthodoxy in, 25–26; publication in, 135; as a social force, 15

scientists: careers, 135; commercial requirements of, 123; as dreamers, 305–7; professional ethos of, 135; responsibilities of, 123–24

Sea Around Us, The (Carson), 200

selection: levels of, 243–44, 245; multilevel, 298. *See also* natural selection

self-deception, dreaming and, 306

Selfish Gene, The (Dawkins), 242–43, 297

"selfish genes," 117, 294

self-sacrifices, 242

sensationalism, 154–55

septage lagoons, 169

septic tanks, septage from, 169

Sethe, Anna, 37–38, 46; death of, 40–41; marriage to Haeckel, 40. See also *Discomedusa Desmonema Annasethe*

Seung, Sebastian, 7

sewage treatment, 168–69

Sherrington, Charles Scott, 258

Siberia, woolly mammoths and, 148–49

Sierra Club, 202

signal processing, 234

Silent Spring (Carson), 196; DDT and, 204; public impacts of, 205; views of morality and, 202

Sillén, Lars, 278–79

Silverman, Robert, 232

Simberloff, Daniel, 183

simplicity, Hubbell on, 184

siphonophores, 44, 45

Skinner, Mary Scott, 198

Sloan, Daniel, 299

Sober, Elliot, 243, 244–45; *Unto Others*, 244–45

social behavior, natural selection and, 294

social behaviors, evolution of, 242

Social Contract, The (Ardrey), 246

social insect communities, 65n25

social networks, Internet-based, 296

social organisms, religion and, 247

social organizations, evolutionary theory and, 241

Sociobiology (E. O. Wilson), 243, 294

solar aquatic systems, 163, 172

solar energy, 165

space programs, 150–51

speciation: definition of, 185; symbiotic bacteria and, 288

species: abundance distributions, 180*f*; collaboration among, 57; competition among, 53; cycles of disturbances, 193n16; dominance-diversity curves, 180*f*; environmental constraints on, 25; hybridization techniques, 148; intra-species conflicts, 66n33; island-mainland model, 178–79; limiting similarity, 179; model organisms, 133–34; niches, 176; organic change within, 30–31; time of fixation of, 185